AF358486

Innovative Solutions for Concrete Applications

Innovative Solutions for Concrete Applications

Guest Editor

Patricia Kara De Maeijer

Basel • Beijing • Wuhan • Barcelona • Belgrade • Novi Sad • Cluj • Manchester

Guest Editor
Patricia Kara De Maeijer
Faculty of Civil and
Mechanical Engineering
Riga Technical University
Riga
Latvia

Editorial Office
MDPI AG
Grosspeteranlage 5
4052 Basel, Switzerland

This is a reprint of the Special Issue, published open access by the journal *Infrastructures* (ISSN 2412-3811), freely accessible at: www.mdpi.com/journal/infrastructures/special_issues/concrete.

For citation purposes, cite each article independently as indicated on the article page online and using the guide below:

Lastname, A.A.; Lastname, B.B. Article Title. *Journal Name* **Year**, *Volume Number*, Page Range.

ISBN 978-3-7258-3628-4 (Hbk)
ISBN 978-3-7258-3627-7 (PDF)
https://doi.org/10.3390/books978-3-7258-3627-7

Cover image courtesy of Patricia Kara De Maeijer
The Antwerp Port House–The headquarters of the Port of Antwerp-Bruges, Belgium.

Contents

About the Editor

Patricia Kara De Maeijer

Patricia Kara De Maeijer is a civil engineer whose research is focused on the pioneering advancements in concrete and alkali-activated materials technologies. Her expertise encompasses the application of recycling industrial waste and by-products to create sustainable construction materials, focusing on reducing the environmental footprint of the construction industry, as well as the study of asphalt, bitumen, structural health monitoring, and the application of life cycle assessment and life cycle cost analysis. As a Senior Member of RILEM (*International Union of Laboratories and Experts in Construction Materials, Systems and Structures*), Patricia actively contributes to several RILEM Technical Committees. Over the past five years, she has led the *Expert Group RRT1* within RILEM *TC 294-MPA*, where her team focused on developing mix designs and evaluating the mechanical properties of ground granulated blast furnace slag-based alkali-activated concrete. Patricia's academic contributions are extensive, with over 100 scientific publications and three patents to her name. She has been involved in multiple industrial research projects at Riga Technical University (Latvia) and the University of Antwerp (Belgium). Additionally, she serves as a peer reviewer for various scientific journals and as a Guest Editor for *Infrastructures*, a journal published by MDPI.

Editorial

Innovative Solutions for Concrete Applications

Patricia Kara De Maeijer

Faculty of Civil and Mechanical Engineering, Riga Technical University, LV-1048 Riga, Latvia; pkdm.rtu@gmail.com

Abstract: Concrete, having evolved over the last 2000 years, is integral to modern infrastructure, with continuous innovations aiming to address sustainability challenges. From Roman concrete mixes to the invention of Portland cement (PC), concrete has evolved to meet growing infrastructure demands. As urbanization and energy consumption increase, the construction industry is focusing on high-performance materials, recycling, and minimizing harmful substances. Research on sustainable concrete alternatives shows promising reductions in global warming potential and other environmental impacts compared to traditional PC. However, challenges such as higher material costs and performance limitations remain. Alternatives such as alkali-activated concrete (AAC), self-healing concrete, and bacterial concrete (BC) have emerged in response to environmental concerns, along with fiber-reinforced AAC, waste-based concrete composites, and the reuse of construction and demolition waste (CDW), further enhancing sustainability. Foamed concrete, with its lightweight and insulating properties, offers additional potential for reducing environmental impact due to its ability to incorporate recycled materials and reduce raw material consumption. Technologies like three-dimensional concrete printing (3DCP) are improving resource efficiency and reducing carbon footprints while also lowering labor and material waste. However, concerns regarding cost-effectiveness and social sustainability persist. Overall, continued innovation is the key to balancing performance, cost, and sustainability in the development of concrete and to meet the growing demands of global infrastructure.

Keywords: alkali-activated concrete (AAC); bacterial concrete (BC); fiber-reinforced alkali-activated concrete (FRAAC); crumb rubber concrete (CRC); steel dust-based concrete; foamed concrete; construction and demolition waste (CDW); life cycle assessment (LCA); life cycle cost analysis (LCCA); three-dimensional concrete printing (3DCP)

Received: 28 February 2025
Accepted: 5 March 2025
Published: 10 March 2025

Citation: Patricia Kara De Maeijer. Innovative Solutions for Concrete Applications. *Infrastructures* **2025**, *10*, 59. https://doi.org/10.3390/infrastructures10030059

Concrete, having evolved over the last 2000 years, is the second most used material on the Earth. The Romans first combined lime, pozzolan, and aggregate to create a durable material, and in 1756, John Smeaton used a similar mixture to rebuild the Eddystone Lighthouse. In 1824, Joseph Aspdin patented Portland cement (PC), which spread worldwide, with annual production now around 4.2 billion tons. Over time, alternatives like alkali-activated concrete (AAC) have emerged, particularly in response to cement shortages during the 20th century. As the demand for more resilient and sustainable infrastructure grows, prioritizing investments in both new projects and the maintenance of existing structures is crucial. Research programs have focused on key performance indicators (KPIs) such as RAMSSHE€P—Reliability, Availability, Maintainability, Safety, Security, Health, Environment, Economics, and Politics—to guide these efforts [1]. In this context, innovative concrete solutions are being explored to ensure that infrastructure meets these objectives of durability and sustainability. As the construction industry faces growing demand for infrastructure, research on sustainable alternatives to traditional PC has intensified, leading to materials like "green concrete". Research highlights the potential for a more sustainable built environment through innovative concrete formulations, recycling efforts, and advanced technologies. Several promising solutions were showcased at the IOCI2022 conference [1], with some of them being additionally highlighted in this editorial.

Twenty-five years ago, innovations began revolutionizing concrete technology with breakthroughs like self-healing concrete, which was first pioneered by researchers in the Netherlands through bacterial concrete (BC)—a type of concrete that uses biological processes to repair cracks. Meanwhile, researchers around the globe have developed various other self-healing methods, including those using chemical reactions and encapsulated healing agents. For instance, Snoeck [2] examined the self-healing capabilities of cement-based samples after ten years of maturation under different conditions. The results showed that cracks can still heal, primarily through the formation of calcium carbonate crystals, contributing to a partial recovery in mechanical properties. Samples with superabsorbent polymers (SAPs) demonstrated better healing compared to reference samples, even after a decade, highlighting their potential as a sustainable solution for cement composites. In addition to self-healing concrete, other innovations such as AAC present opportunities to reduce environmental impact. The construction industry is also exploring the use of waste materials like crumb rubber (CR) and developing non-traditional supplementary cementitious materials (SCMs) to minimize the carbon footprint and optimize resource use in construction projects. Furthermore, three-dimensional concrete printing (3DCP) has also reshaped concrete construction by enabling designs that traditional methods cannot achieve. Urbanization and demographic growth have significantly increased global energy consumption, with the construction sector being a major contributor due to its energy-intensive processes and environmental impact. The extraction and processing of raw materials for construction lead to high energy use and carbon emissions. In response, architectural design and civil engineering are focusing on using high-performance materials while minimizing toxic or harmful ones. There is also a strong emphasis on reducing energy consumption throughout a building's life cycle. One promising solution is reusing construction and demolition waste (CDW), which, for example, accounts for over 30% of waste in the EU.

The study by Ramagiri et al. [3] evaluated the environmental impact and cost of three sustainable concrete mixes: AAC with natural aggregates, AAC with recycled coarse aggregates (RCA), and BC—compared to Portland cement concrete (PCC). Using a life cycle assessment (LCA), this study revealed that AAC mixes were most affected by transportation and sodium silicate, while BC mixes had the highest impacts compared to PC, nutrient broth and coarse aggregates. PCC had a 1.4–2 times higher global warming potential (GWP) than the other mixes. BC, without nutrients, had the lowest environmental impact overall, except for GWP. AAC mixes were 98.8–159.1% more expensive, while BC mixes were 21.8–54.3% costlier than PCC. AAC, with a 0.7 activator modulus, showed the lowest environmental impact, while AAC, with RCA, had higher impacts. This LCA and life cycle cost analysis (LCCA) provided insights into sustainable concrete alternatives in the industry.

Another study by Ramagiri et al. [4] explored AAC made from fly ash and slag, focusing on its high-temperature performance, microstructural changes, and environmental impact. The research examined how different fly ash–slag ratios, activator moduli (Ms), and temperature exposures affect AAC's microstructure and mechanical properties. High temperatures caused the formation of crystalline phases like gehlenite, akermanite, and nepheline, which were linked to improved compressive and bond strengths. Ground granulated blast-furnace slag (GGBFS)-based AAC resulted in a 151.8–339.7% increase in 28-day compressive strength. The optimal mix was found to be a fly ash–slag ratio of

70:30 and Ms of 1.4. However, exposure to temperatures above 760 °C led to a significant drop in compressive strength for mixes with higher slag content due to pore pressure build-up. LCA revealed that transportation and sodium silicate production had the highest environmental impact, accounting for 45.5–48.2% and 26.7–35.6% of the total emissions, respectively. Compared to PC, AAC demonstrated lower global warming potential, though its manufacturing process still contributed to higher impacts in ecotoxicity, eutrophication, and ozone depletion. This study also found that the cost of AAC was competitive, ranging from INR 16,532 to INR 17,265 (EUR 200–210) per m^3, and that implementing a carbon tax would raise costs by 18.4% for AAC compared to an 81.7% increase for PC, underscoring AAC's potential as a sustainable alternative.

The study by Ganapathi Chottemada et al. [5] evaluated the environmental impacts of AAC with different precursors using a cradle-to-grave approach. It compared the impacts of AAC mixes with 28-day compressive strengths ranging from 35 to 55 MPa, selecting the most sustainable option for further assessment with fiber reinforcement. The results showed that PCC has an 86% and 34% higher environmental impact on ecosystem quality and human health, respectively, compared to AAC. Sodium silicate in AAC production accounted for 30–50% of its total environmental impact. Among fiber-reinforced AAC (FRAAC) mixes, glass fibers increased the environmental impact more than steel or polypropylene fibers. The FS50 AAC mix, containing 50% fly ash and 50% GGBFS, had the lowest environmental impact. LCCA showed that AAC is 132% more expensive than PCC, primarily due to sodium silicate. However, polypropylene fibers result in the lowest production cost among FRAAC mixes. This comprehensive LCA highlighted the environmental and economic implications of using FRAAC, offering valuable insights for policymakers and the construction industry to adopt more sustainable alternatives, particularly in regions with similar economic and climatic conditions to the Indian subcontinent.

The study by Tahwia et al. [6] aimed to develop a more sustainable and eco-friendly engineered geopolymer composite (EGC) by using common, low-cost pozzolanic waste materials—rice husk ash (RHA), granite waste powder (GWP), and volcanic pumice powder (VPP)—as partial replacements (10–50%) of GGBFS. These materials were chosen due to their high content of aluminum and silicon, which are key for geopolymer formation. The research focused on evaluating how these waste materials affected the workability, mechanical properties, durability, and microstructure of EGC. The results showed that using RHA and GWP as 50% replacements for GGBFS reduced workability by up to 23% and 31%, respectively, whereas VPP increased workability by up to 38.5%. Among the different mixes, the optimal performance was found with 30% RHA, 20% GWP, and 10% VPP, which achieved the highest compressive, tensile, and flexural strengths and the best residual compressive strength when exposed to elevated temperatures. In contrast, water absorption and porosity increased significantly with higher amounts of RHA and GWP, while VPP showed the opposite effect, decreasing both water absorption and porosity. When exposed to high temperatures, the optimal mixes (RHA-30, GWP-20, VPP-10) retained strength better than other mixes at 200 °C. However, at 400 °C and 600 °C, all mixes experienced substantial strength loss. This suggests that while these alternative materials improve some properties, they may not perform as well in high-temperature applications compared to conventional EGC. Scanning electron microscopy analysis indicated that the VPP mix had the densest matrix, while RHA and GWP mixes showed more porosity. The study suggested further research on the durability of EGC and further investigation of long-term strength development for RHA-based EGC.

The review by Kara De Maeijer et al. [7] highlighted 30 years of research on CR in concrete, identifying up-to-date key barriers in the construction industry, such as high recycling costs, reduced mechanical properties, limited research on environmental risks, and recyclability concerns. Improving the effectiveness of CR particles through surface treatments and optimized concrete mix designs could provide significant benefits as a replacement for natural aggregates. However, the application of CR in concrete is often region-dependent and may be limited to environmental concerns. A few promising key points were highlighted. Workability can be improved with the addition of admixtures like superplasticizers. CR also lowers concrete density, making it ideal for lightweight applications. Pre-treating CR enhances the bond at the interfacial transition zone (ITZ), mitigating strength loss. In general, the cementitious materials surrounding CR effectively confine the trace metals or volatile organics that exist in rubber particles. The optimal CR replacement is 10–15% for fine aggregates and 5% for coarse aggregates. Crumb rubber concrete (CRC) shows improved resistance to freeze–thaw cycles, chloride penetration, acid resistance, and abrasion, though it is more susceptible to sulfate attacks. Additionally, CR improves vibration and moisture absorption, making it useful for dynamic structures like railway sleepers, bridges, and seismic-prone structures. However, further cost-effective studies are needed to support its broader use in the construction industry.

A major challenge in the construction industry is the absence of scientifically supported, versatile building materials that perform effectively in aggressive environments, such as those exposed to chloride attack. The study by Shcherban' et al. [8] compared the durability of conventional and variotropic concrete mixes modified with microsilica under cyclic chloride exposure using three different production methods: vibrating, centrifuging, and vibro-centrifuging. The results showed that vibro-centrifuged concrete exhibited the highest resistance to chloride attack, with a significantly lower decrease in compressive strength compared to vibrated (87%) and centrifuged concrete (24%). Adding 2–6% microsilica improved the concrete's resistance, with the best results seen at 4%, reducing strength loss by 45–55% after 90 wet–dry cycles. The combination of vibro-centrifuging and microsilica led to an 188% decrease in strength loss as a result of cyclic chloride attack, significantly improving concrete durability. The authors concluded that variotropic concrete, especially when combined with microsilica, offers superior resistance to chloride attack. These findings are valuable for designing more durable infrastructure, and it was indicated that further research will focus on full-scale tests of reinforced concrete structures exposed to chloride attack, with potential adjustments based on the technological capabilities of production plants.

The study by Jahami et al. [9] investigated the effects of replacing cement with steel dust in reinforced concrete beams, focusing on workability, mechanical properties, and durability. Steel dust was added at 0%, 10%, 20%, and 30% replacement levels, with a constant water–cement ratio of 0.55. The results showed that increasing steel dust content reduced workability and density and significantly affected the elasticity modulus. At 10% replacement, compressive, tensile splitting, and flexural strengths improved but declined with higher steel dust levels. The best results in terms of ductility and maximum load were seen at the 10% replacement level, where ductility increased by 13% and load capacity improved by 5%. However, at 30% replacement, ductility and load capacity dropped significantly. The SD_{10} beam (with 10% of steel dust) exhibited fewer microcracks and better deformation resistance, sustaining a 20 kN load without yielding, while other beams with higher steel dust cement replacement levels resulted in premature failure. This suggests that steel dust can enhance concrete properties at moderate levels, but caution is needed for higher replacements, as they can weaken the material. However, further research is necessary to assess the long-term durability and corrosion potential of steel dust-based concrete.

A vast number of concrete structures are approaching the end of their expected service life, creating a growing need for repairs. Reinforcement corrosion is a major cause of damage, leading to cracking and concrete spalling, which require effective repair techniques. As Europe shifts to a circular economy, considering both environmental impact and life cycle costs in repair decisions is crucial. The review by Renne et al. [10] highlighted the gaps in the existing literature, comparing repair methods in terms of the differences in structures, damage causes, repair techniques, estimated and expected life span assumptions, etc. Although refurbishment is often seen as more environmentally beneficial than new construction, economic outcomes vary. Preventive maintenance is typically more cost-effective over the long-term, while curative repairs may be better for short-term life extension. In terms of specific repair methods, low-labor options like patch repair are favored for short-term fixes, though they may not always be the most economical for longer-term extensions. To determine the most sustainable concrete repair, LCA and LCCA should be performed, considering life cycle perspectives for optimal environmental and economic outcomes. While sustainability frameworks exist, no research compares repairs using LCA and LCCA with all five EN1504-9 [11] repair principles related to reinforcement corrosion. Further research is needed on the leaching behavior of concrete with sacrificial galvanic anodes. Service life prediction should be more integrated into LCA and LCCA for accurate service life assessments.

The interface between old and new concrete is critical in construction, but it often remains a weak point despite various bonding treatments. Traditional design methods overlook the complexity of this interface, leading to uncertainties in structural performance. The study by Zhang and Lu [12] introduced a novel framework using X-ray computed tomography and finite element-based numerical homogenization to quantify the anisotropic properties of the old–new concrete interface. The analysis showed that the interface has significantly lower stiffness and greater anisotropy compared to non-interface regions due to microcracks and voids. This study identified the "weakest vectors" for normal and shear stresses, revealing an orthogonal relationship between them with slight local deviations. Cosine similarity analysis demonstrated more consistent directional features at the interface, further highlighting its heterogeneous nature. These findings challenge traditional design assumptions and provide crucial insights for improving concrete structure rehabilitation and design.

The growing issue of construction waste, which is costly to dispose of, can be mitigated by reusing it in concrete production. This approach not only reduces waste disposal but also offers a sustainable solution for the depletion of natural concrete resources. Pervious concrete, which can contain up to 80% coarse aggregates, is a promising medium for recycling construction waste. The study by Sangthongtong et al. [13] explored the mechanical properties of pervious concrete made with both natural and recycled aggregates, with recycled aggregates enhanced by natural fibers from sackcloth. The research involved 45 samples, focusing on air void ratios and aggregate sizes. The results showed that increasing the air void ratio led to a 40–60% decrease in compressive strength, regardless of aggregate type or size. The permeability of pervious concrete remained unaffected by the type of aggregate, while the temperature increased as the air void ratio rose. Specifically, for small-size aggregates with 10% designed air voids, the permeability was 0.705 cm/s for both natural and recycled aggregates with sackcloth.

The review by Suarez-Riera et al. [14] explored various strategies for producing more sustainable cement and concrete, with an emphasis on leveraging CDW as a valuable resource. Sustainable improvements in cement-based materials can be pursued in two main ways: first, by using eco-friendly cements that have a lower environmental footprint than traditional materials, and second, by fully exploiting CDW as aggregates while also improving the properties of recycled aggregates to enhance their performance. These efforts are central to the work of architects and engineers striving to create high-performance, eco-friendly construction materials. The EU's ongoing support of CDW recovery strategies further emphasizes the importance of managing construction waste in a way that benefits both the environment and the economy. A key advancement in sustainable cement is green cement, which lowers energy use and carbon emissions compared to traditional PC. The use of crystallizing agents in concrete also boosts durability and self-healing properties and reduces maintenance. These innovations improve both the environmental impact and lifespan of cement materials. By incorporating CDW, green cement, and crystallizing agents into construction, the industry can make strides toward sustainability. Ongoing research and collaboration are needed to refine these strategies and ensure broad adoption for a more sustainable built environment.

Foamed concrete, also known as cellular concrete or aerated concrete, is an innovative solution that has gained significant attention in the construction industry due to its unique properties and wide range of applications. The study by Markin et al. [15] examined how different foamed concrete (FC) production methods—cavitation disintegration (CD) and turbulent mixing (TM)—affect pore size distribution, compressive strength, and water absorption. Six FC mixes with densities ranging from 820 to 1480 kg/m^3 and compressive strengths up to 47 MPa were prepared and tested at 7, 28, and 180 days. Digital image correlation was used to analyze pore structure, revealing that the production method significantly influences pore formation, which in turn affects strength and water absorption. CD promoted a finer, more uniform pore structure, improving compressive strength. TM was effective in producing low-density FC with a lower amount of foaming agents. This study found that porosity directly impacted water absorption, though pore shape and distribution also played a role. At 28 days, compressive strength ranged from 9.4 to 47.4 MPa and continued to improve with curing time, reaching 53 MPa at 180 days. The findings highlighted the importance of choosing production methods that optimize pore structure for stronger, more durable FC. It was indicated that future research should refine these methods, expand the dataset, and use advanced imaging techniques, such as micro-computed tomography (micro-CT), to better understand how pore structure affects material performance.

With significant growth since 2014, 3DCP has gained attention for its potential to reduce the construction industry's carbon footprint, though its impact on social sustainability and project success is often overlooked. The study by Shivendra et al. [16] examined how strategic decisions influence the balance between economic, environmental, and social sustainability in 3DCP adoption. Interviews with 20 Indian industry leaders revealed that companies invest in 3DCP primarily for automation and workforce development rather than for environmental benefits alone. A key barrier to wider adoption is the lack of government incentives for sustainable practices. This study identified five strategies companies use to promote sustainability through 3DCP and suggested government measures to accelerate its adoption. Additionally, 3DCP's environmental impact, especially regarding raw materials, requires more assessment to explore alternative combinations that could further reduce its footprint. Design factors like modularity and structure thickness also need to be studied for their influence on circularity and overall performance. The social sustainability of 3DCP is complex. While it reduces reliance on seasonal labor, it could also reduce opportunities for low-skilled workers. The potential job displacement due to automation needs further exploration to understand its effects on the workforce. Finally, 3DCP's cost-effectiveness remains unclear. Technoeconomic models that consider cost, quality, labor, and maintenance factors are needed, and studying cost trends in similar industries could help predict future expenses as 3DCP technology evolves.

Concrete is vital for infrastructure like bridges, tunnels, and power plants, which consume large amounts of the material. As infrastructure demand increases and sustainability concerns rise, alternative SCMs are needed. The industry is also turning to automated methods like 3DCP to address labor shortages. The study by Hanžič et al. [17] explored using oil shale ash (OSA) as an SCM in 3DCP and investigated collision milling as a pre-treatment. The research found that OSA from flue gases, particularly OSA-Ees(nid), showed the most promise due to its smooth, globular particles and high active β-calcium silicate content. Concrete with this ash achieved a 56-day compressive strength of 60 MPa, similar to conventional concrete. Collision milling, while effective for reducing particle size in bottom ash, did not significantly improve the performance of flue gas-derived ashes. Milling was only beneficial for bottom ash, which was too coarse for direct use as an SCM. Although milling increased the reactivity of the ash, it did not lead to significant improvements in concrete strength, except for bottom ash. Printability tests showed no major differences between untreated and milled OSA-Ees(nid) in terms of yield stress and buildability. This study concluded that OSA-Ees(nid) is the most suitable ash for 3D-printable concrete, and further research should focus on optimizing its use. The results highlighted the potential for non-traditional SCMs and digital fabrication methods in addressing sustainability, efficiency, and labor challenges in large-scale infrastructure projects.

This editorial highlighted key themes such as sustainability, performance, durability, cost, environmental impact, recycling, and construction waste management. It also explored emerging techniques and innovations, including AAC, BC, FRAAC, CRC, steel dust-based concrete, foamed concrete, and the use of CDW. In conclusion, sustainable concrete alternatives, innovative materials, and improved recycling and mixing techniques offer significant opportunities to reduce the environmental impact of concrete. However, challenges related to cost, long-term durability, and the application of LCA and LCCA still need to be addressed through further research. Furthermore, technologies such as 3DCP present advanced opportunities for innovation in the construction industry.

Acknowledgments: The Guest Editor would like to thank all Authors and RILEM colleagues who supported, participated, and contributed to this Special Issue titled "Innovative Solutions for Concrete Applications" and took part in the online conference IOCI2022. Heartfelt and sincere thanks are extended to Sharon Fan, Pedro Arias-Sánchez, Naibing Peng, the entire *Infrastructures* MDPI Editorial Team, Academic Editors, and Reviewers for their support and immense efforts in the efficient processing of each manuscript. Your dedication and expertise have been truly invaluable.

Conflicts of Interest: The Author declare no conflict of interests.

References

1. Martínez-Sánchez, J.; Kara De Maeijer, P.; Lo Presti, D.; Lee, H.; Sánchez Rodríguez, A.; Liberati, F.; Soilán, M. The 1st International Online Conference on Infrastructures—IOCI 2022. *Eng. Proc.* **2022**, *17*, 100. [CrossRef]
2. Snoeck, D. Autogenous Healing in 10-Years Aged Cementitious Composites Using Microfibers and Superabsorbent Polymers. *Infrastructures* **2022**, *7*, 129. [CrossRef]
3. Ramagiri, K.K.; Chintha, R.; Bandlamudi, R.K.; Kara De Maeijer, P.; Kar, A. Cradle-to-Gate Life Cycle and Economic Assessment of Sustainable Concrete Mixes—Alkali-Activated Concrete (AAC) and Bacterial Concrete (BC). *Infrastructures* **2021**, *6*, 104. [CrossRef]
4. Ramagiri, K.K.; Kara De Maeijer, P.; Kar, A. High-Temperature, Bond, and Environmental Impact Assessment of Alkali-Activated Concrete (AAC). *Infrastructures* **2022**, *7*, 119. [CrossRef]
5. Ganapathi Chottemada, P.; Kar, A.; Kara De Maeijer, P. Environmental Impact Analysis of Alkali-Activated Concrete with Fiber Reinforcement. *Infrastructures* **2023**, *8*, 68. [CrossRef]
6. Tahwia, A.M.; Aldulaimi, D.S.; Abdellatief, M.; Youssf, O. Physical, Mechanical and Durability Properties of Eco-Friendly Engineered Geopolymer Composites. *Infrastructures* **2024**, *9*, 191. [CrossRef]
7. Kara De Maeijer, P.; Craeye, B.; Blom, J.; Bervoets, L. Crumb Rubber in Concrete—The Barriers for Application in the Construction Industry. *Infrastructures* **2021**, *6*, 116. [CrossRef]
8. Shcherban', E.M.; Stel'makh, S.A.; Beskopylny, A.N.; Mailyan, L.R.; Meskhi, B.; Varavka, V.; Chernil'nik, A.; Elshaeva, D.; Ananova, O. The Influence of Recipe-Technological Factors on the Resistance to Chloride Attack of Variotropic and Conventional Concrete. *Infrastructures* **2023**, *8*, 108. [CrossRef]

9.	Jahami, A.; Younes, H.; Khatib, J. Enhancing Reinforced Concrete Beams: Investigating Steel Dust as a Cement Substitute. *Infrastructures* **2023**, *8*, 157. [CrossRef]

10.	Renne, N.; Kara De Maeijer, P.; Craeye, B.; Buyle, M.; Audenaert, A. Sustainable Assessment of Concrete Repairs through Life Cycle Assessment (LCA) and Life Cycle Cost Analysis (LCCA). *Infrastructures* **2022**, *7*, 128. [CrossRef]

11.	*EN 1504-9*; Products and Systems for the Protection and Repair of Concrete Structures—Definitions, Requirements, Quality Control and Evaluation of Conformity—Part 9: General Principles for the Use of Products and Systems. European Committee for Standardization: Brussels, Belgium, 2008.

12.	Zhang, G.; Lu, Y. Quantifying Anisotropic Properties of Old–New Concrete Interfaces Using X-Ray Computed Tomography and Homogenization. *Infrastructures* **2025**, *10*, 20. [CrossRef]

13.	Sangthongtong, A.; Semvimol, N.; Rungratanaubon, T.; Duangmal, K.; Joyklad, P. Mechanical Properties of Pervious Recycled Aggregate Concrete Reinforced with Sackcloth Fibers (SF). *Infrastructures* **2023**, *8*, 38. [CrossRef]

14.	Suarez-Riera, D.; Restuccia, L.; Falliano, D.; Ferro, G.A.; Tuliani, J.-M.; Pavese, M.; Lavagna, L. An Overview of Methods to Enhance the Environmental Performance of Cement-Based Materials. *Infrastructures* **2024**, *9*, 94. [CrossRef]

15.	Markin, S.; Sahmenko, G.; Korjakins, A.; Mechtcherine, V. The Impact of Production Techniques on Pore Size Distribution in High-Strength Foam Concrete. *Infrastructures* **2025**, *10*, 14. [CrossRef]

16.	Shivendra, B.T.; Shahaji; Sharath Chandra, S.; Singh, A.K.; Kumar, R.; Kumar, N.; Tantri, A.; Naganna, S.R. A Path towards SDGs: Investigation of the Challenges in Adopting 3D Concrete Printing in India. *Infrastructures* **2024**, *9*, 166. [CrossRef]

17.	Hanžič, L.; Štefančič, M.; Šter, K.; Zalar Serjun, V.; Šinka, M.; Sapata, A.; Šahmenko, G.; Šerelis, E.; Migliniece, B.; Korat Bensa, L. Collision Milling of Oil Shale Ash as Constituent Pretreatment in Concrete 3D Printing. *Infrastructures* **2025**, *10*, 18. [CrossRef]

 *infrastructures*

Article

Cradle-to-Gate Life Cycle and Economic Assessment of Sustainable Concrete Mixes—Alkali-Activated Concrete (AAC) and Bacterial Concrete (BC)

Kruthi Kiran Ramagiri, Ravali Chintha, Radha Kiranmaye Bandlamudi, Patricia Kara De Maeijer and Arkamitra Kar

[1] Department of Civil Engineering, Birla Institute of Technology and Science-Pilani, Hyderabad Campus, Hyderabad 500078, Telangana, India; h20191430099@hyderabad.bits-pilani.ac.in (R.C.); p20180462@hyderabad.bits-pilani.ac.in (R.K.B.); arkamitra.kar@hyderabad.bits-pilani.ac.in (A.K.)

[2] EMIB, Faculty of Applied Engineering, University of Antwerp, Groenenborgerlaan 171, 2020 Antwerp, Belgium; patricija.karademaeijer@uantwerpen.be

* Correspondence: p20170008@hyderabad.bits-pilani.ac.in

Abstract: The negative environmental impacts associated with the usage of Portland cement (PC) in concrete induced intensive research into finding sustainable alternative concrete mixes to obtain "green concrete". Since the principal aim of developing such mixes is to reduce the environmental impact, it is imperative to conduct a comprehensive life cycle assessment (LCA). This paper examines three different types of sustainable concrete mixes, viz., alkali-activated concrete (AAC) with natural coarse aggregates, AAC with recycled coarse aggregates (RCA), and bacterial concrete (BC). A detailed environmental impact assessment of AAC with natural coarse aggregates, AAC with RCA, and BC is performed through a cradle-to-gate LCA using openLCA v.1.10.3 and compared versus PC concrete (PCC) of equivalent strength. The results show that transportation and sodium silicate in AAC mixes and PC in BC mixes contribute the most to the environmental impact. The global warming potential (GWP) of PCC is 1.4–2 times higher than other mixes. Bacterial concrete without nutrients had the lowest environmental impact of all the evaluated mixes on all damage categories, both at the midpoint (except GWP) and endpoint assessment levels. AAC and BC mixes are more expensive than PCC by 98.8–159.1% and 21.8–54.3%, respectively.

Keywords: life cycle assessment (LCA); bacterial concrete (BC); environmental impact assessment; alkali-activated concrete (AAC); alkali-activated binder (AAB); recycled coarse aggregates

Citation: Kruthi Kiran Ramagiri, Ravali Chintha, Radha Kiranmaye Bandlamudi, Patricia Kara De Maeijer and Arkamitra Kar Cradle-to-Gate Life Cycle and Economic Assessment of Sustainable Concrete Mixes— Alkali-Activated Concrete (AAC) and Bacterial Concrete (BC). *Infrastructures* **2021**, *6*, 104. https://doi.org/10.3390/infrastructures6070104

Academic Editor: Ali Behnood

Received: 28 June 2021
Accepted: 13 July 2021
Published: 15 July 2021

Publisher's Note: MDPI stays neutral with regard to jurisdictional claims in published maps and institutional affiliations.

1. Introduction

To mitigate the worst health impacts of climate change, global annual greenhouse gas (GHG) emissions must be halved by 2030 and attain net-zero by 2050 [1] or incur the marginal expense of negative emissions. This expense could be in the range of USD 100–300/t CO_2-Eq. depending on the cost of biomass or direct CO_2 capture through carbon capture and storage [2]. The global energy-associated CO_2 emissions from the building sector alone accounted for 38% of the total emissions in 2019 [3]. The Paris Agreement, an international legal treaty adopted by 196 states, aims to limit global warming to 1.5 °C compared to pre-industrial levels [4]. The International Energy Agency (IEA) estimates that to achieve net-zero emissions by 2050, the direct and indirect CO_2 emissions from the building sector should decline by 50% and 60%, respectively, by 2030 [3]. Portland cement (PC) is the primary contributor of emissions in the building sector, accounting for approximately 14% of non-energy use CO_2 emissions, owing to its significant use and innate characteristics of its manufacturing process [5]. Energy conservation, carbon extraction, and the use of alternative materials are some of the strategies for reducing CO_2 pollution associated with the use of PC, which were lately introduced [6].

Recent research reports the use of materials such as calcined clay, marble dust, and granite dust as PC replacements for improving the sustainability of concrete production [7,8]. Utilizing industrial wastes, bio-wastes, and agro-wastes with alternative activations such as alkali-activation and carbonation has been performed in various applications [9–11]. Alkali-activated binders (AABs) have shown significant potential in replacing PC in concrete mixes [12–14]. AAB concrete, viz., alkali-activated concrete (AAC), produced by the activation of aluminosilicate-rich industrial wastes with an alkaline activator, exhibits superior mechanical and durability performance and a lower carbon footprint than PC concrete (PCC) [15]. The commonly used precursors in AAC are fly ash and ground granulated blast furnace slag (GGBFS, further referred to as slag). There is extensive literature available on fly ash-based AAC compared to slag-based AAC [16]. Furthermore, previous research shows that slag can be effectively used as a precursor when used in combination with other alternative binders such as fly ash or glass powder [11]. However, the latest tendency of the application of alternative binders has to also be investigated in depth in regard to the environmental impact.

In addition to the emissions associated with binders used in concrete, the use of natural aggregates (NA) (usually limestone or granite) results in resource scarcity, deterring its sustainable development. NA can be replaced with recycled aggregates, resulting in the twofold advantage of minimizing the extraction of non-renewable sources and environmental impact from disposed of resources. Construction and demolition waste (CDW) generated in India is estimated to be about 100 million tonnes yearly. The utilization of CDW in India constitutes only 10–30% of the generation. In the European Union (EU), for example, 850 million tonnes are generated per year and 70% are set to be recycled by 2020 according to the Waste Framework Directive [17]. Usually, CDW is primarily used for land-leveling and backfilling projects, while the remaining is disposed of in landfills [18]. CDW from building-derived materials (BDM) can be a potential partial replacement of NA in PC and AAC [19,20]. The effect of incorporating recycled coarse aggregates (RCA) on the mechanical properties and durability of PCC has been extensively reported in the literature [21–23]. The lower replacement levels of about 20% of NA with RCA did not alter the mechanical performance of PCC [22]. However, a further increase in the replacement percentage reduced the mechanical strength. This reduction in strength is attributed to increased porosity and a weak interfacial transition zone (ITZ) between aggregates and the matrix [24]. AAB concrete exhibited a higher potential for the replacement of NA with RCA. The change in compressive strength is reported in the range of 20% when 50% of NA is replaced with RCA [25]. Due to the increased porosity, the addition of RA has a negative impact on the durability of PCC. However, because of the higher alkaline environment of the alkali-activated systems, recycled aggregates could be more suitable for use [25]. It is evident from the existing literature that alternative binders and RCA could be used as a complete or partial substitute for their conventional counterparts.

However, the main rationale for their use is to improve sustainable development, which necessitates a rigorous environmental assessment to determine their efficacy. A few studies reported that replacing PCC with AAC can reduce the emissions associated with climate change by 9–80% compared to PCC [26,27]. This significant disparity in emissions' mitigation by AAC is due to differences in the types of precursors and activators, mix proportions, transit distances, and the type of production and sources of raw materials. There have been relatively few investigations on the environmental effects of RCA inclusion in AAC. The treatment procedures used to improve the quality of RCA can result in additional environmental burdens. Since the adhered matrix is the phase increasing porosity, treatment approaches are commonly used to extract it from the surface of RCA. The treatment approaches include either dissolving RCA in a solvent or imposing internal tension at their ITZ to isolate them from the adhered hardened binder. The treatment processes used in previous studies on AAC are ultrasonic cleaning [28], pre-saturation [29], carbon dioxide sequestration (CS) [30], nitric acid dissolution [29], thermal expansion [31], the freeze–thaw method [32], microwave heating [33], mechanical grinding [34], and the heating and rubbing method [35]. Most of the above-mentioned treatment processes are energy and/or resource-intensive. Using RCA treated through CS improved the compressive strength by approximately nine times due to more stable polymorphs. Furthermore, CS-treated RCA can reduce GHG emissions by 50% compared to NA cured under ambient conditions [30]; however, they increased the cost/m^3 of concrete owing to the higher cost of RCA. Furthermore, to promote in situ applications of AAC with RCA, a detailed environmental analysis is necessary. As a preliminary step, the present study aims to evaluate the environmental and economic impact of untreated (virgin) BDM used as coarse aggregates in AAC.

The other aspect of concrete structures that can incur high cost is their repair and maintenance. The proclivity of concrete for cracking is the primary factor compromising its structural integrity, serviceability, and durability. Conventional rehabilitation and maintenance are effective strategies for prolonging the service life of concrete structures. However, the repair and maintenance cost of concrete cracking can be 84% to 125% higher than the actual construction cost/m^3 [36]. The USA spends USD 266.5 billion on infrastructure maintenance and repair, accounting for 65% of all public infrastructure expenditures [37]. Infrastructure assessment, repair, and maintenance account for nearly 33% and 45% of the annual civil engineering budgets in The Netherlands and the United Kingdom, respectively [38,39]. According to recent estimates, India spends about 1.08% of its annual GDP on infrastructure repair and maintenance [40]. Furthermore, depending on the location of the crack, its extent, and the service needs for infrastructures such as highways, there might be situations where repair is impractical. Self-healing concrete can be very beneficial in these circumstances by automatically healing fractures without the need for external intervention [41].

Self-healing can be achieved through two different mechanisms, viz., autogenous and autonomous healing. Autogenous healing refers to reactions and/or processes originating from the cementitious system, such as calcium hydroxide carbonation, crack blockage caused by impurities in water and loose concrete particles, and ongoing hydration of cement [42], whereas autonomous healing comprises the utilization of additions/methods such as bacteria, shape memory alloy capsules, and electrodeposition technology. Autonomous healing via the addition of mineral-depositing bacteria is one of the most efficient and potentially sustainable methodologies [43].

The bacteria added along with the nutrients to the concrete mix is activated when it encounters the moisture through the newly formed cracks and begins the precipitation of calcium carbonate through metabolism. There are three major metabolic pathways to precipitate calcium carbonate, viz., the hydrolysis of urea, the oxidation of organic compounds, and denitrification depending on the type of bacteria [43–45]. Different ureolytic bacteria such as *Sporosarcina ureae*, *Bacillus sphaericus*, *Bacillus megaterium*, *Proteus vulgaris*, *Proteus mirabilis*, *Bacillus subtilis*, and *Sporosarcina pasteurii* have been used in self-healing concrete [46]. Denitrifying bacteria such as *Pseudomonas aeruginosa* and *Diaphorobacter nitroreducens* and aerobic heterotrophic bacteria (which produces calcium carbonate by the oxidation of organic compounds) such as *Bacillus pseudoformus* and *Bacillus cohnii* are also used in successfully developing self-healing behavior in concrete [45,46].

Bacterial concrete is reported to enhance compressive strength by approximately 40% under simulated cracks by healing them [47]. There is an optimum concentration of bacteria that shows a positive effect on the compressive strength of bacterial concrete with fly ash, and it is reported to be 10^5 cells/mL for *Sporoscarcina pasteurii* bacteria [48]. Studies show that the direct inclusion of bacteria, that is, without immobilization or encapsulation, improved compressive strength in the range of 15–32% [49,50], flexural strength by 11% [49], and split tensile strength by 14–16% [49,50]. The immobilization and encapsulation of bacteria are known to improve the self-healing efficiency of concrete [51]. Limestone powder [52], iron oxide nano-sized particles [53], crushed brick aggregate [54], graphite nanoplatelets [55], expanded perlite [56], and porous ceramsite particles [57] are used for the immobilization of bacteria in concrete. Though efficient, the shortcoming of protective materials is their varying efficiencies depending on the type. Encapsulation resulted in a healing rate of 70 to 100% for cracks in the range of 0.3 mm [58]. Typically used Ca-precursors for aerobic respiration in bacterial concrete are calcium acetate, calcium lactate, calcium nitrate, and calcium formate [59]. Calcium nitrate acts as an accelerator when added to concrete, reducing the setting times, accelerating hydration, and decreasing compressive strength [59,60]. It is reported that calcium lactate used in the range of 1–2% by mass of cement improved its compressive strength; however, there are contradictory studies on its effect on setting times [59,61]. Calcium formate and calcium acetate are also found to be suitable Ca-precursors based on compressive strength investigations [59]. It is evident that there is extensive literature reported on the mechanical performance of bacterial concrete. However, to the best of the authors' knowledge, there are no studies on the environmental assessment of bacterial concrete.

The present study performs a comprehensive life cycle assessment of three different sustainable concrete mixes: alkali-activated concrete (AAC) with natural coarse aggregates, (2) AAC with recycled coarse aggregates, and bacterial concrete (BC) with and without Ca-precursors. AAC with natural coarse aggregates, AAC with RCA, and BC are evaluated using a cradle-to-gate life cycle assessment (LCA) and compared to PCC of equal strength. The influence of mix proportions on the environmental impact of three distinct AAC mixes and two distinct BC mixes is investigated. The life cycle impact assessment is performed both at midpoint and endpoint damage categories using ReCiPe 2016 methodology. Simple cost analysis, including electricity, water tariffs and transportation charges in the Indian context, is presented.

2. Materials and Methods

2.1. Materials

The precursors considered in the present study are class F fly ash, slag, and Portland cement (PC). Class F fly ash utilized in this investigation is obtained from the National Thermal Power Corporation in Ramagundam, India. Slag is acquired from JSW Ltd. in Vijayanagar, India. In the current investigation, Type I Portland cement of 53 grade complying to the standards IS [62] and ASTM [63] requirements is used. The properties of the precursors are provided in Table 1.

Table 1. Specifications of precursors.

Specification	Fly Ash	Slag	PC
CaO (%)	3.80	37.63	65.23
SiO_2 (%)	48.81	34.81	18.64
Al_2O_3 (%)	31.40	17.92	5.72
MgO (%)	0.70	7.80	0.85
SO_3 (%)	0.91	0.20	2.34
Fe_2O_3 (%)	7.85	0.66	4.54
TiO_2 (%)	2.93	-	0.5
K_2O (%)	1.52	-	0.59
Na_2O (%)	1.04	-	-
MnO (%)	-	0.21	-
LOI (%)	3.00	1.41	1.69
Strength activity index (%)	96.46	114.46	-
d_{50} (μm)	51.90	13.93	-
Blaine fineness (m^2/kg) *	327.00	386.00	285.00
Specific gravity	2.06	2.71	-

* Supplied by manufacturer.

Locally available river sand and crushed rock fines (CRF) are used as fine aggregates in the present study. Crushed granite with a nominal maximum size of 10 mm is used as coarse aggregate. River sand and crushed granite aggregates comply with standard specifications of IS 383:2016 [64]. For recycled aggregate concrete, tested concrete specimens are crushed through a jaw crusher and sieved to obtain aggregates with a nominal maximum size of 10 mm. The obtained recycled coarse aggregates are used without any treatment in their virgin form. The properties of aggregates are presented in Table 2. Figure 1 illustrates the gradation curve of river sand and CRF.

Table 2. Physical properties of aggregates.

Aggregate	Specific Gravity	Water Absorption (%)	Fineness Modulus
Natural coarse aggregate	2.72	0.10	-
River sand	2.65	0.50	3.62
Crushed rock fines	2.65	0.36	2.89

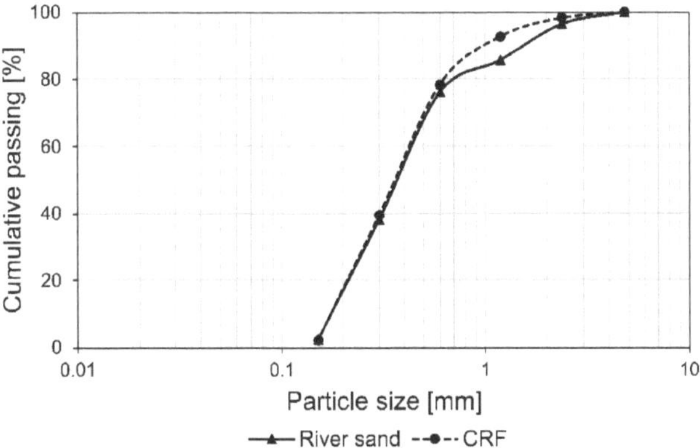

Figure 1. Grading curves of fine aggregates.

As activators, sodium silicate (29.4% SiO_2, 14.7% Na_2O, and 55.9% H_2O) and sodium hydroxide (food-grade, 99% purity) from Hychem Laboratories are utilized in the preparation of AAC mixes.

A denitrifying bacterium *Pseudomonas putida* GAP P-45 purchased from the Microbial Type Culture Collection and Gene Bank (MTCC) Chandigarh is used in BC mixes. Commercially available nutrient broth with 15 g peptone, 3 g yeast extract, 1 g glucose, and 6 g NaCl acquired from Hychem Laboratories is used as the culture medium to cultivate the bacteria. Calcium formate and calcium nitrate obtained from Krishna Chemicals are used as nutrients for bacteria.

Polycarboxylic ether (PCE)-based superplasticizer supplied by BASF is used. Regular tap water free from any deleterious materials is used in concrete mix preparation.

2.2. Mix Proportions

In the present study, three different types of concrete mixes are prepared: AAC, BC, and PCC. Furthermore, by varying mix proportions, 3 AAC mixes and 2 BC mixes are investigated. The details of the mixes are provided in Table 3.

The mixes are selected in such a way that they have equivalent or comparable compressive strength. AAC mixes are denominated based on their activator modulus (ratio of SiO_2/Na_2O in the activating solution). AAC-0.7 has only fly ash as a precursor with an activator modulus of 0.7. In contrast, AAC-1 and AAC-1.4 have fly ash and slag as precursors with activator modulus of 1 and 1.4, respectively. AAC-R has a similar mix proportion to AAC-1.4, with 50% of natural coarse aggregates replaced with RCA. The water content is reduced to ensure sufficient strength development, and hence the workability is improved by PCE-based SP.

Table 3. Mix proportions of AAC, BC and PCC, in [kg/m^3].

Mix	AAC-0.7	AAC-1	AAC-1.4	AAC-R	BC-N	BC	PCC
Fly ash	425.00	280.00	280.00	280.00	115.00	115.00	-
Slag	-	120.00	120.00	120.00	-	-	-
PC	-	-	-	-	335.00	335.00	450.00
Sodium silicate	70.13	115.07	129.43	129.00	-	-	-
Sodium hydroxide	25.08	24.93	10.57	10.60	-	-	-
CRF	-	-	-	-	750.00	750.00	-
River sand	676.49	651.00	651.00	653.00	-	-	623.00
Calcium nitrate	-	-	-	-	4.78	-	-
Calcium formate	-	-	-	-	11.96	-	-
Coarse aggregate	1014.74	1209.00	1209.00	645.00	1020.00	1020.00	1084.00
RCA	-	-	-	645.00	-	-	-
Nutrient broth	-	-	-	-	0.67	0.67	-
Water	68.06	85.37	77.38	67.60	157.00	157.00	150.00
SP	-	-	-	3.14	2.00	2.00	-
Density	2280	2485	2477	2553	2396	2380	2307

PC—Portland cement; CRF—Crushed rock fines; RCA—Recycled coarse aggregate; SP—Superplasticizer.

2.3. Methods

2.3.1. Preparation of Alkali-Activated Concrete Specimens

The activator with a combination of sodium silicate solution and sodium hydroxide pellets is prepared a day before the casting in order to allow sufficient time for heat dissipation. The mixing procedure is commenced with batching, followed by blending of dry materials, viz., coarse aggregates, fine aggregates, and precursors (fly ash and/or slag) in the order mentioned. The uniform blending of dry ingredients is followed by the gradual addition of activator with simultaneous mixing. A measured quantity of additional water is then added and mixed to obtain a homogenous mixture. Specimens are cast in three approximately equal layers, and each layer is compacted using a needle vibrator. The surface of the specimens is finished using a trowel. Molds are then sealed with a plastic sheet to minimize moisture loss. Throughout this study, all the AAC specimens are allowed to cure for at least 24 h in the molds.

The time for demolding the specimens is varied depending on the fly ash content in the mix, owing to its slow reactivity. All the specimens are demolded after gaining sufficient strength to prevent any damage. The AAC-0.7 specimens are demolded at the age of 7 days, as they exhibited the slowest final setting times. AAC-1, AAC-1.4, and AAC-R specimens are demolded at the age of 2 days. On demolding, the specimens are cured until 7-day age underwater and then under ambient laboratory conditions until the commencement of tests. The average ambient temperature was 25 °C, and the relative humidity was in the range of 60–75% in Hyderabad during the experimental activities. This curing regime is selected based on the recommendations of previous research on blended AAC [65,66].

2.3.2. Preparation of Bacterial Concrete Specimens
Bacterial Growth Conditions and Gram Staining

The culture medium and all the related equipment are sterilized in an autoclave at 121 °C for 30 min. The sterilized culture medium is then inoculated with the activated bacterial strain and then placed in an incubator operating at 170 rpm and 37 °C for 24 h. The bacterium is further sub-cultured before being inoculated into the sterilized nutrient broth solution and cultured for 24 h under optimal growth conditions. During this time, the bacterial cell concentration is ensured through the optical density/absorbance at 600 nm (OD$_{600}$) of the culture broth at regular intervals.

Gram staining is a technique used to distinguish between Gram-positive and Gram-negative bacteria based on the physical and chemical characteristics of their cell walls. The current study utilizes a Gram-negative bacterial strain, and this test is performed to ensure that the culture did not include any Gram-positive or other microorganisms. Gram staining is dependent on the bacteria's ability to preserve their original color, which is dependent on the cell wall structure. Gram-positive bacteria are characterized by periwinkle color, whereas Gram-negative are characterized through amaranth color [67]. The test procedure commenced by placing the bacteria from the prepared culture on a clean glass slide with a sterile loop. Following this, the slide is heat-fixed by running it over the flame multiple times while ensuring maintenance of the appropriate temperature range. Care is taken to avoid excessive heating as it can result in staining abnormalities and disrupt the morphology of cells. This is followed by four steps of staining. In the first step, the slide is flooded with crystal violet ($C_{25}N_3H_{30}Cl$) for 60 s and then washed with tap water. Step 2 entails exposing the slide to Gram's iodine for 90 s to bind and encapsulate the crystal violet in the cell, followed by washing the slide with tap water. The third step is to destain the smear with 95% ethanol until the thinnest sections of the smear become colorless before rinsing it with water. The fourth step includes flooding the slide with safranin (pink color) for 60 s and rinsing with tap water. Then, the slide is allowed to air-dry. The recorded micrograph of Gram-negative bacteria is presented in Figure 2.

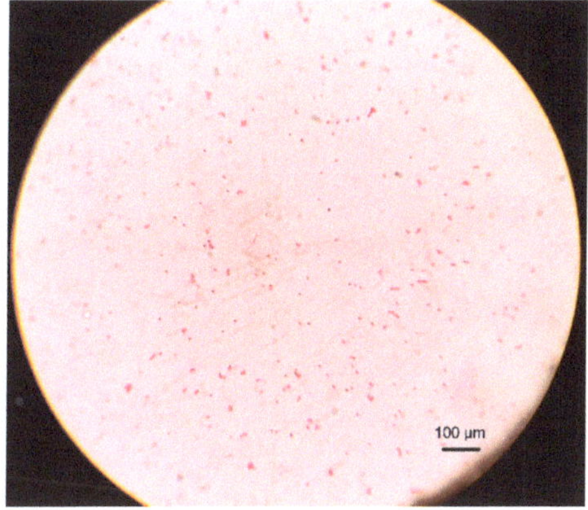

Figure 2. Gram staining of *Pseudomonas putida*.

Casting

The coarse aggregates, CRF, PC, and fly ash are dry mixed, followed by addition of nutrients (calcium nitrate and calcium formate) The required quantities of nutrients are added in case of BC-N mix, while a part of the water is replaced by bacterial solution with the proviso that the water-to-cement ratio is maintained. The specimens are cast following a similar procedure as AAC mixes and cured in water for 28 days until testing.

The preparation of PCC specimens is similar to BC mixes, except for the mixing water free from bacteria and nutrients.

2.3.3. Compressive Strength

The test procedure complies with ASTM standard [68] regulations and is performed on a HEICO compression testing machine (CTM) with a capacity of 2000 kN. The specimens are ensured to surface dry before testing. The specimens are tested in a load control set-up at a loading rate of 0.25 ± 0.05 MPa/s.

2.3.4. Life Cycle Assessment (LCA)
Goal and Scope

The primary aim of this study is to evaluate and compare the environmental impact of preparing different sustainable concrete mixes. The environmental impacts are evaluated for a functional unit of 1 m³ of concrete with equivalent 28-day compressive strengths in the range of 40–57 MPa. The mix selection also entails an equivalent curing regime, viz., curing under ambient conditions (20–27 °C) and relative humidity of 60–100%. Seven concrete mixes are evaluated for their environmental impact using cradle-to-gate life cycle assessment (LCA). These seven mixes are: (i) AAC with fly ash as a precursor, activator modulus of 0.7 and a water–solids ratio of 0.2; (ii) AAC with fly ash and slag as precursors, activator modulus of 1 and a water-to-solids ratio of 0.3; (iii) AAC with fly ash and slag as precursors, activator modulus of 1.4 and a water-to-solids ratio of 0.3; (iv) AAC-1.4 with equal proportions of natural and recycled coarse aggregates; (v) BC with nutrients, calcium nitrate and calcium formate; (vi) BC without nutrients, calcium nitrate and calcium formate; (vii) PCC.

In this study, the cradle-to-gate LCA technique is used, including estimations of all emissions and energy consumption from raw material acquisition through concrete preparation. As the major goal of this study is to compare different sustainable concrete mixes, the consumption and disposal phases are not studied, and the environmental effect from these phases is presumed to be similar. The analysis is performed using openLCA v.1.10.3. The system boundary for the mixes investigated in this study from cradle-to-gate is presented in Figure 3.

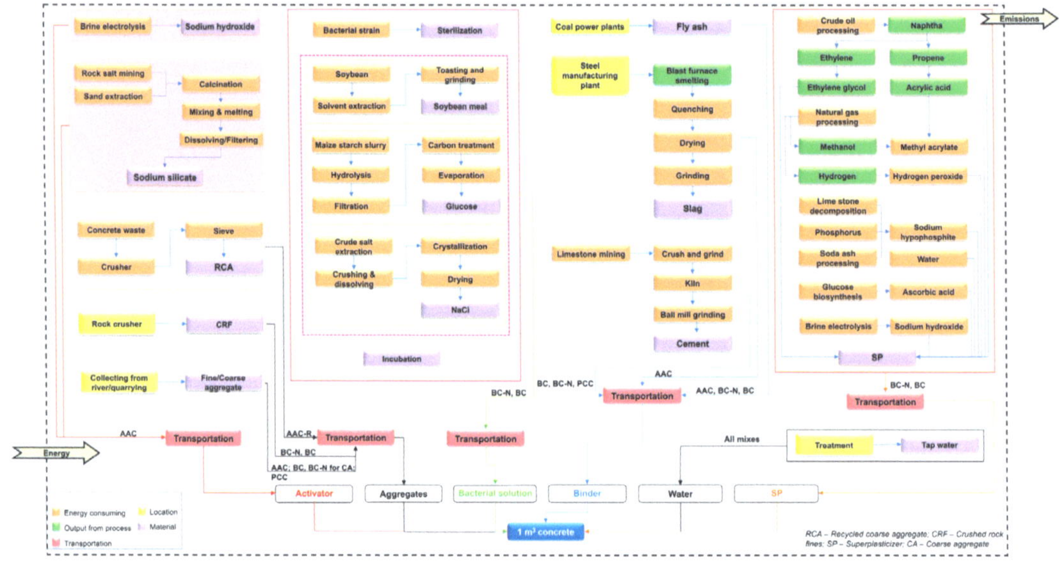

Figure 3. System boundary for cradle-to-gate LCA.

Life Cycle Inventory Analysis

The required data to perform LCA are categorized into two groups: the data associated with emissions and energy consumptions of materials and transportation specified in the system boundary, and those related to electricity consumption associated with RCA and sterilization and incubation associated with preparing the bacterial solution. The emission and energy consumption data for the materials available in the Ecoinvent database (v. 3.7.1) are used [69]. The electricity consumption is calculated based on the specification of the equipment used and the duration or quantity of material processed depending on its energy consumption. The calculated electricity consumption for the services used in the present study is listed in Table 4. The energy consumption per unit quantity (jaw crusher) or time (autoclave and incubator) provided by the manufacturer in the datasheet is used.

Table 4. Energy consumption data of services.

Service	Electricity Consumption (kWh)
Jaw crusher	1.61
Autoclave	1.00
Incubator	4.08

The data for yeast extract are not available in the Ecoinvent database; hence, an alternative for yeast extract production is used in the present study. Previous studies using yeast extract as one of the ingredients show that soybean meal is a potential alternative compared to other alternatives such as fish-powder waste, white gluten waste, perilla meal, sesame meal, and wheat bran [70,71]. Furthermore, the energy consumption per unit (kg) production of soybean meal and yeast extract are approximately equivalent [72,73]. Therefore, soybean meal is used to simulate the energy consumption by yeast extract production in the present study. The environmental burden associated with peptone production is not considered in this study as per the recommendations in the existing literature [74]. Table 5 presents the freight distances and unit cost of raw materials used in the current study in INR. The transportation distances are calculated from the source of their manufacturing to the laboratory where specimens are prepared.

Table 5. Freight distances and unit cost of raw materials and services.

Raw Material	Distance (km)	Cost/kg (Including Freight, INR)
Fly ash	202.0	4.8
Slag	438.0	6.6
PC	231.0	8.3
Sodium silicate	22.8	80.2
Sodium hydroxide	28.4	106.2
CRF	24.2	0.7
River sand	217.0	1.5
Calcium nitrate	22.8	210.0
Calcium formate	22.8	96.0
Coarse aggregate	24.2	1.8
RCA	-	-
Nutrient broth	29.2	2800.0
Water	-	0.035
SP	39.7	232.5
Electricity	-	6.7

The cost of electricity and water is obtained by the guidelines issued by Telangana State Electricity Regulatory Commission, an Indian government regulatory body. The tariff charge per unit of energy consumption (kWh) for industries or commercial use is INR 6.70 [75].

Life Cycle Impact Assessment

The environmental impact of a product is quantified through life cycle impact assessment (LCIA) [76]. ReCiPe 2016, following a hierarchist perspective, is used in the present study. Eleven midpoint indicators are evaluated in this study, which includes global warming potential (GWP), fossil depletion potential (FDP), freshwater ecotoxicity (FETP), freshwater eutrophication (FEP), human toxicity potential (HTP), marine ecotoxicity potential (METP), marine eutrophication potential (MEP), ozone depletion potential (ODP), photochemical oxidant formation potential (POFP), terrestrial acidification potential (TAP), and terrestrial ecotoxicity potential. The endpoint assessment includes three areas of protection, viz., quality of ecosystem, human health, and resource depletion. The openLCA v 1.10.3, an open-source software developed by Hildenbrand et al. [77], is used to evaluate the impacts for each product within the system boundary.

3. Results

3.1. Compressive Strength

The compressive strength of different mixes used in the present study is illustrated in Figure 4. The overall standard deviation in the compressive strength of different mixes is 5.50. Hence, the selected mixes are categorized as having equivalent or comparable compressive strength.

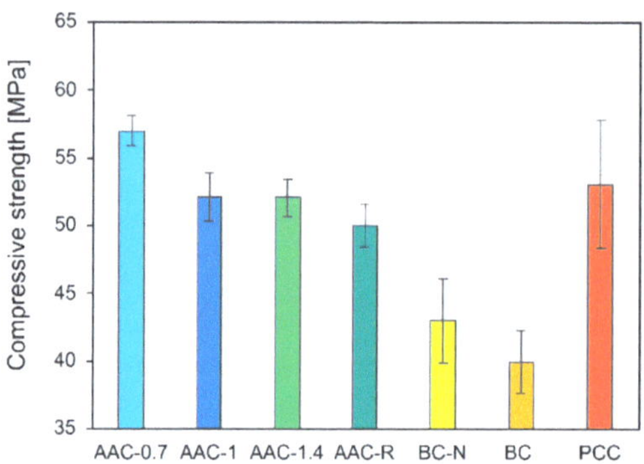

Figure 4. Compressive strength test results.

3.2. Life Cycle Assessment

3.2.1. Midpoint Assessment

Figure 5 presents the results from midpoint impact assessment using ReCiPe 2016 v1.13. It is observed that of all the materials and services considered for four AAC mixes in the present study, transportation followed by sodium silicate contributed the most to climate change. The contributions from transportation and sodium silicate for different AAC mixes vary in the range of 46–53% and 33–21%, respectively. A lower contribution from sodium silicate is identified in AAC-0.7, owing to a lower quantity of sodium silicate compared to the other three AAC (AAC-1, AAC-1.4, and AAC-R) mixes. Transportation entails the combustion of fossil fuels, which results in the release of GHG emissions to the environment. The primary pollutants released due to the combustion of fuels are GHG emissions (CO_2, CH_4, N_2O), ozone precursors (CO, NO_x, non-methane volatile organic compounds (NMVOCs)), particulate matter, and toxic materials (furans and dioxins) [78].

In addition to climate change, transportation is the primary contributor to fossil depletion (51–57%), marine eutrophication (46–52%), photochemical oxidant formation (61–67%), terrestrial acidification (47–53%), and terrestrial ecotoxicity (66–73%) in AAC mixes. Fossil depletion due to transportation is attributed to the use of fuels. In the case of marine eutrophication, excess availability of nitrogen and/or phosphorus can increase phytoplankton production, disrupt the energy balance of the marine ecosystem, and result in the acidification of the ocean [79]. Fossil fuel combustion during transportation releases NO_x in addition to other pollutants. If released in higher quantities, NO_x can contribute to the formation of smog and acid rain, subsequently redepositing to the land and/or aquatic ecosystems [80]. The higher contribution of transportation to photochemical oxidant formation is due to the release of NMVOCs. These pollutants, released due to the partial combustion of fuel, react with nitrogen oxide in the presence of light, resulting in the formation of photochemical ozone [81]. Acidification is caused by air pollutants such as SO_2, NH_3, and NO_x acidifying rivers/streams and soil. These pollutants are also released during the combustion of fossil fuels during transportation. Acidification aggravates trace metal mobilization and leaching in soil, causing harm to aquatic and terrestrial animals and plants by disrupting their energy balance [82]. Terrestrial ecotoxicity is ascribed to the release of heavy metals by road transportation. It is reported in the literature that one of the significant sources of heavy metal emissions to the environment is road transportation. The major sources of heavy metal emissions from road transportation include three sources of tailpipe emissions: (i) the attrition of engines or after-treatment systems, (ii) lubricants, and (iii) fuel, and non-tailpipe emissions from the attrition of tires or brakes, and road abrasion. In the present investigation, it is identified that petrol (*petrol, low sulfur: ecoinvent 3.7.1*) is the primary source of these trace metal emissions to the environment. The primary trace metals due to road transportation include arsenic (As), lead (Pb), selenium (Se), chromium, copper (Cu), cadmium (Cd), zinc (Zn), and mercury (Hg) [83].

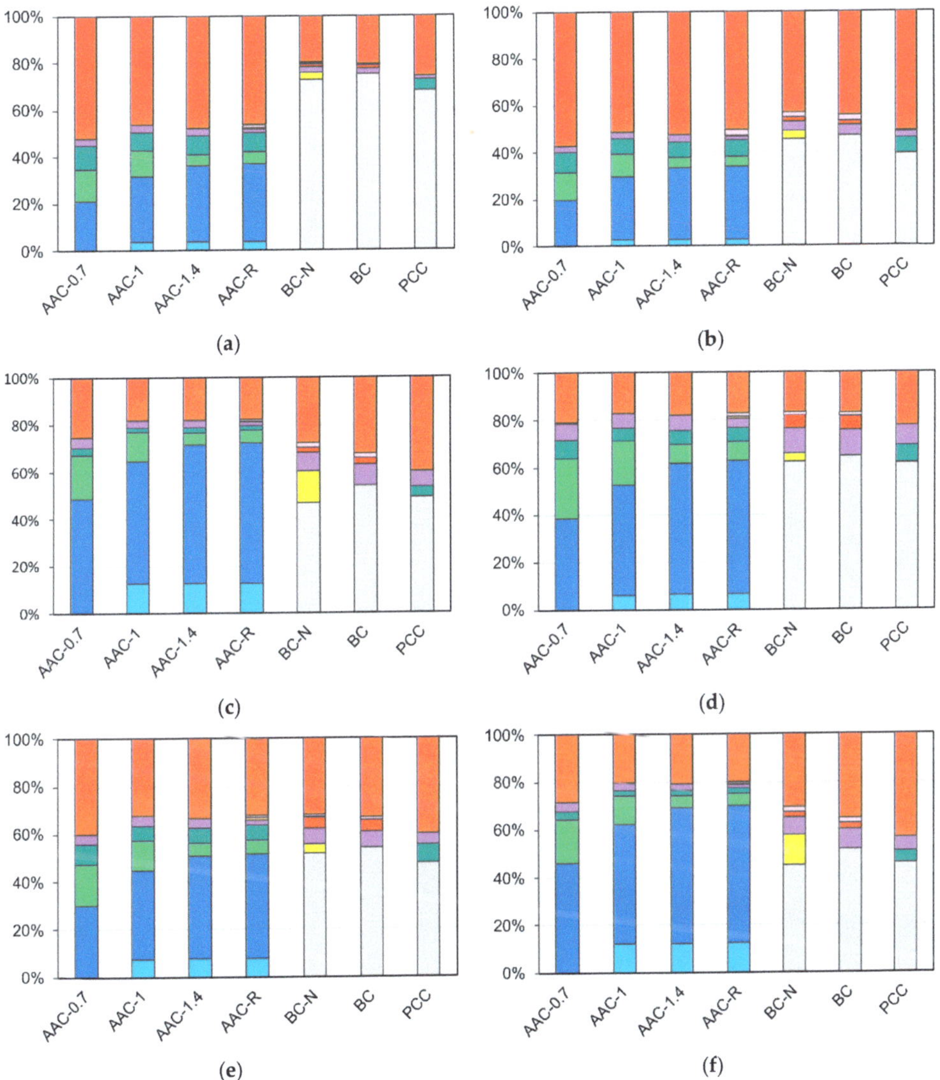

Figure 5. *Cont.*

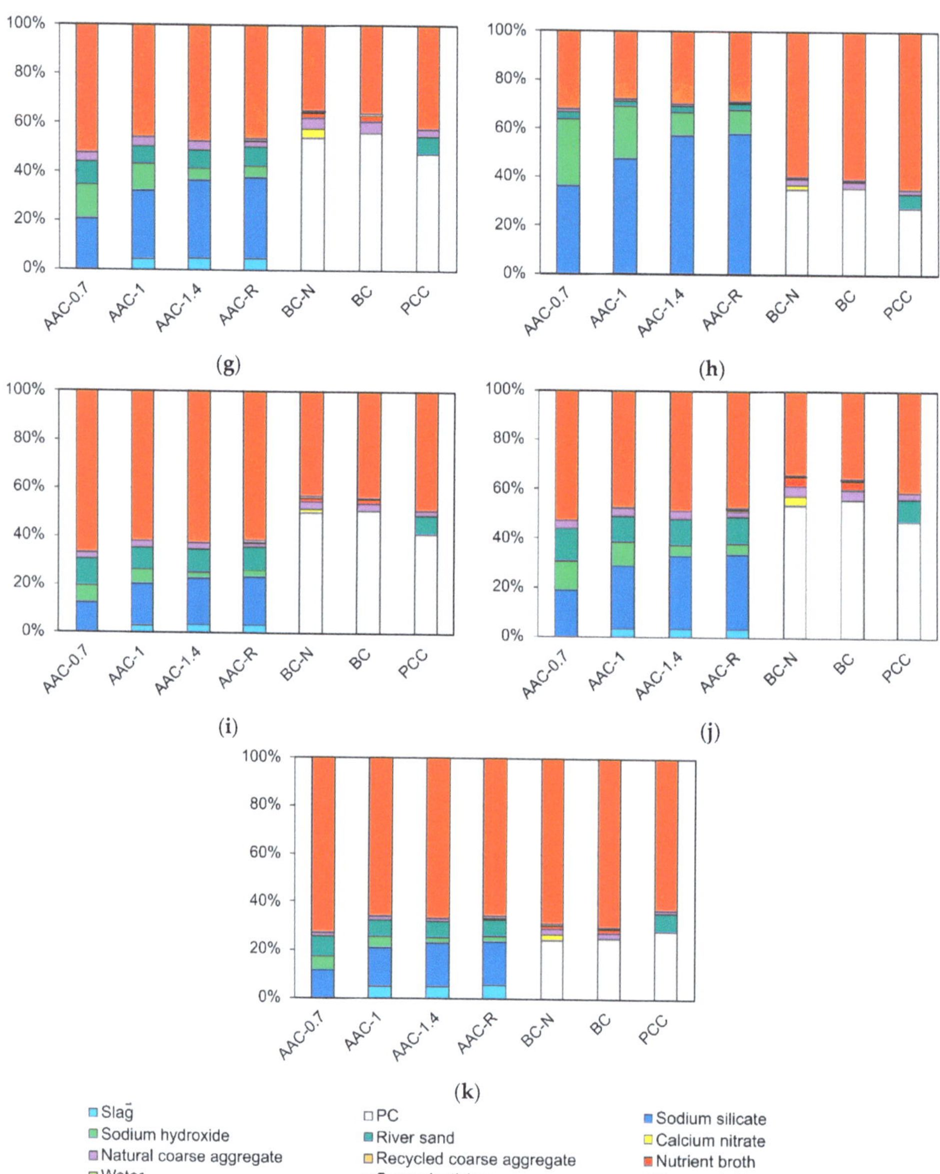

Figure 5. Environmental impacts (%) of different concrete mixes on midpoint damage categories: (**a**) Climate change, (**b**) Fossil depletion, (**c**) Freshwater ecotoxicity, (**d**) Freshwater eutrophication, (**e**) Human toxicity, (**f**) Marine ecotoxicity, (**g**) Marine eutrophication, (**h**) Ozone depletion, (**i**) Photochemical oxidant formation, (**j**) Terrestrial acidification, and (**k**) Terrestrial ecotoxicity.

The other significant contributor to the environmental impact of AAC mixes is sodium silicate, as identified from Figure 5. The contribution of sodium silicate varies in the range of 49–59% for freshwater ecotoxicity, 39–56% for freshwater eutrophication, 30–44% for human toxicity, 46–57% for marine ecotoxicity, and 36–57% for ozone depletion. The hydrothermal dissolution of silica sand in sodium hydroxide solution used to produce sodium silicate is considered in the present study. The procedure is carried out in an autoclave, followed by filtration. The production process of sodium silicate entails substantial energy consumption, which is reported as approximately 1.5 kWh/kg for the hydrothermal process [84]. The principal cause of the aforementioned environmental impacts by sodium silicate is electricity consumption during the manufacturing process of sodium silicate and sodium hydroxide (utilized as one of the raw materials for sodium silicate manufacture). In the present study, electricity generated by coal combustion is selected. Coal-fired electricity generation plants release trace metals, SO_2, NO_2, and particulate matter ($PM_{2.5}$) [85]. The principal source for ecotoxicity (freshwater and marine) is trace metals comprising Hg, As, Se, Pb, Cd, and Cr, of which Hg is of critical concern. Coal-fired power plants account for more than one-third of all mercury emissions attributed to human activities [86]. The release of SO_2, NO_2, and $PM_{2.5}$, in addition to trace metals, affects human health. The ozone depletion contributed by sodium silicate is also due to electricity consumption. Mining, plant operation, and the emission of halons 1211, 1301 used as fire suppressants and coolants in gas pipeline distribution all contribute to ozone depletion from coal-fired power plants [87].

PC is the principal contribution from the mixes BC-N, BC, and PC to climate change (68–75%), fossil depletion (39–47%), freshwater ecotoxicity (46–54%), freshwater eutrophication (62–65%), human toxicity (47–54%), marine ecotoxicity (46–52%), marine eutrophication (48–56%), photochemical oxidant formation (41–51%), and terrestrial acidification (47–56%). The calcination phase of PC production results in about 0.525 kg/kg of CO_2 emissions. However, depending on the clinker/PC ratio, the emissions ratio can vary between 0.5 and 0.95, and the remainder is attributed to fuel and electricity consumption [88]. The fossil depletion by PC production can be attributed to the utilization of natural resources as raw materials and fuel consumption. The emissions contributing to ecotoxicity (freshwater and marine) and human toxicity are trace metals, volatile organic compounds (VOCs), dioxins, and particulate matter. Raw materials and fuel are sources for trace metals, and an incomplete combustion of fuel results in the emission of VOCs. The sources of particulate matter include raw material crushing, grinding and drying facilities, clinker combustion process, PC grinding, and the dispatch of PC [89]. Marine eutrophication is primarily caused by NO_x emitted during the production of PC.

However, transportation is responsible for most of the ozone depletion (61–65%) and terrestrial ecotoxicity (63–70%) resulting from the preparation of BC-N, BC, and PC mixes. The reasoning behind the contribution of transportation to ozone depletion and terrestrial ecotoxicity is similar to that of AAC mixes. Fuel combustion also releases ozone precursors, NO_x, CO, and hydrocarbons. The following series of reaction equations elucidate the conversion of ozone precursors to ozone.

$$NO_2 + h\nu \rightarrow NO + O$$

$$O + O_2 \rightarrow O_3$$

$$NO + HO_2 \rightarrow NO_2 + OH$$

The availability of H_2O is from the oxidation of CO or hydrocarbons.

$$CO + OH \rightarrow CO_2 + H$$

$$H + O_2 \rightarrow HO_2$$

Figure 6 presents the normalized midpoint impact indicators for all the mixes. The ReCiPe normalization factors are provided in Table 6.

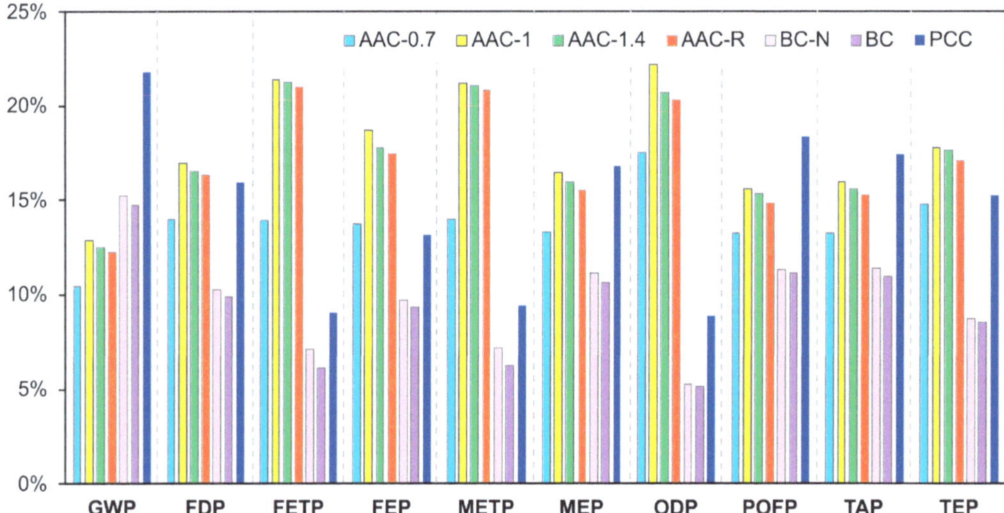

Figure 6. Normalized impacts of different concrete mixes on midpoint damage categories.

Table 6. Normalization scores for ReCiPe 2016 methodology [90].

Indicator	ReCiPe Midpoint (H)
GWP	5.51×10^{13}
FDP	3.93×10^{12}
FETP	1.74×10^{11}
FEP	4.48×10^{9}
HTP-cancer	7.10×10^{10}
HTP-no cancer	2.16×10^{14}
METP	3.00×10^{11}
MEP	3.18×10^{10}
ODP	4.14×10^{8}
POFP	1.42×10^{11}
TAP	2.83×10^{11}
TETP	1.05×10^{14}

The overall contribution of individual blends to the impact categories can be effectively interpreted using Figure 6. The GWP of PC is approximately twice that of AAC-0.7 and 1.4 times that of BC-N. AAC-0.7 has lower environmental impacts among the AAC mixes across all the impact categories. This is attributed to a lower content of alkaline activators, viz., sodium silicate and sodium hydroxide used in AAC-0.7 compared to other AAC mixes. AAC-1, AAC-1.4, and AAC-R show the highest environmental impacts for all the damage categories, excluding POFP and TAP. This is ascribed to the manufacturing process of sodium silicate and sodium hydroxide. Furthermore, there is negligible variance in the impacts exhibited by these three mixes due to the usage of an equivalent number of activators. BC-N and BC have lower environmental impacts for all the damage categories, excluding GWP. The higher GWP of BC-N and BC is attributed to the use of PC as the binder. The lower impacts for all the other damage categories are attributed to two factors: the usage of a lower PC content (335 kg/m^3) compared to PCC (450 kg/m^3) and the complete replacement of silica sand with CRF as fine aggregates. The CRF are not allotted any environmental burden in the present study and are considered as waste generated from the stone grinding. Furthermore, the usage of CRF eliminates the environmental impacts associated with the mining of sand from natural reserves.

PCC has the highest POFP and TAP in addition to GWP. The calcination phase in the production of PC is the primary supplier of CO_2 emissions and consequently to climate change. A reported quantity of 0.525 kg CO_2/kg of clinker is produced during the decarbonization of limestone in the kiln. The fuel and electricity consumption result in 0.335 kg CO_2 and 0.05 kg CO_2 emissions per kg of cement [91]. The VOCs emitted during incomplete combustion contribute to photochemical oxidant formation, whereas NO_x, SO_x and their associated compounds contribute to terrestrial acidification. Approximately 0.0023–0.138 kg VOCs/tonne of clinker are emitted due to the incomplete combustion of fuel during PC production [89]. NO_x and other nitrogen compounds are released when a fuel nitrogen combines with oxygen in the flame, or due to a combination of ambient nitrogen and oxygen in the combustion air. The nitrogen-related emissions are in the range of 0.33–4.67 kg/tonne of clinker. The source of SO_x and other sulfur compounds is fuel and raw materials with a high volatile sulfur content, and their emissions are in the range of 11.12 kg/tonne of clinker.

3.2.2. Endpoint Assessment

Figure 7 presents the endpoint assessment of different AAC, BC and PCC mixes used in the present study.

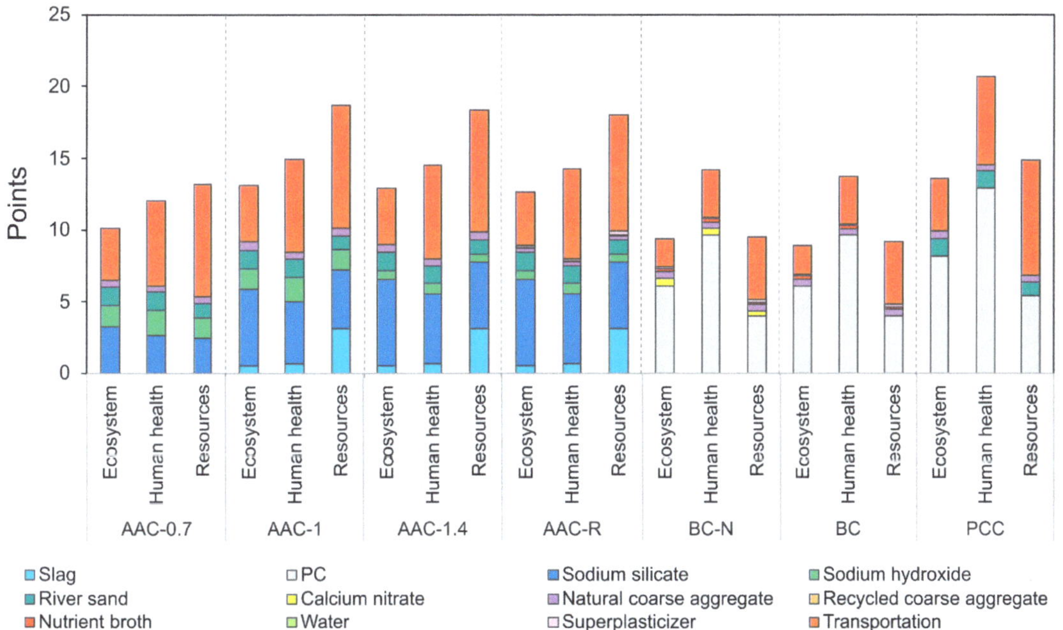

Figure 7. Endpoint environmental impact assessment of different concrete mixes.

The production of all the AAC and BC mixes has a lower negative impact on the quality of the ecosystem than PCC. The ecosystem quality value of AAC mixes ranges between 10.12 (AAC-0.7) and 13.09 (AAC-1). In contrast, PCC has a value of 13.59, which is 25.6% and 3.6% higher than AAC-0.7 and AAC-1, respectively. The emissions from the production process of PC have a detrimental effect on ecosystem quality, whereas in AAC mixes, sodium silicate followed by transportation (service) contribute adversely to the ecosystem quality. Of all the mixes, BC exhibited a lower impact on the quality of the ecosystem, with an endpoint value of 8.91, which is 34.4% lower than PC. The raw materials and/or services contributing to the GWP, FEP, FETP, TETP, TAP, and METP are associated with ecosystem quality.

PCC has the highest impact on human health with an endpoint value of 20.69, followed by AAC-1 with an endpoint value of 14.96. As evident from Figure 7, about 62.5% of the total impact on human health is caused by the use of PC and 29.7% by transportation in the PCC mix. In AAC-1, 29.2% of the overall impact on human health is contributed by sodium silicate and about 43.8% by transportation. BC has the lowest impact on human health of all the mixes, with an endpoint value of 13.7, which is 33.8% lower than the PCC mix. The raw materials and/or services considered in the present study and discussed in the midpoint assessment contributing to particulate matter, HTP, and GWP are associated with adverse effects on human health.

AAC mixes show the highest resource depletion compared to other mixes. AAC-1 has the highest resource depletion value of 18.66, of which 45.8% is contributed by transportation and 21.8% by sodium silicate. The endpoint values of AAC-1 and AAC-1.4 differ only by 1.5%; however, this difference is attributed to sodium hydroxide. The electrolytic phase of the chloralkali process using diaphragm technology to produce sodium hydroxide is reported to consume 2.97 kWh/kg of electricity [92]. This electricity consumption by the manufacturing process of sodium hydroxide is approximately thrice as high as PC and twice as high as sodium silicate production. When individually considered, the environmental impact due to the production of sodium hydroxide is higher than sodium silicate. However, due to its usage in lower quantities in the present study, the effect of sodium hydroxide on different damage categories is minimal. BC has the lowest resource depletion score of 9.14, which is approximately half of ACC-1.

When an individual type of concrete mix is considered, AAC-0.7 has the lowest environmental impact on all three areas of protection: ecosystem quality, human health, and resource depletion of all the AAC mixes. This is attributed to the lower quantity of sodium silicate used in AAC-0.7 compared to other AAC mixes. In the case of BC mixes, due to eliminating environmental impacts associated with the use of calcium nitrate and calcium formate, BC exhibited lower environmental impacts on all three areas of protection. When all the mixes in the present study are considered, BC exhibits the lowest environmental impact. Despite the similar binder used in BC and PCC mixes, the lower environmental impacts by BC are attributed to the lower PC content, the complete replacement of sand with CRF, and the transportation effect. Though in the case of BC the number of raw materials is higher compared to PCC, the overall quantity multiplied by a distance given as input to transportation is approximately half (45.8% lower) of the PCC mix.

3.3. Economic Assessment

A sustainable concrete mix should also be economically viable to promote its in situ applications and commercial use. Therefore, a cost analysis is performed for the mixes investigated in the present study and the results are presented in Figure 8.

The cost of AAC mixes ranging between INR 13172 and INR 17168 is 98.8–159.1% higher than PCC. BC mixes have an overall cost of INR 8074–10226, which is approximately 21.8–54.3% higher than PCC. Sodium silicate used in AAC contributes to 42.7–62.7%, lowest in AAC-0.7 and highest for AAC-1.4. Though the unit cost of sodium hydroxide is high, the higher contribution from sodium silicate is attributed to its higher proportion in these mixes. Analogously, owing to higher quantities of sodium hydroxide in AAC-0.7 and AAC-1, its contribution to the total cost is 20.2% and 15.4%, respectively. The contribution from PC ranging between 27.2% and 34.4% is highest in BC-N and BC mixes. The coarse aggregates and nutrient broth result in 17.9–22.7% and 18.3–23.2%, respectively. Since the overall cost of BC mix is lower than BC-N (by 21.0%) owing to the omission of nutrients, the contribution from coarse aggregates and nutrient broth is higher in the BC mix. Calcium nitrate and calcium formate together account for 21.0% of the overall cost of the combination, with approximately equal individual contributions. The cost of water and electricity is negligible compared to raw materials and transportation.

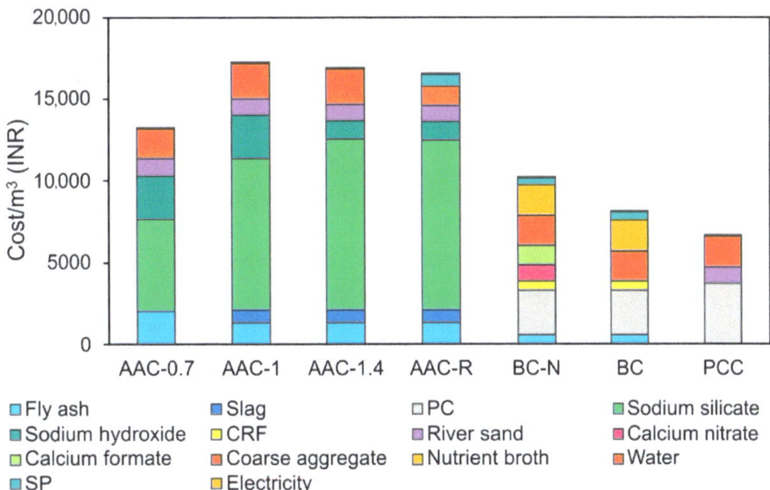

Figure 8. Cost estimate of different concrete mixes.

4. Conclusions

This study performs a comprehensive life cycle assessment of three different sustainable concrete mixes: (1) alkali-activated concrete (AAC) with natural coarse aggregates, (2) AAC with recycled coarse aggregates, and (3) bacterial concrete (BC). All the selected mixes are compared for their environmental impact versus Portland cement concrete (PCC) of equivalent strength. The environmental implications of three different AAC mixes with natural coarse aggregates and two different BC mixes (with and without nutrients) are evaluated as a function of mix proportions. A simplified cost analysis is performed in the Indian context. The findings are summarized as follows:

- For different AAC mixtures, contributions from transportation and sodium silicate are the highest for different midpoint damage categories.
- Portland cement is the principal contributor from the BC and PCC mixes to various midpoint damage categories.
- PCC has a GWP that is almost twice that of AAC with an activator modulus of 0.7 (AAC-0.7) and 1.4 times that of BC with nutrients.
- Owing to lower quantities of alkaline activators and only fly ash as a precursor, AAC-0.7 has the lowest environmental impact of all AAC mixes. The electricity consumption by recycled coarse aggregates (RCA) increased the environmental impact caused by AAC with RCA.
- PCC and AAC with an activator modulus of 1 (AAC-1) have the most detrimental effect on the quality of ecosystems and human health. AAC-1 has the highest resource depletion value (18.66) of all the mixes.
- BC (bacterial concrete without nutrients) has the lowest environmental effect of the evaluated mixes for all midpoint damage categories except GWP and endpoint damage categories. BC has the lowest resource depletion value (9.14), which is almost half of ACC-1.
- The cost of AAC mixes is 98.8–159.1%, and the cost of BC mixes is 21.8–54.3% higher than PCC.
- Sodium silicate in AAC mixes, PC, coarse aggregates, and nutrient broth in BC mixes have the highest contribution to the total cost.

The present study provides a detailed LCA of bacterial concrete, which is not reported in the existing literature. A comprehensive comparison of the environmental impact of different concrete mixes at midpoint and endpoint damage levels and a cost analysis would assist the industry and policymakers in identifying a viable, sustainable alternative.

Infrastructures **2021**, *6*, 104

Author Contributions: Conceptualization, K.K.R. and A.K.; methodology, K.K.R. and R.C.; validation, K.K.R., R.C. and R.K.B.; formal analysis, K.K.R., R.C., and A.K.; investigation, K.K.R., R.C. and R.K.B.; data curation, K.K.R., R.C., R.K.B., and A.K.; writing—original draft preparation, K.K.R. and A.K.; writing—review and editing, P.K.D.M.; supervision, A.K. All authors have read and agreed to the published version of the manuscript.

Funding: This research received no external funding.

Institutional Review Board Statement: This study does not involve any studies on humans or animals.

Informed Consent Statement: Not applicable.

Data Availability Statement: All data, models, and code generated or used during the study appear in the submitted article. Some or all data, models, or code that support the findings of this study is available from the corresponding author upon reasonable request.

Conflicts of Interest: The authors declare no conflict of interest.

References

1. Rogelj, J.; Shindell, D.; Jiang, K.; Fifita, S.; Forster, P.; Ginzburg, V.; Handa, C.; Kheshgi, H.; Kobayashi, S.; Kriegler, E.; et al. *Global Warming of 1.5 °C—Chapter 2 Mitigation Pathways Compatible with 1.5 °C in the Context of Sustainable Development*; Intergovernmental Panel on Climate Change: Geneva, Switzerland, 2018; pp. 93–174.
2. Keith, D.W.; Holmes, G.; Angelo, D.S.; Heidel, K. A process for capturing CO_2 from the atmosphere. *Joule* **2018**, *2*, 1573–1594. [CrossRef]
3. UNEP. Building Sector Emissions Hit Record High, but Low-Carbon Pandemic Recovery Can Help Transform Sector. Available online: https://www.unep.org/news-and-stories/press-release/building-sector-emissions-hit-record-high-low-carbon-pandemic (accessed on 15 June 2021).
4. UNFCCC. *Adoption of the Paris Agreement*; United Nations Framework Convention on Climate Change: Geneva, Switzerland, 2015. Available online: https://unfccc.int/sites/default/files/english_paris_agreement.pdf (accessed on 15 June 2021).
5. Huang, L.; Krigsvoll, G.; Johansen, F.; Liu, Y.; Zhang, X. Carbon emission of global construction sector. *Renew. Sustain. Energy Rev.* **2018**, *81*, 1906–1916. [CrossRef]
6. Benhelal, E.; Zahedi, G.; Shamsaei, E.; Bahadori, A. Global strategies and potentials to curb CO_2 emissions in cement industry. *J. Clean. Prod.* **2013**, *51*, 142–161. [CrossRef]
7. Danish, A.; Mosaberpanah, M.A.; Salim, M.U.; Fediuk, R.; Rashid, M.F.; Waqas, R.M. Reusing marble and granite dust as cement replacement in cementitious composites: A review on sustainability benefits and critical challenges. *J. Build. Eng.* **2021**, *44*, 102600. [CrossRef]
8. Rakhimova, N.R.; Rakhimov, R.; Morozov, V.; Eskin, A. Calcined low-grade clays as sources for zeolite containing material. *Period. Polytech. Civ. Eng.* **2021**, *65*, 204–214. [CrossRef]
9. Barbuta, M.; Bucur, R.D.; Cimpeanu, S.M.; Paraschiv, G.; Bucur, D. *Wastes in Building Materials Industry*; IntechOpen: London, UK, 2015; Volume 1, pp. 81–99. [CrossRef]
10. Shao, Y.; Mahoutian, M.; Ghouleh, Z. Carbonation-activated steel slag binder as alternative cementing material. *ACI Spec. Publ.* **2017**, *320*, 21.1–21.14.
11. Marvila, M.T.; Azevedo, A.R.G.D.; Matos, P.R.D.; Monteiro, S.N.; Vieira, C.M.F. Rheological and the fresh state properties of alkali-activated mortars by blast furnace slag. *Materials* **2021**, *14*, 2069. [CrossRef] [PubMed]
12. Ramagiri, K.K.; Kar, A. Effect of high-temperature on the microstructure of alkali-activated binder. *Mater. Today* **2020**, *28*, 1123–1129. [CrossRef]
13. Ramagiri, K.K.; Patil, S.; Mundra, H.; Kar, A. Laboratory investigations on the effects of acid attack on concrete containing portland cement partially replaced with ambient-cured alkali-activated binders. *Adv. Concr. Constr.* **2020**, *10*, 221–236. [CrossRef]
14. Ramagiri, K.K.; Chauhan, D.R.; Gupta, S.; Kar, A.; Adak, D.; Mukherjee, A. High-temperature performance of ambient-cured alkali-activated binder concrete. *Innov. Infrastruct. Solut.* **2021**, *6*, 71. [CrossRef]
15. Torres-Carrasco, M.; Puertas, F. La activación alcalina de diferentes aluminosilicatos como una alternativa al Cemento Portland: Cementos activados alcalinamente o geopolímeros. *Rev. Ing. Constr.* **2017**, *32*, 5–12. [CrossRef]
16. Provis, J.L.; Bernal, S.A. Geopolymers and related alkali-activated materials. *Annu. Rev. Mater. Sci.* **2014**, *44*, 299–327. [CrossRef]
17. Villoria Sáez, P.; Osmani, M. A diagnosis of construction and demolition waste generation and recovery practice in the European Union. *J. Clean. Prod.* **2019**, *241*, 118400. [CrossRef]
18. NITI Aayog. *Resource Efficiency and Circular Economy-Current Status and Way Forward*; National Institution for Transforming India: New Delhi, India, 2019.
19. Akhtar, A.; Sarmah, A.K. Construction and demolition waste generation and properties of recycled aggregate concrete: A global perspective. *J. Clean. Prod.* **2018**, *186*, 262–281. [CrossRef]

20. Pyngrope, M.; Hossiney, N.; Chen, Y.; Thejas, H.K.; Sarath, C.K.; Alex, J.; Lakshmish Kumar, S. Properties of alkali-activated concrete (AAC) incorporating demolished building waste (DBW) as aggregates. *Cogent Eng.* **2021**, *8*, 1870791. [CrossRef]
21. Xiao, J.Z.; Li, J.B. Study on relationships between strength indexes of recycled concrete. *J. Build. Mater.* **2005**, *2*, 197–201. (In Chinese) [CrossRef]
22. Li, X. Recycling and reuse of waste concrete in China: Part I. Material behaviour of recycled aggregate concrete. *Resour. Conserv. Recycl.* **2008**, *53*, 36–44. [CrossRef]
23. Kwan, W.H.; Ramli, M.; Kam, K.J.; Sulieman, M.Z. Influence of the amount of recycled coarse aggregate in concrete design and durability properties. *Constr. Build. Mater.* **2012**, *26*, 565–573. [CrossRef]
24. Liu, Q.; Xiao, J.; Sun, Z. Experimental study on the failure mechanism of recycled concrete. *Cem. Concr. Res.* **2011**, *41*, 1050–1057. [CrossRef]
25. Abdollahnejad, Z.; Mastali, M.; Falah, M.; Luukkonen, T.; Mazari, M.; Illikainen, M. Construction and Demolition Waste as Recycled Aggregates in Alkali-Activated Concretes. *Materials* **2019**, *12*, 4016. [CrossRef]
26. Turner, L.K.; Collins, F.G. Carbon dioxide equivalent (CO_2-e) emissions: A comparison between geopolymer and OPC cement concrete. *Constr. Build. Mater.* **2013**, *43*, 125–130. [CrossRef]
27. Fořt, J.; Vejmelková, E.; Koňáková, D.; Alblová, N.; Čáchová, M.; Keppert, M.; Rovnaníková, P.; Černý, R. Application of waste brick powder in alkali activated aluminosilicates: Functional and environmental aspects. *J. Clean. Prod.* **2018**, *194*, 714–725. [CrossRef]
28. Katz, A. Properties of concrete made with recycled aggregate from partially hydrated old concrete. *Cem. Concr. Res.* **2003**, *33*, 703–711. [CrossRef]
29. Movassaghi, R. Durability of Reinforced Concrete Incorporating Recycled Concrete as Aggregate (RCA). Master's Thesis, University of Waterloo, Waterloo, ON, Canada, 2006. Available online: http://hdl.handle.net/10012/2884 (accessed on 1 June 2021).
30. Mastali, M.; Abdollahnejad, Z.; Pacheco-Torgal, F. 15—Carbon dioxide sequestration of fly ash alkaline-based mortars containing recycled aggregates and reinforced by hemp fibers: Mechanical properties and numerical simulation with a finite element method. In *Carbon Dioxide Sequestration in Cementitious Construction Materials*; Woodhead Publishing: New Delhi, India, 2018; pp. 373–391. [CrossRef]
31. de Juan, M.S.; Gutiérrez, P.A. Study on the influence of attached mortar content on the properties of recycled concrete aggregate. *Constr. Build. Mater.* **2009**, *23*, 872–877. [CrossRef]
32. Abbas, A.; Fathifazl, G.; Burkan Isgor, O.; Razaqpur, A.; Fournier, B.; Foo, S. Proposed Method for Determining the Residual Mortar Content of Recycled Concrete Aggregates. *J. ASTM Int.* **2007**, *5*, 1–12. [CrossRef]
33. Akbarnezhad, A.; Ong, K.C.G.; Zhang, M.H.; Tam, C.T.; Foo, T.W.J. Microwave-assisted beneficiation of recycled concrete aggregates. *Constr. Build. Mater.* **2011**, *25*, 3469–3479. [CrossRef]
34. Yanagibashi, K.; Inoue, K.; Seko, S.; Tsuji, D. A study on cyclic use of aggregate for structural concrete. In Proceedings of the SB05: The 2005 World Sustainable Building Conference, Tokyo, Japan, 27–29 September; pp. 9–11.
35. Shima, H.; Tateyashiki, H.; Matsuhashi, R.; Yoshida, Y. An Advanced Concrete Recycling Technology and its Applicability Assessment through Input-Output Analysis. *J. Adv. Concr. Technol.* **2005**, *3*, 53–67. [CrossRef]
36. Sidiq, A.; Setunge, S.; Gravina, R.J.; Giustozzi, F. Self-repairing cement mortars with microcapsules: A microstructural evaluation approach. *Constr. Build. Mater.* **2020**, *232*, 117239. [CrossRef]
37. Kane, J.W.; Tomer, A. Shifting into an Era of Repair: Us Infrastructure Spending Trends. Brookings. Available online: https://www.brookings.edu/research/shifting-into-an-era-of-repair-us-infrastructure-spending-trends/ (accessed on 15 June 2021).
38. Alves, L.; Alves, L.; Mello, M.; Barros, S.D. Characterization of bioconcrete and the properties for self-healing. *Proceedings* **2019**, *38*, 4. [CrossRef]
39. ONS. Construction Output in Great Britain: May 2016, Office for National Statistics. Available online: https://www.ons.gov.uk/businessindustryandtrade/constructionindustry/bulletins/constructionoutputingreatbritain/may2016 (accessed on 1 June 2021).
40. Best, R.D. Global Investments on the Construction and Maintenance of Infrastructure as Share of GDP in 2018. Available online: https://www.statista.com/statistics/566787/average-yearly-expenditure-on-economic-infrastructure-as-percent-of-gdp-worldwide-by-country/ (accessed on 15 June 2021).
41. Wu, M.; Johannesson, B.; Geiker, M. A review: Self-healing in cementitious materials and engineered cementitious composite as a self-healing material. *Constr. Build. Mater.* **2012**, *28*, 571–583. [CrossRef]
42. Termkhajornkit, P.; Nawa, T.; Yamashiro, Y.; Saito, T. Self-healing ability of fly ash–cement systems. *Cem. Concr. Compos.* **2009**, *31*, 195–203. [CrossRef]
43. Jonkers, H.M.; Thijssen, A.; Muyzer, G.; Copuroglu, O.; Schlangen, E. Application of bacteria as self-healing agent for the development of sustainable concrete. *Ecol. Eng.* **2010**, *36*, 230–235. [CrossRef]
44. De Muynck, W.; De Belie, N.; Verstraete, W. Microbial carbonate precipitation in construction materials: A review. *Ecol. Eng.* **2010**, *36*, 118–136. [CrossRef]
45. Erşan, Y.Ç.; Hernandez-Sanabria, E.; Boon, N.; de Belie, N. Enhanced crack closure performance of microbial mortar through nitrate reduction. *Cem. Concr. Compos.* **2016**, *70*, 159–170. [CrossRef]

46. Zhang, W.; Zheng, Q.; Ashour, A.; Han, B. Self-healing cement concrete composites for resilient infrastructures: A review. *Compos. B. Eng.* **2020**, *189*, 107892. [CrossRef]

47. Achal, V.; Mukerjee, A.; Sudhakara Reddy, M. Biogenic treatment improves the durability and remediates the cracks of concrete structures. *Constr. Build. Mater.* **2013**, *48*, 1–5. [CrossRef]

48. Chahal, N.; Siddique, R.; Rajor, A. Influence of bacteria on the compressive strength, water absorption and rapid chloride permeability of fly ash concrete. *Constr. Build. Mater.* **2012**, *28*, 351–356. [CrossRef]

49. Sonali Sri Durga, C.; Ruben, N.; Sri Rama Chand, M.; Venkatesh, C. Performance studies on rate of self healing in bio concrete. *Mater. Today* **2020**, *27*, 158–162. [CrossRef]

50. Jena, S.; Basa, B.; Panda, K.C.; Sahoo, N.K. Impact of Bacillus subtilis bacterium on the properties of concrete. *Mater. Today* **2020**, *32*, 651–656. [CrossRef]

51. Khaliq, W.; Ehsan, M.B. Crack healing in concrete using various bio influenced self-healing techniques. *Constr. Build. Mater.* **2016**, *102*, 349–357. [CrossRef]

52. Shaheen, N.; Khushnood, R.A.; Ud din, S. Bioimmobilized limestone powder for autonomous healing of cementitious systems: A feasibility study. *Adv. Mater. Sci. Eng.* **2018**, *2018*, 7049121. [CrossRef]

53. Khushnood, R.A.; ud din, S.; Shaheen, N.; Ahmad, S.; Zarrar, F. Bio-inspired self-healing cementitious mortar using Bacillus subtilis immobilized on nano-/micro-additives. *J. Intell. Mater. Syst. Struct.* **2018**, *30*, 3–15. [CrossRef]

54. Manzur, T.; Rahman, F.; Afroz, S.; Huq, R.S.; Efaz, I.H. Potential of a microbiologically induced calcite precipitation process for durability enhancement of masonry aggregate concrete. *J. Mater. Civ. Eng.* **2017**, *29*, 04016290. [CrossRef]

55. Papanikolaou, I.; Arena, N.; Al-Tabbaa, A. Graphene nanoplatelet reinforced concrete for self-sensing structures—A lifecycle assessment perspective. *J. Clean. Prod.* **2019**, *240*, 118202. [CrossRef]

56. Alazhari, M.; Sharma, T.; Heath, A.; Cooper, R.; Paine, K. Application of expanded perlite encapsulated bacteria and growth media for self-healing concrete. *Constr. Build. Mater.* **2018**, *160*, 610–619. [CrossRef]

57. Xu, J.; Wang, X.; Zuo, J.; Liu, X. Self-Healing of Concrete Cracks by Ceramsite-Loaded Microorganisms. *Adv. Mater. Sci. Eng.* **2018**, *2018*, 5153041. [CrossRef]

58. Wang, J.; Van Tittelboom, K.; De Belie, N.; Verstraete, W. Use of silica gel or polyurethane immobilized bacteria for self-healing concrete. *Constr. Build. Mater.* **2012**, *26*, 532–540. [CrossRef]

59. Luo, M.; Qian, C. Influences of bacteria-based self-healing agents on cementitious materials hydration kinetics and compressive strength. *Constr. Build. Mater.* **2016**, *121*, 659–663. [CrossRef]

60. Wang, J.Y.; Soens, H.; Verstraete, W.; De Belie, N. Self-healing concrete by use of microencapsulated bacterial spores. *Cem. Concr. Res.* **2014**, *56*, 139–152. [CrossRef]

61. Paine, K. Bacteria-based self-healing concrete: Effects of environment, exposure and crack size. In Proceedings of the RILEM Conference on Microorganisms-Cementitious Materials Interactions, Delft, The Netherlands, 23 June 2016; Wiktor, V., Jonkers, H., Bertron, A., Eds.; RILEM Publications S.A.R.L.: Paris, France, 2016.

62. BIS. *IS 12269 Ordinary Portland Cement, 53 Grade—Specification*; Bureau of Indian Standards: New Delhi, India, 2013.

63. ASTM. *C150/C150M Standard Specification for Portland Cement*; ASTM International: West Conshohocken, PA, USA, 2020.

64. BIS. *IS 383 Coarse and Fine Aggregate for Concrete—Specification (Third Revision)*; Bureau of Indian Standards: New Delhi, India, 2016.

65. El-Hassan, H.; Ismail, N. Effect of process parameters on the performance of fly ash/GGBS blended geopolymer composites. *J. Sustain. Cem.-Based Mater.* **2018**, *7*, 122–140. [CrossRef]

66. El-Hassan, H.; Shehab, E.; Al-Sallamin, A. Influence of different curing regimes on the performance and microstructure of alkali-activated slag concrete. *J. Mater. Civ. Eng.* **2018**, *30*, 04018230. [CrossRef]

67. Salmasi, F.; Mostofinejad, D. Investigating the effects of bacterial activity on compressive strength and durability of natural lightweight aggregate concrete reinforced with steel fibers. *Constr. Build. Mater.* **2020**, *251*, 119032. [CrossRef]

68. ASTM. *C39/C39M-18 Standard Test Method for Compressive Strength of Cylindrical Concrete Specimens*; ASTM International: West Conshohocken, PA, USA, 2018.

69. Moreno Ruiz, E.; Valsasina, L.; FitzGerald, D.; Symeonidis, A.; Turner, D.; Müller, J.; Minas, N.; Bourgault, G.; Vadenbo, C.; Ioannidou, D.; et al. *Documentation of Changes Implemented in Ecoinvent Database v3. 7 & v3. 7.1*; Ecoinvent Association: Zürich, Switzerland, 2020. Available online: https://www.ecoinvent.org/files/change_report_v3_7_1_20201217.pdf (accessed on 15 June 2021).

70. Sahnoun, M.; Kriaa, M.; Elgharbi, F.; Ayadi, D.-Z.; Bejar, S.; Kammoun, R. Aspergillus oryzae S2 alpha-amylase production under solid state fermentation: Optimization of culture conditions. *Int. J. Biol. Macromol.* **2015**, *75*, 73–80. [CrossRef] [PubMed]

71. Hong, M.; Jang, I.; Son, Y.; Yi, C.; Park, W. Agricultural by-products and oyster shell as alternative nutrient sources for microbial sealing of early age cracks in mortar. *AMB Express* **2021**, *11*, 11. [CrossRef]

72. Kumar, L.R.; Yellapu, S.K.; Zhang, X.; Tyagi, R.D. Energy balance for biodiesel production processes using microbial oil and scum. *Bioresour. Technol.* **2019**, *272*, 379–388. [CrossRef]

73. Hill, J.; Nelson, E.; Tilman, D.; Polasky, S.; Tiffany, D. Environmental, economic, and energetic costs and benefits of biodiesel and ethanol biofuels. *Proc. Natl. Acad. Sci. USA* **2006**, *103*, 11206. [CrossRef]

74. Myhr, A.; Røyne, F.; Brandtsegg, A.S.; Bjerkseter, C.; Throne-Holst, H.; Borch, A.; Wentzel, A.; Røyne, A. Towards a low CO_2 emission building material employing bacterial metabolism (2/2): Prospects for global warming potential reduction in the concrete industry. *PLoS ONE* **2019**, *14*, e0208643. [CrossRef] [PubMed]
75. TSERC. Retail Supply Tariff Order for FY 2018-19—Salient Features. Available online: https://tserc.gov.in/file_upload/uploads/ Tariff%20Orders/Tariff%20Schedule/Tariff%20Schedule%20for%20FY2018-19.pdf (accessed on 15 June 2021).
76. BIS. *IS/ISO 14040 Environmental Management: Life Cycle Assessment*; Requirements and Guidelines; Bureau of Indian Standards: New Delhi, India, 2006.
77. Hildenbrand, J.; Srocka, M.; Siroth, A. Why We Started the Development of OpenLCA. Available online: https://www.openlca. org/the-idea/ (accessed on 26 June 2021).
78. Ntziachristos, L.; Samaras, Z. *EMEP/EEA Emission Inventory Guidebook 2009*; European Environment Agency (EEA): Copenhagen, Denmark, 2009.
79. Jessen, C.; Roder, C.; Villa Lizcano, J.F.; Voolstra, C.R.; Wild, C. In-situ effects of simulated overfishing and eutrophication on benthic coral reef algae growth, succession, and composition in the central Red Sea. *PLoS ONE* **2013**, *8*, e66992. [CrossRef]
80. Selman, M.; Greenhalgh, S. Eutrophication: Sources and Drivers of Nutrient Pollution. WRI Policy Note, Water Quality: Utrophication and Hypoxia. Available online: https://files.wri.org/d8/s3fs-public/pdf/eutrophication_sources_and_drivers. pdf (accessed on 26 June 2021).
81. Hauschild, M.; Potting, J. Spatial Differentiation in Life Cycle Impact Assessment-The EDIP2003 Methodology. Available online: https://lca-center.dk/wp-content/uploads/2015/08/Spatial-differentiation-in-life-cycle-impact-assessment-the-EDIP2 003-methodology.pdf (accessed on 26 June 2021).
82. Kim, T.H.; Chae, C.U. Environmental impact analysis of acidification and eutrophication due to emissions from the production of concrete. *Sustainability* **2016**, *8*, 578. [CrossRef]
83. Pulles, T.; Denier van der Gon, H.; Appelman, W.; Verheul, M. Emission factors for heavy metals from diesel and petrol used in European vehicles. *Atmos. Environ.* **2012**, *61*, 641–651. [CrossRef]
84. Fawer, M.; Concannon, M.; Rieber, W. Life cycle inventories for the production of sodium silicates. *Int. J. Life Cycle Assess.* **1999**, *4*, 207. [CrossRef]
85. Koiwanit, J.; Manuilova, A.; Chan, C.; Wilson, M.; Tontiwachwuthikul, P. A comparative study of human health impacts due to heavy metal emissions from a conventional lignite coal-fired electricity generation station, with post-combustion, and oxy-fuel combustion capture technologies. In *Greenhouse Gases: Selected Case Studies*; Manning, A., Ed.; IntechOpen: London, UK, 2016. [CrossRef]
86. Castleden, W.M.; Shearman, D.; Crisp, G.; Finch, P. The mining and burning of coal: Effects on health and the environment. *Med. J. Aust.* **2011**, *195*, 333–335. [CrossRef]
87. Atilgan, B.; Azapagic, A. Life cycle environmental impacts of electricity from fossil fuels in Turkey. *J. Clean. Prod.* **2015**, *106*, 555–564. [CrossRef]
88. Worrell, E.; Price, L.; Martin, N.; Hendriks, C.; Meida, L.O. Carbon dioxide emissions from the global cement industry. *Annu. Rev. Environ. Resour.* **2001**, *26*, 303–329. [CrossRef]
89. Schorcht, F.; Kourti, I.; Scalet, B.M.; Roudier, S.; Sancho, L.D. *Best Available Techniques (BAT) Reference Document for the Production of Cement, Lime and Magnesium Oxide*; European Commission Joint Research Centre Institute for Prospective Technological Studies: Seville, Spain, 2013.
90. Huijbregts, M.A.; Steinmann, Z.J.; Elshout, P.M.; Stam, G.; Verones, F.; Vieira, M.D.M.; Hollander, A.; Zijp, M.; van Zelm, R. *ReCiPe2016. A Harmonized Life Cycle Impact Assessment Method at Midpoint and Endpoint Level. Report I: Characterization*; RIVM Report 2016-0104; National Institute for Human Health and the Environment: Bilthoven, The Netherlands, 2016. Available online: https://www.rivm.nl/bibliotheek/rapporten/2016-0104.pdf (accessed on 15 June 2021).
91. Bosoaga, A.; Masek, O.; Oakey, J.E. CO_2 capture technologies for cement industry. *Energy Procedia* **2009**, *1*, 133–140. [CrossRef]
92. EuroChlor. The European chlor-alkali industry: An Electricity Intensive Sector Exposed to Carbon Leakage. Brussels, Belgium. 2010. Available online: https://www.eurochlor.org/wp-content/uploads/2019/04/3-2-the_european_chlor-alkali_industry_ -_an_electricity_intensive_sector_exposed_to_carbon_leakage.pdf (accessed on 26 June 2021).

 *infrastructures*

Article

Physical, Mechanical and Durability Properties of Eco-Friendly Engineered Geopolymer Composites

Ahmed M. Tahwia, Duaa S. Aldulaimi, Mohamed Abdellatief and Osama Youssf

1 Structural Engineering Department, Faculty of Engineering, Mansoura University, Mansoura 35516, Egypt; duaa.salah@std.mans.edu.eg (D.S.A.); osama.youssf@mymail.unisa.edu.au (O.Y.)

2 Department of Civil Engineering, Higher Future Institute of Engineering and Technology, Mansoura 35516, Egypt; besttime703@std.mans.edu.eg

* Correspondence: atahwia@mans.edu.eg

Abstract: Engineered geopolymer composite (EGC) is a high-performance material with enhanced mechanical and durability capabilities. Ground granulated blast furnace slag (GGBFS) and silica fume (SF) are common binder materials in producing EGC. However, due to the scarcity and high cost of these materials in some countries, sustainable alternatives are needed. This research focused on producing eco-friendly EGC made of cheaper and more common pozzolanic waste materials that are rich in aluminum and silicon. Rice husk ash (RHA), granite waste powder (GWP), and volcanic pumice powder (VPP) were used as partial substitutions (10–50%) of GGBFS in EGC. The effects of these wastes on workability, unit weight, compressive strength, tensile strength, flexural strength, water absorption, and porosity of EGC were examined. The residual compressive strength of the proposed EGC mixtures at high elevated temperatures (200, 400, and 600 °C) was also evaluated. Additionally, scanning electron microscope (SEM) was employed to analyze the EGC microstructure characteristics. The experimental results demonstrated that replacing GGBFS with RHA and GWP at high replacement ratios decreased EGC workability by up to 23.1% and 30.8%, respectively, while 50% VPP improved EGC workability by up to 38.5%. EGC mixtures made with 30% RHA, 20% GWP, or 10% VPP showed the optimal results in which they exhibited the highest compressive, tensile, and flexural strengths, as well as the highest residual compressive strength when exposed to high elevated temperatures. The water absorption and porosity increased by up to 106.1% and 75.1%, respectively, when using RHA; increased by up to 23.2% and 18.6%, respectively, when using GWP; and decreased by up to 24.7% and 22.6%, respectively, when using VPP in EGC.

Keywords: eco-friendly; engineered geopolymer composite; rice husk ash; granite waste powder; volcanic pumice powder

Citation: Ahmed M. Tahwia, Duaa S. Aldulaimi, Mohamed Abdellatief and Osama Youssf. Physical, Mechanical and Durability Properties of Eco-Friendly Engineered Geopolymer Composites. *Infrastructures* **2024**, 9, 191. https://doi.org/10.3390/infrastructures9110191

Academic Editors: Patricia Kara De Maeijer and Pedro Arias-Sánchez

Received: 29 September 2024
Revised: 18 October 2024
Accepted: 24 October 2024
Published: 25 October 2024

1. Introduction

Concrete is the most utilized material for construction in the world, with a tensile strain capacity of about 0.01%. It is strong in compression but weak in tension [1]. Since the 1960s, a variety of studies have concentrated on developing fiber-reinforced cementitious composites in an attempt to counteract this weakness [2]. Early in the 1990s, researchers developed engineered cementitious composites (ECCs), which have exceptional tensile ductility and multiple cracking characteristics that can overcome the brittleness of regular cementitious materials [1]. However, the typical ECC mix usually requires a cement amount that is two to three times more than that required in conventional concrete [3]. The production of ECC releases 8% of world carbon dioxide (CO_2) emissions [4,5]. As a result, in recent decades, a number of studies have concentrated on the production of greener and more sustainable ECCs.

Industrial byproducts that are alkali-activated to create geopolymers are thought to be a viable substitute for Portland cement because of their superior technical qualities and reduced carbon emissions [6,7]. Geopolymer has been recognized as one of the most promising new green cementitious materials. It is environmentally and economically friendly and has a short setting time, early strength, high-temperature tolerance, and good corrosion resistance [8,9]. Natural minerals or industrial wastes like fly ash, metakaolin, ground granulated blast furnace slag (GGBFS), and rice husk ash (RHA) are typically precursors of geopolymers. These materials are rich in aluminum and silicon and can chemically react with alkaline activators like sodium hydroxide (SH), sodium silicate (SS), and sodium carbonate [10,11]. Given the potential benefit of utilizing geopolymers in the field of concrete construction, numerous attempts have been made to replace Portland cement in ECC with geopolymers to create engineered geopolymer composites (EGCs). EGC is a fiber-reinforced geopolymer composite that is highly ductile, environmentally friendly, and sustainable [12]. Because of their exceptional strength, durability in hard environmental conditions, and friendliness to the environment, these composites are becoming more and more popular in the infrastructure and construction sectors. The EGCs were able to show enhanced mechanical characteristics under dynamic stress but similar tension and flexural characteristics under static loads, thus rendering it a more sustainable material than ECC [12].

Several previous studies have been conducted to develop more sustainable EGCs by applying longer curing periods and higher curing temperatures or utilizing various kinds of alkaline activators and precursors. Wang et al. [13] studied the utilization of recycled concrete powder (RCP) and recycled fine sand (RFS) and found that the addition of RCP slightly improved the flowability. While the inclusion of RFS had the opposite effect, the decrease in flowability induced by RFS was minimized by the combined use of RCP and RFS; this also applied to the setting time and water absorption.

Nematollahi et al. [14] investigated how strain hardening performance for polyvinyl alcohol fiber (PVA) EGC was affected by alkali activator solutions. They tried four types of alkali activators, namely a potassium-based alkali activator liquid, two varieties of SS liquids, and a lime-based activator in powder form. Their results showed that utilizing an SS-based activator and 8M SH liquid with a mixing ratio (SS/SH) of two was extremely advantageous in terms of cost reduction and it showed an EGC tensile strength of 4.7 MPa and compressive strength of 60 MPa. In addition, this alkali activator enhanced the fracture toughness and other matrix fracture parameters. Other researchers studied the impact of various factors on EGC behavior, including the matrix design [15], the combination of superplasticizers and activators [16], the curing temperature [17], and various fly ash and slag ratios.

Yazan and Dai [18] evaluated the impact of fiber hybridization on the tensile characteristics and microstructure of EGCs that were cured at room temperature. The research team discovered that while the EGC hybrid composites' compressive strength, fracture properties, relative slump, and dry density all increased with the addition of steel fiber, the multiple cracking actions and tensile strain capacity were significantly reduced. Additionally, even though the fiber bridging stress and fracture properties were improved, the addition of micro-silica sand dramatically decreased the multiple cracking behaviors and the tensile ductility of EGC composites. Ling et al. [19] reported the effect of slag content on the mechanical properties and bond strength of fly ash EGC. They evaluated the compressive strength, pullout bond strength, flexural strength, tensile properties, and elastic modulus of EGCs when substituting fly ash binder by slag at several weight ratios (0%, 10%, 20%, and 30%). The findings showed that fly ash EGC without slag had low fracture toughness and tensile strain ductility, whereas all EGCs made with slag had superior compressive strength, elastic modulus, initial cracking strength, and fiber bridging stress. Moreover, under uniaxial tensile and flexural loadings, many tiny cracks were seen in all EGC specimens (with and without slag). As the slag percentage of EGCs grew, the number of cracks decreased and their spacing increased. Deb et al. [20] investigated whether the slag and activator contents affect the workability and strength characteristics of fly ash-based geopolymer concrete. They discovered that at all ages up to 180 days, the compressive strength of the slag-fly ash geopolymer increased with increasing slag content. Ahmed et al. [21] studied replacing slag with recycled brick waste powder (RBWP) and found that the amount of RBWP in the EGC mixture progressively raised water absorption, which reached 304% when slag was completely substituted. The porous surface of unreacted RBWP may be the cause of the increase in water absorption in the mixture of EGC with a high RBWP concentration. Wu et al. [22] studied the incorporation of recycled concrete powder (RCP) and recycled paste powder (RPP) as a substitute for FA-slag EGC and found that the water-permeable porosity and water absorption were increased by incorporating RCP up to 100% as a substitute for FA-slag. The substitution of RCP for FA-slag increases the porosity and associated water transport in EGC because it has low alkali-activated activity and substantial inert components. A noteworthy increase in water absorption and water-permeable porosity is seen, particularly for the fully recycled powder EGC manufactured with 100% RCP. Also, it was found that the water absorption and water-permeable porosity of EGC are not considerably increased when 50% or 100% of FA-slag is replaced with RPP; however, these parameters are greatly increased when 75–100% of FA-slag is replaced with RPP.

With the significant scientific development that has occurred in slag geopolymer composites, the market price of slag has gradually increased. As a result, industrial wastes that are high in silicon and aluminum have turned into a hotspot for research. Therefore, a variety of alternative materials are needed to replace slag in EGC. Rice husk ash (RHA) is a dark powder made by burning rice husks that is a non-environmentally friendly material [23]. It has been reported that using RHA in cementitious composites improved their mechanical properties and reduced their water absorption [24]. Zhao et al. [8] studied the effect of RHA at high content (50–80%) on concrete compressive strength, flexural strength, moisture absorption coefficient and water absorption, and found that 60% RHA and 40% slag can be used together as raw materials to obtain the preferred mechanical properties. Granite waste powder (GWP) is produced during the processing and cutting of granite stone. Using GWP in the construction industry can address many environmental issues related to waste disposal and serve as a tool for resource conservation. It has been shown that GWP can be used as a binder in cementitious composites technology. Filling up the gaps in the cementitious composites' matrix with GWP significantly improves the volume porosity and increases the composites' durability [25]. The demand for natural resources can be decreased by substituting GWP for natural materials like cement and sand. Additionally, industries can save money on raw material costs and develop their circular economy by using GWP to replace a portion of their material requirements [26]. Shilar et al. [27] investigated the optimum percentage of GWP as a binder, studied the effect of the molarity-to-binder ratio on the mechanical properties of geopolymer concrete, and concluded that at 16 M, the maximum compressive, tensile, and flexural strengths were obtained with GWP of 20% content. Volcanic pumice powder (VPP) is a fine, lightweight, and porous material derived from volcanic pumice, which is a type of volcanic rock. Because of its potential as a sustainable substitute for conventional building materials, it has attracted interest in the field of geopolymer research. VPP can improve the mechanical characteristics of geopolymer formulations, giving the material more strength and endurance. Since VPP is a naturally occurring substance that can be sourced responsibly, it is also environmentally beneficial. Because of its abundance and lightweight nature, it presents a promising alternative to the environmentally friendly building techniques [28]. Kabay et al. [29] evaluated the physical, mechanical and microstructural properties of geopolymer pastes and mortars manufactured from VPP and GGBFS with sodium hydroxide, potassium hydroxide, and sodium silicate solution. Their results revealed that the ratio of VPP and the activator concentration have an important role in the compressive strength of pastes.

This research focused on producing more sustainable and eco-friendly EGC made of cheaper and more common pozzolanic waste materials, namely RHA, GWP, and VPP. These waste materials were used as partial substitutions (10–50%) of GGBFS in EGC. The effects of these wastes on workability, unit weight, compressive strength, tensile strength, flexural strength, water absorption, and porosity of EGC were examined through sixteen proposed mixtures. The residual compressive strength of the proposed EGC mixtures when exposed to high elevated temperatures (200, 400, and 600 °C) and the microstructure of selected mixtures were also evaluated.

2. Experimental Procedures

2.1. Raw Materials and Mix Proportions

A total of sixteen mixtures were designed and tested in this study. The precursor materials utilized in the control EGC mixture of this research included GGBFS and silica fume (SF). The specific gravity of GGBFS and SF were 2.87 and 2.15, respectively. The slag was partially replaced by RHA, GWP, or VPP at 10–50% by volume; while SF was used with constant content in all mixtures. The specific gravities of RHA, GWP and VPP were 1.95, 2.8, and 2.17, respectively. The chemical compositions of all binder materials used are shown in Table 1. A constant content of sand was used as the fine aggregate in all mixtures. The sand had a specific gravity of 2.65, a fineness modulus of 2.20, and a unit weight of 1420 kg/m^3.

A mix of SH and SS solutions was used as the alkaline activator of EGC mixtures. The SH solution with specific gravity of 2.13 was prepared with a molar concentration of 14 M using SH pellets of 97–98% purity locally purchased and dissolved in portable water (dissolving 560 gm of SH pellets in every 1 L of water) and then left in room temperature for 2 h. The prepared SH solution was then mixed with the SS solution (specific gravity of 1.5) for 30 min using a mechanical liquid stirrer. The alkaline activator was then covered and left for 24 h before using in EGC mixing. The weight mixing ratio of SS to SH was kept constant at a value of 2.0.

To control the workability of the proposed EGC mixtures, extra water with 5% binder and superplasticizer (SP) with 1.5% binder were added. The SP had a commercial name of "Sika ViscoCrete ®–3425" type F with specific gravity of 1.15 according to ASTM C494/C494M−17 [30]. Curved wave polypropylene (PP) macro fibers, as shown in Figure 1, with a specific gravity of 0.91 were used to reinforce EGC mixtures. Table 2 lists the characteristics of PP fibers, as provided by the supplier.

The proportions of GGBFS and SF in control EGC precursor materials were 85% and 15%, respectively, and the ratio of the alkaline solution to precursor materials was held constant at 0.4. The volume content of PP fiber remained at 2%. All materials used and the mixing method are shown in Figure 2. Table 3 lists the proportions of the designed EGC mixtures. In this table, EGC refers to the control mix, and RHA, GWP, and VPP refer to mixes including rice husk ash, granite waste powder, and volcanic pumice powder, respectively, as partial replacements of slag followed by the replacement ratio by volume. The replacements in the designed mixes have been carried out by volume to ensure constant absolute volume (1 m^3) for all comparable mixes.

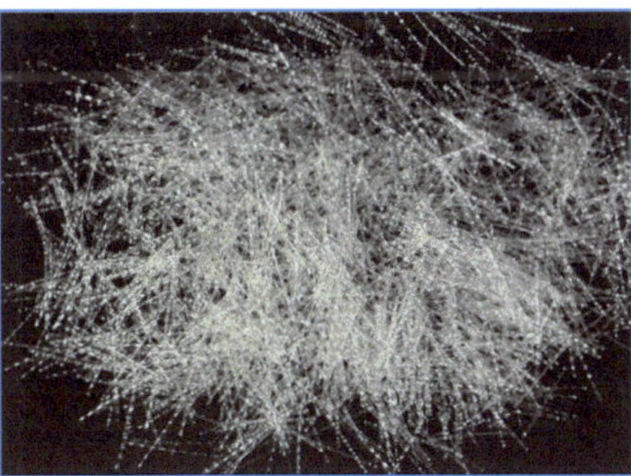

Figure 1. Polypropylene fibers used.

Table 1. Chemical composition of geopolymer precursors.

Geopolymer Precursors	SiO_2 (%)	CaO (%)	Al_2O_3 (%)	Fe_2O_3 (%)	SO_3 (%)	MgO (%)	Na_2O (%)	K_2O (%)	SrO (%)	TiO_2 (%)	P_2O_5 (%)	Mn_2O_3 (%)	LOI (%)
GGBFS	39.8	41.1	11.4	0.4	1.7	3.5	0.4	0.3	0.8	0.6	0.01	0.0	0.8
SF	97.5	0.3	0.2	0.3	0.0	0.6	0.5	0.6	0.0	0.0	0.0	0.0	5.1
RHA	90.14	0.97	3.88	1.15	0.05	0.26	1.39	1.24	0.0	0.0	0.91	0.0	6.13
GWP	69.7	3.6	13.03	3.04	0.28	0.79	3.66	5.29	0.0	0.42	0.09	0.0	0.65
VPP	49.4	7.76	16.87	14.04	0.54	6.01	2.21	2.01	0.0	0.0	0.72	0.19	2.21

Table 2. Physical and mechanical characteristics of macro-PP fibers.

Property	Value
Compressive Strength	550 MPa
Tensile Strength	560 MPa
Modulus of Elasticity	5 GPa
Crack Elongation (%)	$\geq$15
Melting Point	160–170 °C
Length	30 mm
Thickness	0.3 mm
Aspect Ratio	100
Density	910 kg/m^3

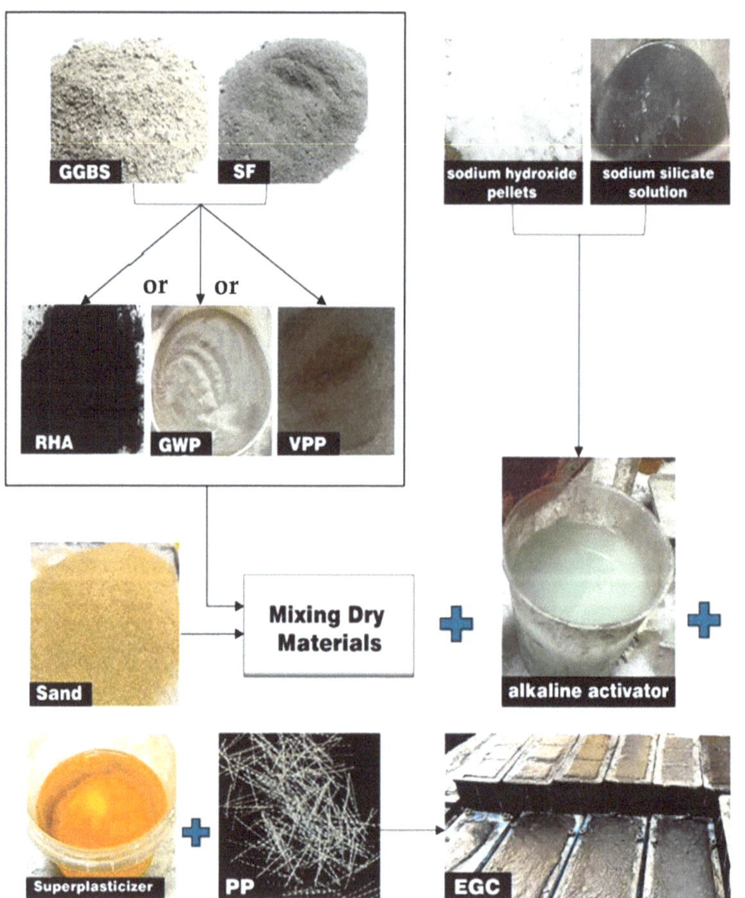

Figure 2. Materials and mixing method used in this study.

Table 3. Mix proportions of EGC (kg/m^3).

Mix ID	Precursors					Sand	AAS		SP	Extra Water	PP
	GGBS	SF	RHA	GWP	VPP		SS	SH			
EGC	996.0	118	0	0	0	570	297	148.5	16.71	55.7	18
RHA-10	896.4	118	67.6	0	0	570	297	148.5	16.71	55.7	18
RHA-20	796.8	118	135.3	0	0	570	297	148.5	16.71	55.7	18
RHA-30	697.2	118	203.0	0	0	570	297	148.5	16.71	55.7	18
RHA-40	597.6	118	270.6	0	0	570	297	148.5	16.71	55.7	18
RHA-50	498.0	118	338.3	0	0	570	297	148.5	16.71	55.7	18
GWP-10	896.4	118	0	97.1	0	570	297	148.5	16.71	55.7	18
GWP-20	796.8	118	0	194.3	0	570	297	148.5	16.71	55.7	18
GWP-30	697.2	118	0	291.5	0	570	297	148.5	16.71	55.7	18
GWP-40	597.6	118	0	388.6	0	570	297	148.5	16.71	55.7	18
GWP-50	498.0	118	0	485.8	0	570	297	148.5	16.71	55.7	18
VPP-10	896.4	118	0	0	75.3	570	297	148.5	16.71	55.7	18
VPP-20	796.8	118	0	0	150.6	570	297	148.5	16.71	55.7	18
VPP-30	697.2	118	0	0	225.9	570	297	148.5	16.71	55.7	18
VPP-40	597.6	118	0	0	301.2	570	297	148.5	16.71	55.7	18
VPP-50	498.0	118	0	0	376.5	570	297	148.5	16.71	55.7	18

2.2. Mixing and Curing

All the solid materials (GGBS, SF, sand, and RHA/GWP/VPP) were mixed dry for two minutes. After that, the alkaline activator solution, SP, and extra water were added and mixed for another three minutes. The PP fibers were subsequently added, progressively, evenly distributed, and mixed for another three minutes. The produced EGC was poured into the molds for different tests, and the specimens were subsequently compacted using a vibrating table for ten seconds at a frequency of 3600 vpm. The specimens were kept for 24 h before demolding and cured for another 24 h in an oven at 80 °C, as shown in Figure 3. The specimens were then kept in air after completing the heat-curing process until performing the planned tests. It is worth noting that although heat curing has some practicality and environmental concerns, it is a very effective method of curing for EGC, especially in the very early stages. However, with the recent huge boom in the manufacturing and use of solar cells, more practical and environmentally friendly heat curing of geopolymer concrete and ECC can be applied. Dong et al. [31] have successfully cured geopolymer concrete using the solar curing method and were able to easily reach a 65 °C curing temperature while showing a substantial improvement in the compressive strength.

Figure 3. Heat curing of EGC specimens for 24 h after demolding.

2.3. Test Methods and Specimens Preparation

2.3.1. Workability

The slump and flow diameter tests were conducted for the fresh properties of EGC by taking a sample of fresh EGC immediately after mixing, as shown in Figure 4. The tests were conducted according to AS 1012.3.5 [32]. The mean flow diameter value of EGC was determined by measuring two perpendicular diameters of the material.

Figure 4. Measuring of EGC workability.

2.3.2. Unit Weight

Unit weight tests were conducted for the physical properties of EGC. Prior to conducting the compressive strength test on EGC cubes at 28 days, the unit weight of each EGC mixture was measured by dividing the mass of the cube by the volume.

2.3.3. Mechanical Properties

Compressive Strength

Compressive strength was evaluated in accordance with ASTM C-109 [33] at three EGC ages, namely 7, 28, and 56 days. Nine cubes sized $70 \times 70 \times 70$ mm were cast (three cubes per measurement age), three specimens from each mixture were tested, and the average results were compared.

Tensile Strength

Uniaxial tensile test was performed using dog bone specimens, as shown in Figure 5, in accordance with the guidelines provided by the Japan Society of Civil Engineers [34]. Three specimens from each mix were tested at 28 days, and the average results were compared. Figure 6a describes the uniaxial tensile test setup.

Flexural Strength

A four-point flexural test was carried out at 28 days on EGC plate specimens in accordance with ASTM C1609 [35], as shown in Figure 5. Then, $400 \times 100 \times 20$ mm plates were cast to measure flexural strength (three per mixture). The detailed setup for the four-point flexural tests is shown in Figure 6b. Three specimens were tested, and the average results were compared.

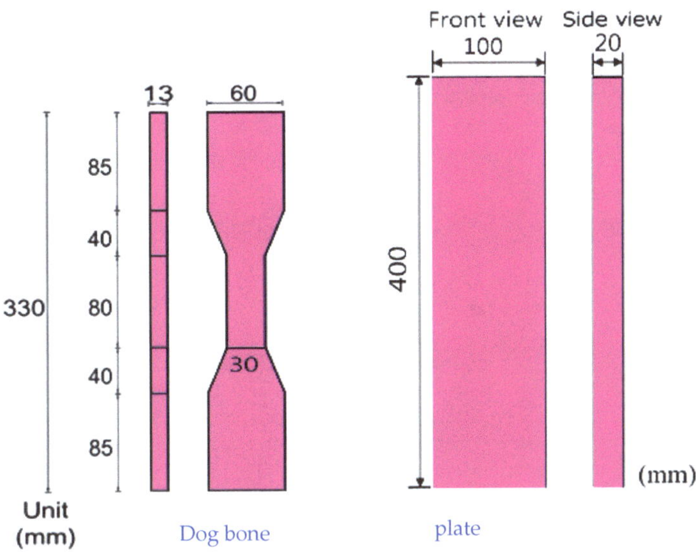

Figure 5. Dimensions of EGC dog-bone and plate specimens.

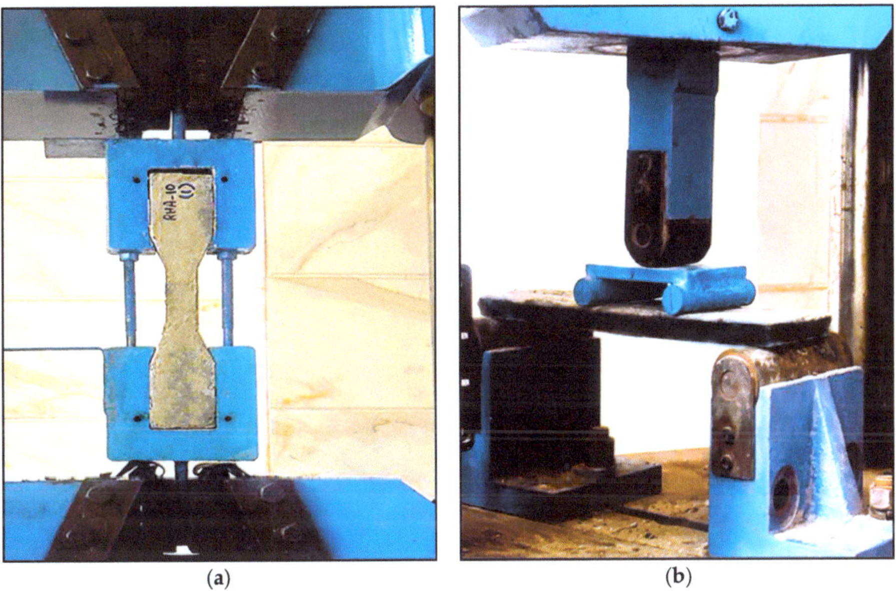

Figure 6. Mechanical test setups of EGC: (**a**) Uniaxial tensile test setup, (**b**) Four-point flexural test setup.

2.3.4. Water Absorption and Apparent Porosity

Water absorption and porosity are important properties that express the durability of the structure. The test was carried out according to ASTM C 948 [36]. At the age of 56, three cube specimens sized $70 \times 70 \times 70$ mm of each mixture were soaked in water for 24 h at a temperature of 21 °C. The weight of the fully saturated specimens was measured as "A". After that, the specimens were removed from the water, the surface was dried with a towel, and the weight of the saturated specimens was measured in air as "B". The specimens were then dried in an oven at a temperature of 100–110 °C for 24 h. After removing them from the oven, the specimens were left to cool down to room temperature, and their weights were measured as "C". The water absorption and apparent porosity values were determined using the following equations:

$$\text{Water absorption } (\%) = \frac{B - C}{C} \times 100 \tag{1}$$

$$\text{Apparent porosity } (\%) = \frac{B - C}{B - A} \times 100 \tag{2}$$

where A = immersed mass (g), B = saturated-surface-dry mass (g), and C = oven-dry mass (g).

2.3.5. Residual Compressive Strength

At 28 days, nine cubes sized $70 \times 70 \times 70$ mm of each mixture were used to gauge the resistance to high temperatures (three per temperature level of 200, 400, and 600 °C), placed in an electric oven with a capacity of 1200 °C, and subjected to different elevated temperatures (200, 400, and 600 °C) for two hours. Following heating, the specimens were allowed to cool down to room temperature for 24 h inside the oven to prevent thermal shock, and the residual compressive strength was then measured.

2.3.6. Microstructure Analysis

Scanning electron microscopy (SEM) analysis was carried out on 1×1 cm cut pieces that were obtained from the core of the fractured cubic specimens following the compressive strength test to investigate the microstructure of selected EGC mixtures. SEM analysis was performed on the control EGC mixture and optimum mixtures that showed the highest 28-day compressive strength, namely EGC, RHA-30, GPW-20, and VPP-10. The microscope used for the SEM analysis was a QUANTA FEG 250 device with an accelerating voltage of 20 kV.

3. Results and Discussions

Table 4 exhibits the outcomes of the measurements carried out on the EGC mixtures, namely workability (slump and flow diameter), unit weight, compressive strength, tensile strength, flexural strength, water absorption, and apparent porosity.

Table 4. EGC tests result.

MIX ID	Slump (mm)	Flow Diameter (mm)	Unit Weight (kg/m³)	Compressive Strength (MPa)			Tensile Strength (MPa)	Flexural Strength (MPa)	Water Absorption (%)	Apparent Porosity (%)
				7 D	28 D	56 D				
EGC	65	135	2128	75	78	79	4.5	6.0	6.88	13.44
RHA-10	62	131	2122	52	57	54	3.5	5.5	7.49	14.42
RHA-20	58	126	2118	50	61	48	3.6	5.7	8.62	16.08
RHA-30	54.5	118	2073	55	65	62	4.5	6.2	8.80	16.50
RHA-40	53	115	2007	49	53	51	4.1	5.0	10.44	18.07
RHA-50	50	110	1950	42	44	43	4.0	4.8	14.18	23.54
GWP-10	85	140	2114	52	55	57	4.0	6.3	6.90	13.50
GWP-20	65	135	2099	53	65	69	5.6	6.9	6.94	13.57
GWP-30	50	130	2093	51	53	55	4.9	6.5	8.09	15.22
GWP-40	48	118	2085	52	54	66	4.4	6.4	8.14	15.45
GWP-50	45	116	2067	43	48	53	3.5	6.2	8.48	15.95
VPP-10	55	132.5	2134	52	60	67	4.4	6.6	6.80	13.40
VPP-20	60	134	2152	51	53	57	4.0	6.0	5.56	10.68
VPP-30	75	140	2166	49	50	55	3.5	5.8	5.32	10.51
VPP-40	80	144	2199	46	47	54	3.1	5.1	5.28	10.48
VPP-50	90	160	2269	41	42	50	2.8	4.8	5.18	10.40

3.1. Workability

Figure 7 shows the results of EGC slump and flow diameter. It was observed that the slump and flow diameter of the control EGC were 65 mm and 135 mm, respectively. When partially replacing the EGC slag with RHA at a rate of 10–50%, it was noted that the higher the replacement rate, the lower the workability of the EGC compared to the control mixture, until it reached the lowest workability in mixture RHA-50. Using 50% RHA decreased the EGC slump by 23% and the flow diameter by 18.5%. The reason for the workability decrease when using RHA is attributed to its high absorption rate and large surface area [37]. This decreased the amount of free water within the EGC matrix, and hence, resulted in lower workability. A similar observation was found by Das et al. [38]. Also, Patel and Shah [39] highlighted that workability decreased with an increase in RHA content for the same reasons mentioned above.

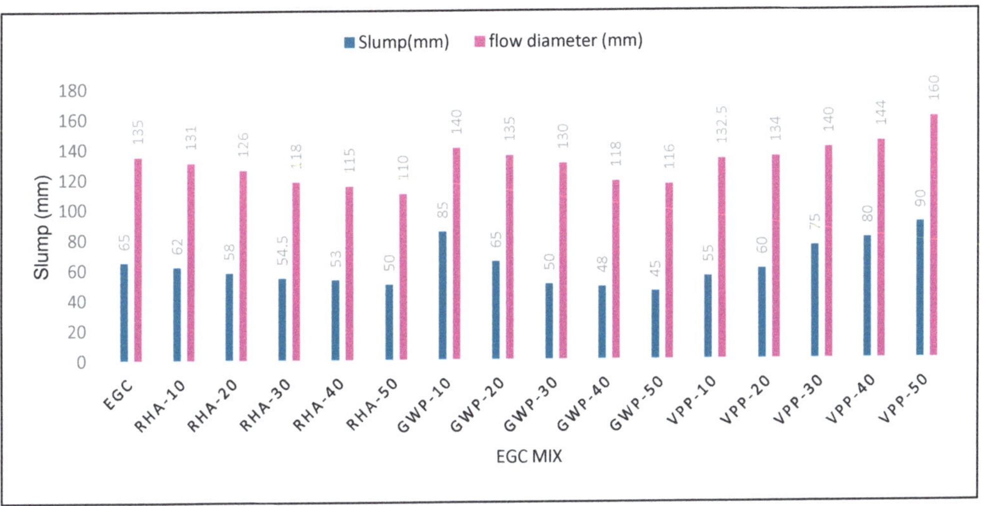

Figure 7. Workability of EGC.

When replacing slag with GWP in the same proportions of 10–50%, an increase in EGC workability was observed at a GWP ratio of 10% (GWP-10 mixture) in which the slump increased by 31% and the flow diameter increased by 4%. At 20% GWP, the workability was similar to that of the control EGC. However, it began to decrease gradually to the lowest value recorded in the GWP-50 mixture. Using 50% GWP decreased the EGC slump by 31% and the flow diameter by 14%. The workability increase at low levels of GWP and decrease at high levels of GWP are due to the irregular shape of GWP particles, which adversely affected the movability of the EGC matrix, and hence, resulted in lower workability. In addition, the high viscosity of the GWP slurry at high contents made the EGC sticky and less workable. Similar results were reported by Shilar et al. [27].

Replacing slag with VPP at the rate of 10% and 20% led to a decrease in EGC workability, compared to the control mixture. At 10% and 20% VPP, the EGC slump decreased by 15% and 8%, respectively, and the flow diameter decreased slightly by 2% and 1%, respectively. Beyond 20% VPP, the workability losses started to recover until it recorded its highest value in mixture VPP-50. Using 50% VPP increased the EGC slump by 38% and the flow diameter by 18%. This might happen because VPP has an uneven, porous particle form and a cellular structure. This may cause the EGC mixture to pack unevenly at low replacement rates, which would make it less workable. Nonetheless, the lubricating action of VPP became more pronounced at high contents, which lessened the particle friction and hence improved the EGC workability. Furthermore, the low weight of VPP could also contribute to the formation of a more cohesive and stable mixture that enhanced the EGC's overall workability.

In conclusion, 10% GWP provides the most promising results in terms of enhanced workability, with a noticeable improvement in slump and flow diameter. While preserving workability is the main priority, RHA should be restricted to lower levels (<50%). Finally, VPP gives a notable improvement in workability compared to the control, even at high replacement levels of 50%.

3.2. Unit Weight

The results of the EGC unit weight tests are plotted in Figure 8. The unit weight of the control mixture was 2128 kg/m^3. By replacing the slag with either RHA or GWP, the unit weight gradually decreased. The EGC unit weight reached its lowest value at a 50% replacement ratio with a value of 1950 kg/m^3 in mixture RHA-50 (8.36% decrease) and a value of 2067 kg/m^3 in mixture GWP-50 (2.86% decrease). This decrease is attributed to the fact that both RHA and GWP have a lower density than the replaced slag. When replacing slag with VPP and when compared to the control mixture, it was found that the EGC unit weight increased with the replacement rate increase because the open-cell structure of VPP allows it to be easily filled and compacted, resulting in reduced porosity in the EGC mixture. This reduction in porosity leads to increased density compared to the EGC control mixture. In addition, the high reactivity of VPP with the alkaline activator contributes to a more compact and cohesive matrix and hence increased unit weight. Using 50% VPP increased the EGC unit weight to 2269 kg/m^3 (6.62% increase).

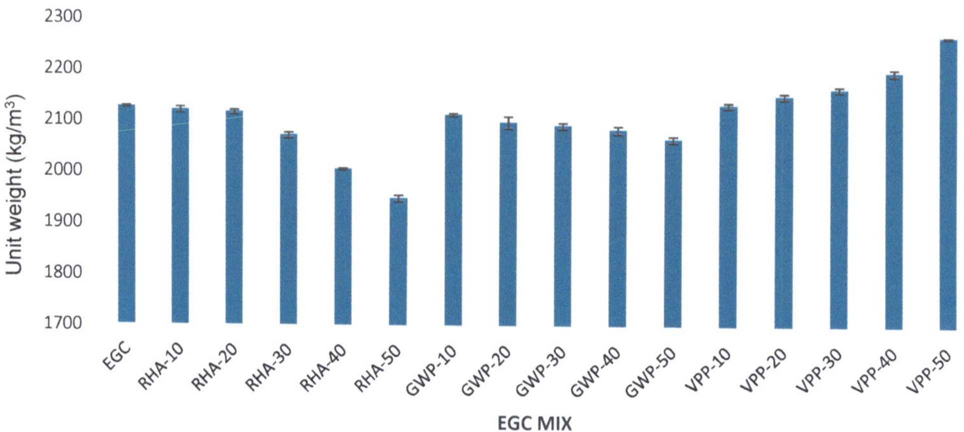

Figure 8. Unit weight of EGC.

In conclusion, unit weights for RHA and GWP drop with rising replacement rates until they achieve their lowest values at a 50% substitution level, whereas unit weights for VPP grow with increasing rates until they achieve their greatest values at a 50% substitution level.

3.3. Compressive Strength

All results are indicated in Table 4 and plotted in Figure 9. The control EGC mixture showed compressive strengths of 75, 78, and 79 MPa at 7, 28, and 56 days, respectively. When replacing slag with RHA/GWP/VPP, it was found that the compressive strength decreased at any given replacement ratio and EGC age compared to the control mixture. For example, at 28 days, replacing EGC slag by 10%, 20%, 30%, 40%, and 50%, decreased its compressive strength by 26.9%, 21.8%, 16.7%, 32.1%, and 43.6%, respectively, when using RHA; by 29.3%, 16.7%, 32.1%, 30.8%, and 38.5%, respectively, when using GWP; and by 23.1%, 32.1%, 35.9%, 39.7%, and 46.2%, respectively, when using VPP. Within each group, the highest EGC compressive strength was shown at 30% RHA (65 MPa), 20% GWP (65 MPa), and 10% VPP (60 MPa), which reflect the optimal ratios to use for these waste materials as an EGC slag partial replacement. The mixtures with a larger slag content had higher compressive strengths because of the development of calcium silicate hydrates (C-S-H) and alumino-silicate hydrate (A-S-H) gel (phases and the density of the microstructure [40,41]. The decrease in EGC compressive strength when using RHA/GWP/VPP is attributed to several reasons. For RHA, increasing the RHA content causes an imbalance between Si and Al, which leads to the formation of low cross-linked aluminosilicates [8,42], which causes a decrease in compressive strength [38]. Zeyad et al. [43] pointed out that the compressive strength was significantly reduced when 15% of RHA was used in place of slag, which potentially led to a higher silica content. According to research of Patel and Shah [39], there is an optimal range of 5% to 10% for substituting RHA with slag. In addition, compared to the geopolymer matrix zero-RHA, it was shown that adding RHA in place of slag at replacement ratios between 15% and 25% resulted in a decrease in compressive strength. According to Kusbiantoro et al. [44], the strength decreased with increasing RHA content because the difference in slag and RHA's degrees of solubility decreases the rate at which aluminosilicate compounds dissolve and condense.

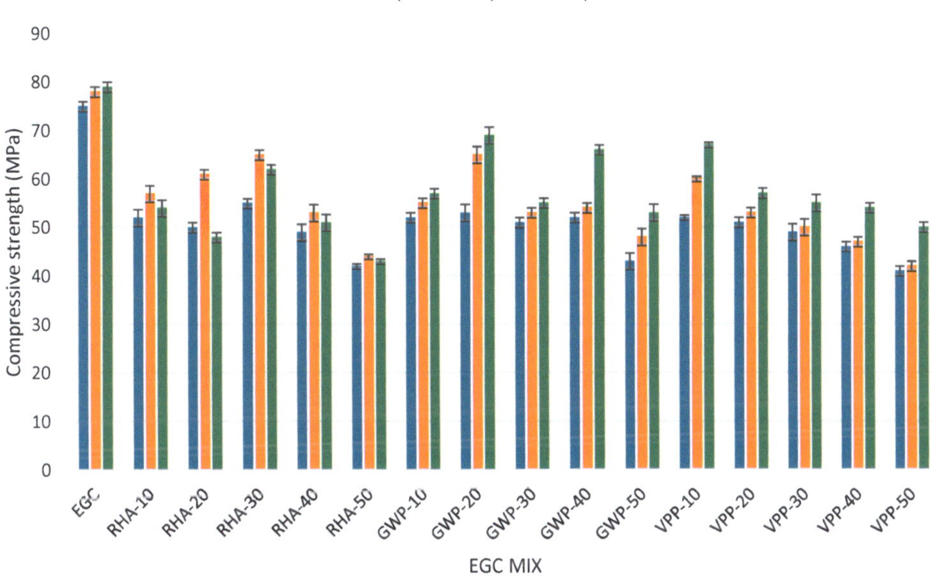

Figure 9. Compressive strength of EGC at different ages.

The excessive incorporation of GWP disrupted the geopolymer matrix formation and weakened the bond between particles, hence resulting in lower strength. Shilar et al. [27] concluded that the maximum compressive strengths were obtained with a GWP content of 20%. Khater et al. [45] concluded that a further increase in the content of granite waste powder of up to 15% causes the created geopolymer structure to deactivate and lose strength. The increased amounts of VPP decreased the compressive strength due to decreased Ca/Si ratio [46].

All EGC mixtures in this study showed common development of their compressive strength when comparing the 7, 28, and 56 days' strengths for each mixture, except for the RHA group. When using RHA as partial slag replacement in EGC, the compressive strength increased from age of 7 days to 28 days by 9.6%, 22%, 18.2%, 8.1%, and 4.7%, respectively, at 10%, 20%, 30%, 40%, and 50% RHA content; however, the corresponding strength of these mixtures decreased from an age of 28 days to 56 days by 5.2%, 21.3%, 4.6%, 3.7%, and 2.2%, respectively. This strength decrease at late EGC ages can be attributed to several reasons. The interaction between RHA and alkaline activators in the EGC mixture may be varied over time, affecting the geopolymerization process and strength development at different ages. It could also be due to moisture absorption, as RHA may have higher moisture absorption properties compared to slag, which may cause changes in the EGC matrix over time, affecting the compressive strength at 56 days. The temperature changes can be another reason why the temperature fluctuations during the curing period can affect the strength development of EGC, leading to differences in compressive strength at different ages. Future research is needed to focus on the RHA-based EGC strength development at late ages.

In conclusion, the optimum ratio regarding strength and material sustainability is 30% RHA replacement and 20% GWP, which maintains enough geopolymer matrix integrity while offering an acceptable degree of strength. Finally, 10% VPP substitution is recommended for higher strength requirements since higher VPP content results in significant strength losses.

3.4. Tensile Strength

As shown in Table 4 and Figure 10, the tensile strength of the EGC control mixture was 4.5 MPa, and when slag was replaced with RHA, the tensile strength ranged from 3.5 to 4.5 MPa. The RHA mixture with a 30% replacement ratio (RHA-30) obtained the highest tensile strength of 4.5 MPa, which is equal to the tensile strength of the control mixture. Beyond 30% RHA, the high substitution of slag decreased the tensile strength. Factors such as poor dispersion of RHA particles, increased porosity, or altered microstructure could lead to a decrease in the tensile strength of RHA mixtures. The tensile strength generally follows a similar trend with compressive strength, decreasing with higher slag replacement ratios. Mehta and Siddique [47] stated that the inclusion of RHA beyond 15% decreased split tensile strength. This was due to the difference in solubility rates of RHA and slag and the presence of unreactive silica in the mixture. Patel and Shah [39] show that when the content of RHA increases, split tensile strength decreases, and at all curing ages and at 100% slag content, the greatest split tensile strength values are obtained at ambient temperature.

When replacing slag with GWP at a rate of 10%, the tensile strength decreased by 11.1%, compared to the control mixture. Beyond that, the tensile strength increased by 24.4% and 8.8% at rates of 20% and 30% GWP, respectively. This is in contrast to what resulted in the compressive strength, in which it decreased at all replacement ratios compared to the control mixture. At 40% and 50% GWP, the EGC tensile strength decreased by 2.2% and 22.2%, respectively. The increase in tensile strength at 20% and 30% GWP may be due to the chemical composition of GWP, which enhances the reactions within the geopolymer matrix, resulting in stronger bond formation and increased tensile strength. Also, these replacement ratios may exhibit better compatibility with PP fibers, which improves the bonding between the fibers and the geopolymer matrix and, hence, provides better tensile strength. Shilar et al. [27] observed that with an increase in the GWP content in the mix, the split tensile strength increases up to 20% GWP; beyond 20%, the split tensile strength decreases.

The tensile strength decreased when VPP was used as a partial replacement of EGC slag by 2.2% to 37.8%, as it showed 4.4 MPa in mixture VPP-10 and 2.8 MPa in mixture VPP-50. The is attributed to the lower adhesion and bonding that might occur between the geopolymer matrix, fibers, and VPP particles, as is the case when slag was present at a high content, leading to weak bonding and decreased tensile strength. There is a clear correlation between compressive and tensile strengths. Both strengths peak at optimum replacement ratios but decline significantly at higher replacement levels.

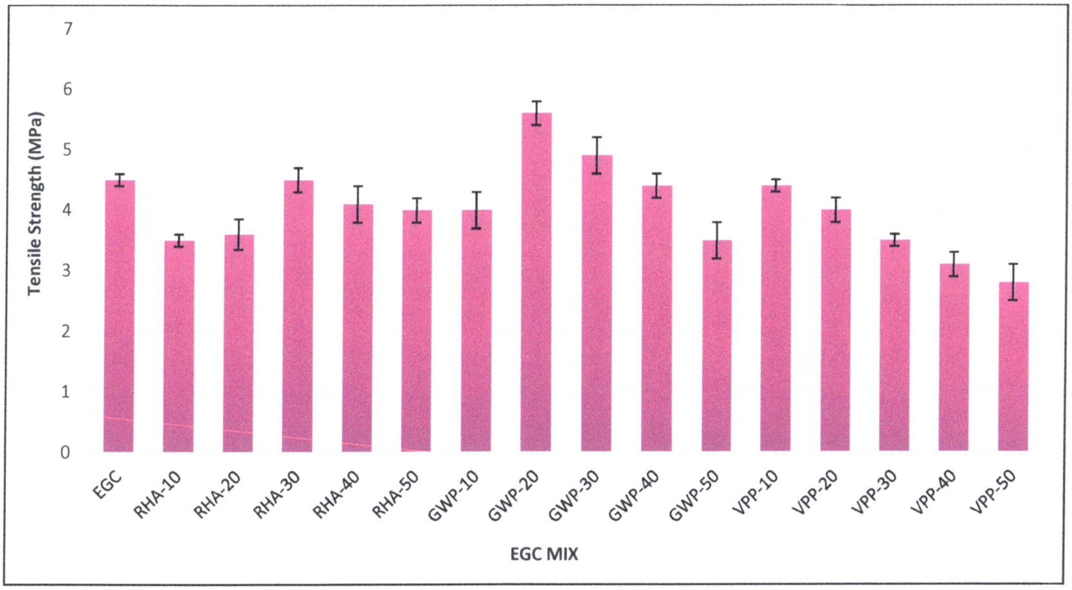

Figure 10. Tensile strength of EGC.

To sum up, according to the tensile strength data, the EGC mixes' tensile performance is greatly impacted when slag is substituted with RHA, GWP, or VPP. A 30% RHA replacement ratio provides an optimal tensile strength equal to the control mixture. The optimum balance is provided by 20% GWP substitution, which increases tensile strength by 24.4%. Since all VPP mixes resulted in a reduction in tensile performance, it is not advised to utilize VPP as a partial substitute for slag in order to increase tensile strength. Replacements of 30% RHA or 20% GWP are advised for situations where tensile strength is crucial since they provide optimum performance while still using sustainable resources.

3.5. Flexural Strength

All results are indicated in Table 4 and plotted in Figure 11.

The flexural strength of the EGC control mixture showed a value of 6 MPa. Using RHA in EGC showed a flexural strength trend similar to that of the corresponding tensile strength due to the strong correlation between them. The flexural strength increased with increasing RHA content until it reached its maximum value at the optimal replacement ratio of 30%, then decreased beyond that. The flexural strength decreased by 8.3% and 5% at 10% and 20% RHA, respectively, and increased by 3.3% at 30% RHA. It again decreased by 16.7% and 20% at 40% and 50% RHA, respectively. When using RHA that has a high SiO_2/Al_2O_3 ratio, the increase in the concentration of amorphous silica in the geopolymer structure may be the reason for the declining flexural strength. This occurrence produces a geopolymer gel with a lower flexural strength that is weaker, less densely bonded, and more porous. Moreover, the dissolution rate and polycondensation of aluminosilicates decreased due to the difference in melting points between GBFS and RHA. As a result, the strength decreases with the increase in RHA amount [43].

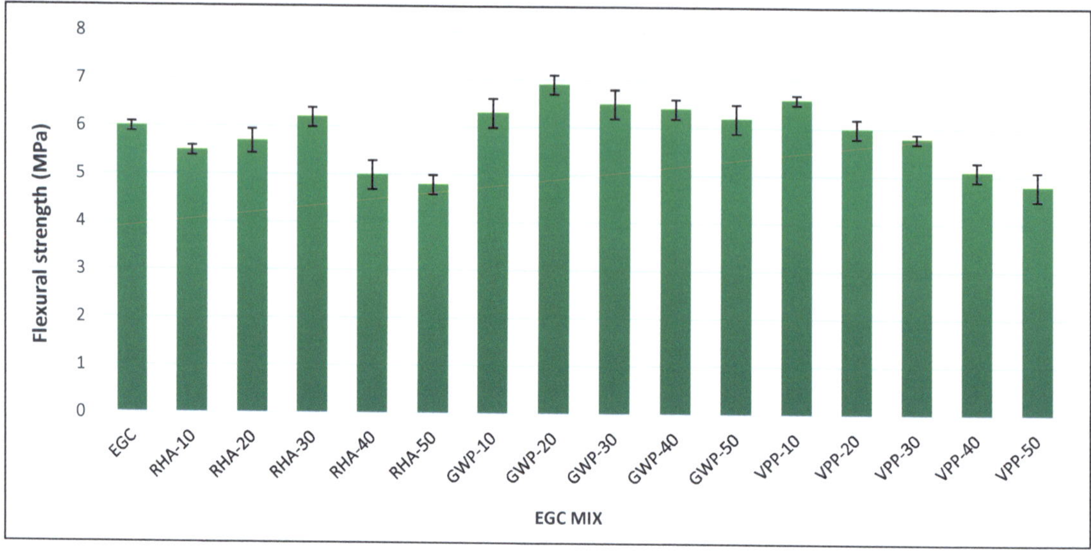

Figure 11. Flexural strength of EGC.

In all GWP mixtures, the flexural strength improved compared to the control mixture, where the maximum flexural strength was achieved at a replacement ratio of 20% with an increase of 15%. This is in contrast to what resulted in the compressive strength, in which it decreased at all replacement ratios compared to the control mixture, as well as the tensile strength, where the increase was at replacement ratios of 20% and 30% only. GWP has pozzolanic properties, reacting with the alkaline activators in the geopolymer matrix to form additional binding phases. This reaction led to the formation of more hydrated products, enhancing the interfacial transition zone between the fibers and the matrix, which improved the load transfer mechanism and flexural strength. In the same direction, Shilar et al. [27] found that the flexural strength increased with GWP levels up to 20%, but beyond 20%, the flexural strength decreased.

The flexural strength at 10% VPP increased by 10% due to the optimum pozzolanic activity, in which VPP actively participates in pozzolanic reactions, resulting in additional C-S-H and other binding phases. These reactions complemented the binding capacity of the geopolymer matrix and enhanced its overall strength. The additional hydrated products formed at this level also improved the adhesion between the fibers and the matrix and enhanced the load transfer mechanism and flexural strength. This is in contrast to what resulted in the compressive and tensile strengths, in which they decreased at all replacement ratios compared to the control mixture. At 20% VPP, the pozzolanic activity of VPP contributed positively to the geopolymer matrix, but the reduced GGBS content began to offset these benefits. GGBS is a major component of the geopolymer, and its low presence means that the matrix cannot benefit much from the improved properties of VPP alone, resulting in no significant gain or loss in flexural strength. As the VPP content increased by 30%, 40%, and 50%, the decrease in GGBS became significant, resulting in a weaker matrix that reduced the load transfer efficiency. This poor interaction resulted in decreased flexural strength.

In conclusion, 30% RHA is the optimum replacement ratio because it maximizes flexural strength; however, performance is adversely affected by larger RHA concentrations. Of all the replacement materials, a 20% GWP replacement ratio provides the maximum flexural strength and the best performance. While a 10% VPP substitution improves flexural strength, greater levels result in a weaker geopolymer matrix and worse performance. A 30% RHA or 20% GWP replacement is advised for increasing flexural strength; GWP exhibits the best increase. To prevent appreciable decreases in matrix strength, VPP usage should be limited to tiny amounts.

3.6. Water Absorption and Apparent Porosity

The results of the water absorption and apparent porosity tests are shown in Table 4 and plotted in Figure 12. The water absorption and apparent porosity of the control EGC mixture were found to be 6.88% and 13.44%, respectively. When slag was replaced by RHA, the water absorption and apparent porosity increased with the increase in the RHA ratio. This is due to the increased porosity of RHA, which tends to have a more porous structure compared to slag. The RHA can absorb and retain water, which contributes to the increased water absorption of the composite. The water absorption of EGC containing RHA ranged from 7.49% to 14.18%, and the apparent porosity ranged from 14.42% to 23.54%. These results are consistent with the study of Zhao, Y. et al. [8] who concluded that RHA increased water absorption and the capillary absorption coefficient of geopolymer paste due to a great number of interrelated pores at a microscopic scale, resulting in RHA being able to absorb a huge amount of water.

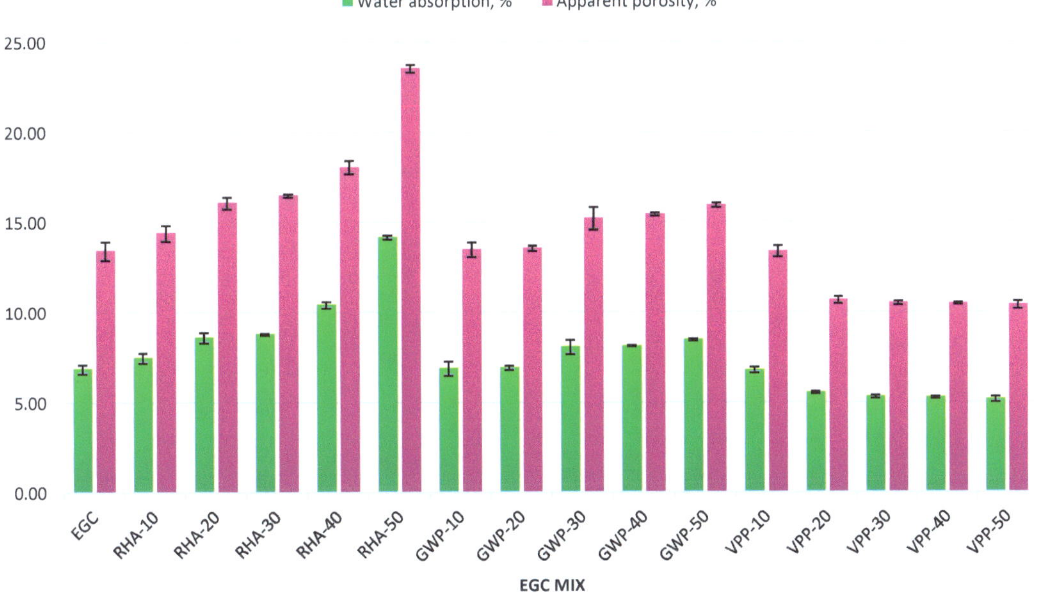

Figure 12. Water absorption and apparent porosity of EGC.

When slag was replaced by GWP, water absorption and apparent porosity increased with increasing GWP content compared to the control EGC mixture until reaching the highest water absorption and porosity of 8.48% and 15.95%, respectively, in the GWP-50 mixture. This increase is due to the larger particle size of GWP compared to slag, which led to less dense packing of particles within the geopolymer matrix, resulting in increased voids and, therefore, increased porosity and water absorption. A similar conclusion was obtained by Shilar et al. [27] in that water absorption increases as the proportion of granite increases.

The use of VPP showed a decrease in water absorption and porosity when compared to the control mixture. By increasing the VPP content, the water absorption and porosity kept decreasing, showing the lowest values of 5.18% and 10.4%, respectively, in the VPP-50 mixture. Since the open cell structure of VPP allows it to be easily filled and compacted, it led to an increase in EGC density, a decrease in its porosity, and a decrease in water absorption. Zeyad et al. [48] showed that a pore's tortuosity increase with increasing volcanic pumice particles can significantly reduce permeability and water absorption. This result is consistent with the unit weight test results discussed in Section 3.2.

The results of water absorption and apparent porosity tests revealed a clear trend in which increasing RHA content consistently resulted in higher water absorption and porosity, highlighting its more porous nature than slag. This porous structure allowed for greater water retention, with the highest values observed at 50% RHA replacement. Similarly, increasing GWP concentration increased water absorption and porosity, resulting in greater voids in the geopolymer matrix. In contrast, VPP responded differently, in which increasing its concentration resulted in a reduction in water absorption and porosity, with the lowest values reported at 50% VPP.

3.7. Residual Compressive Strength

All results are indicated in Figure 13. At 200 °C, the residual compressive strength of the control EGC mixture decreased by 3.8% compared to that measured at room temperature. The irregular shape of the slag particles and its high density, which produces a dense mix, led to the accumulation of vapor pressure that caused crack formations and reduced the transition area between the EGC matrix and the PP fibers. This result is consistent with the unit weight test result discussed in Section 3.2. The mixtures containing the optimum replacement ratios (RHA-30, GWP-20, VPP-10) showed better resistance to high temperatures, as it improved by 4.6%, 20%, and 3.3% for mixtures RHA-30, GWP-20, and VPP-10, respectively, compared with that measured at room temperature. The improvement in strength at 200 °C temperature is the result of the further geopolymerization process between SF and slag in addition to the optimal substitute material ratio with the alkaline solution. The compressive strength decreased by 21.9%, 32.4%,12.2% and 23.1% for mixtures RHA-10, RHA-20, RHA-40 and RHA-50, respectively; by 16.3%, 10.3%,11.8%, and 2.7% for mixtures GWP-10, GWP-30, GWP-40, and GWP-50, respectively; and by 9.4%, 22%, 6.3%, and 13% for mixtures VPP-20, VPP-30, VPP-40, and VPP-50, respectively.

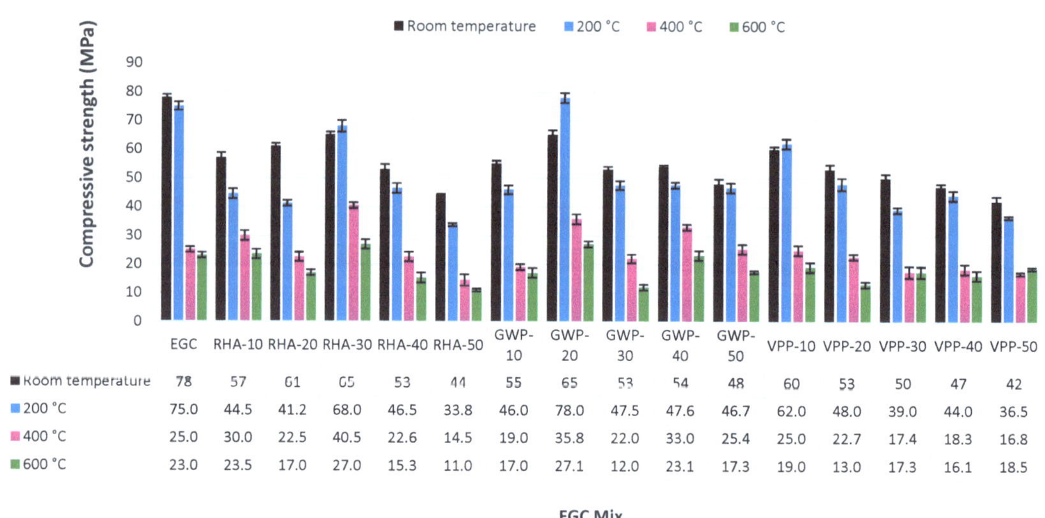

	EGC	RHA-10	RHA-20	RHA-30	RHA-40	RHA-50	GWP-10	GWP-20	GWP-30	GWP-40	GWP-50	VPP-10	VPP-20	VPP-30	VPP-40	VPP-50
Room temperature	78	57	61	65	53	44	55	65	53	54	48	60	53	50	47	42
200 °C	75.0	44.5	41.2	68.0	46.5	33.8	46.0	78.0	47.5	47.6	46.7	62.0	48.0	39.0	44.0	36.5
400 °C	25.0	30.0	22.5	40.5	22.6	14.5	19.0	35.8	22.0	33.0	25.4	25.0	22.7	17.4	18.3	16.8
600 °C	23.0	23.5	17.0	27.0	15.3	11.0	17.0	27.1	12.0	23.1	17.3	19.0	13.0	17.3	16.1	18.5

Figure 13. Residual compressive strength after exposure to high elevated temperatures.

At a temperature of 400 °C, the control mixture EGC witnessed a decrease in compressive strength of 67.9%, while the other mixtures had strength decrease of 47.4%, 63.1%,

37.7%, 57.4%, and 67.0% for mixtures RHA-10, RHA-20, RHA-30, RHA-40, and RHA-50, respectively; of 65.5%, 44.9%, 58.5%, 38.9%, 47.1% for mixtures GWP-10, GWP-20, GWP-30, GWP-40, and GWP-50, respectively; and of 58.3%, 57.2%, 65.2%, 61.1%, and 60% for mixtures VPP-10, VPP-20, VPP-30, VPP-40 and VPP-50, respectively.

At a temperature of 600 °C, the control mixture EGC witnessed a decrease in compressive strength by 70.5%, while the other mixtures had a strength decrease of 58.8%, 72.1%, 58.5%, 71.1%, and 75% for mixtures RHA-10, RHA-20, RHA-30, RHA-40, and RHA-50, respectively, of 69.1%, 58.3%, 77.4%, 57.2%, and 64% for mixtures GWP-10, GWP-20, GWP-30, GWP-40, and GWP-50, respectively; and of 68.3%, 75.5%, 65.4%, 65.7%, and 56% for mixtures VPP-10, VPP-20, VPP-30, VPP-40, and VPP-50, respectively.

The compressive strength decrease under high elevated temperature is attributed to the PP fiber's performance when subjected to high temperature. The melting point of these fibers is between 160 and 170 °C. These fibers start to melt and break down at 200 °C and above, which leaves continuous voids in the EGC matrix and reduces the mechanical characteristics.

There is a clear correlation between compressive strength and residual compressive strength under high temperatures. The mixtures with optimum replacement ratios (RHA-30, GWP-20, VPP-10) not only maintained higher compressive strength at room temperature but also showed superior resistance to strength loss when exposed to high temperatures.

In conclusion, at 200 °C, GWP-20 exhibited the highest increase in compressive strength (20%), followed by RHA-30 (4.6%) and VPP-10 (3.3%). At 400 °C and 600 °C, the compressive strength of all mixtures showed a significant decline. The melting and decomposition of PP fibers above 200 °C are the primary factors behind the reduced compressive strength across all mixtures at elevated temperatures. Among the mixtures, GWP-20 exhibited the smallest strength loss at both 400 °C and 600 °C, indicating better thermal resistance compared to other mixtures.

4. Microstructure Analysis

Figure 14 shows the SEM analysis results of the selected EGC mixes. As shown in Figure 14a, the EGC control mixture presents a compact and dense matrix with fewer microcracks and minimal porosity, in addition to very limited unreacted particles. This denser microstructure is attributed to the effective geopolymerization process. Thus, when the volume and permeability of the voids, pores, and cracks decrease, the compressive strengths of the EGC increase and their durability improves, as observed and discussed in Section 3. PP fibers are embedded within the matrix, enhancing the composite's mechanical properties by bridging microcracks and providing additional tensile strength. The interface between the fibers and the matrix appears to be well-bonded, indicating effective load transfer between the matrix and the fibers. This interaction is crucial for the composite's performance under tensile and flexural loads. Some microcracks are visible in the image, which can be attributed to the curing process or mechanical stresses. These microcracks can propagate under loading conditions, potentially affecting the composite's integrity.

Well-dispersed RHA particles within the matrix are shown in Figure 14b. Compared to the EGC mixture, the RHA-30 mixture exhibits more porosity and microcracks. These voids can lead to higher water absorption and apparent porosity, as evident by the measured relevant characteristics. On the other hand, homogenous microstructure can be seen in Figure 14c for mixture GPW-20, although some regions exhibit higher porosity. The presence of GWP can contribute to this variation in density. The interface between the PP fibers and the geopolymer matrix seems well-integrated, indicating good adhesion that is essential for effective stress transfer between the fibers and the matrix. Some microvoids and pores are also visible in the structure. These can impact the composite's mechanical properties and durability. The inclusion of GWP introduced additional porosity compared to the EGC control mixture. The dense regions in the image likely represent the CSH phase, which is formed by the reaction of GGBS and the alkaline activator. This phase is crucial for the compressive strength of the composite. The presence of GWP can increase the apparent porosity and water absorption of the composite. This is consistent with findings in recent studies [49], where the incorporation of waste powders tends to introduce more porosity due to less efficient packing compared to EGC control mixtures. Figure 14d shows a dense matrix with embedded fibers when VPP was present, which indicates efficient geopolymerization and fiber distribution. The figure also shows that the volcanic pumice powder particles are completely dissolved and have reacted with the activator solution, resulting in the formation of a geopolymer grid.

In conclusion, better mechanical qualities are often associated with dense matrix structures and evenly distributed fibers, whereas increasing porosity tends to decrease strength and durability.

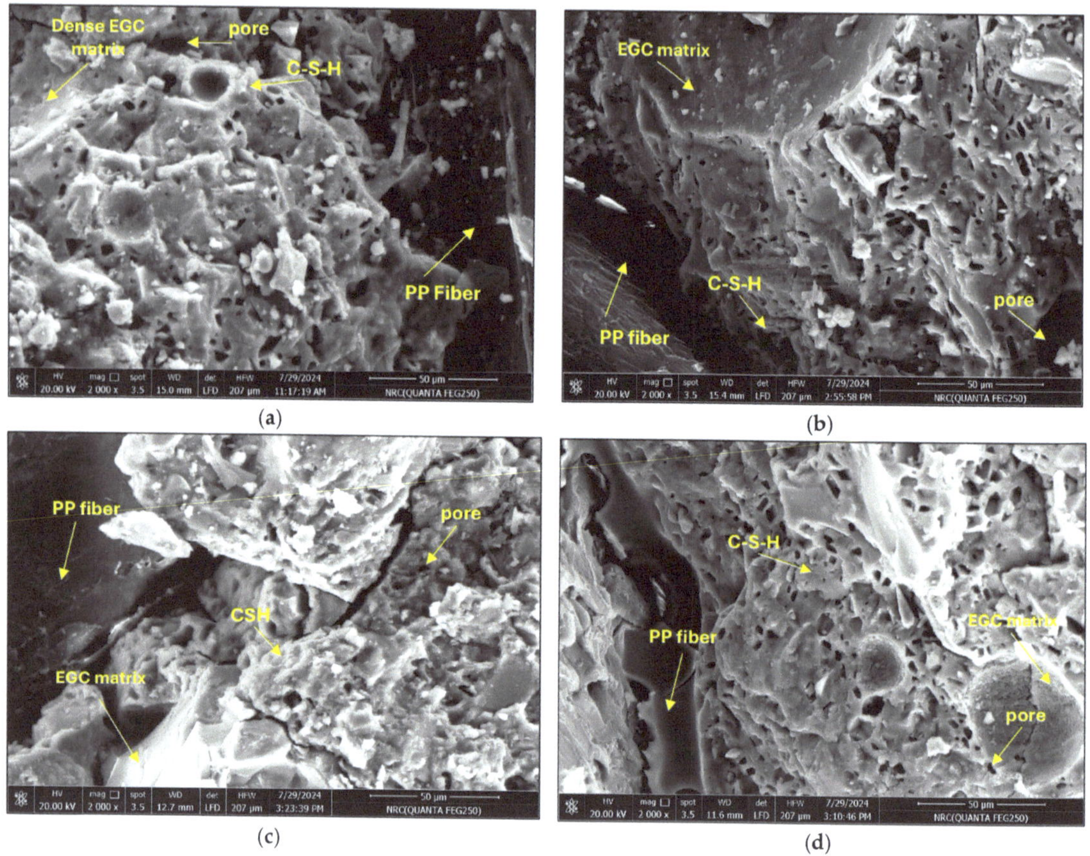

Figure 14. SEM images of EGC mixes: (**a**) EGC-control, (**b**) RHA-30, (**c**) GWP-20, and (**d**) VPP-10.

5. Conclusions

This research focused on producing a more sustainable and eco-friendly EGC made of cheaper and more common pozzolanic waste materials, namely RHA, GWP, and VPP. These waste materials were used as partial substitutions (10–50%) of GGBFS in EGC. The effects of these wastes on the mechanical, durability, and microstructure properties were examined. The following key conclusions can be listed:

1. Using RHA or GWP generally decreased the EGC workability by up to 31%, However, using VPP in EGC increased its workability by up to 38%. The compressive strength decreased by up to 43.6% when using RHA, 38.5% when using GWP, and 46.2% when using VPP compared to the control mixture at 28 days. Optimal replacement ratios were identified as 30% for RHA, 20% for GWP, and 10% for VPP.

2. Increasing GWP content led to higher water absorption and apparent porosity, reaching 8.48% and 15.95%, respectively, at a 50% replacement ratio. In contrast, using VPP reduced water absorption and porosity.

3. At 200 °C, optimal mixtures (RHA-30, GWP-20, VPP-10) showed improved strength, while other mixtures exhibited substantial strength decreases. At 400 °C and 600 °C, all mixtures experienced significant strength reductions. For applications requiring high-temperature resistance, high-melting point refractory fibers used to replace low-melting point polypropylene fibers tend to melt and decompose at elevated temperatures.

4. The SEM analysis of the control EGC mixture exhibited a dense matrix with minimal porosity and well-bonded PP fibers, contributing to enhanced mechanical properties. In the RHA-30 mixture, increased porosity and microcracks were observed. The GWP-20 mixture showed a homogeneous microstructure but with higher porosity. The VPP-10 mixture displayed efficient geopolymerization with a dense matrix and well-distributed fibers.

5. It is recommended for future studies to investigate other durability properties of EGC, such as chloride permeability, sulfuric acid attacks, dry shrinkage, freezing, and corrosion. It is also recommended to focus on investigating the strength development of RHA-based EGC at late ages.

Author Contributions: Conceptualization, A.M.T. and O.Y.; methodology, A.M.T., D.S.A., M.A. and O.Y.; validation, D.S.A. and M.A.; investigation, A.M.T., D.S.A. and O.Y.; data curation, D.S.A.; writing—original draft, D.S.A.; writing—review and editing, A.M.T. and O.Y.; visualization, M.A. and O.Y.; supervision, A.M.T. All authors have read and agreed to the published version of the manuscript.

Funding: This research received no external funding.

Data Availability Statement: The original contributions presented in the study are included in the article. Further inquiries can be directed to the corresponding author.

Conflicts of Interest: The authors declare no conflict of interest.

References

1. Biyani, Y.; Patil, L.G.; Kurhe, C.N. *Engineered Cementitious Composites*; Springer International Publishing: Berlin/Heidelberg, Germany, 2020; Volume 2, ISBN 9783662584378.

2. Stark, J.; Möser, B.; Bellmann, F. *Nucleation and Growth of C-S-H Phases on Mineral Admixtures*; Springer: Berlin/Heidelberg, Germany, 2007; ISBN 9783540724476.

3. Wang, S.; Li, V. Engineered Cementitious Composites with High-Volume Fly Ash. *ACI Mater. J.* **2007**, *104*, 233–241. [CrossRef]

4. Andrew, R.M. Global CO_2 Emissions from Cement Production. *Earth Syst. Sci. Data* **2018**, *10*, 195–217. [CrossRef]

5. Luukkonen, T.; Abdollahnejad, Z.; Yliniemi, J.; Kinnunen, P.; Illikainen, M. One-Part Alkali-Activated Materials: A Review. *Cem. Concr. Res.* **2018**, *103*, 21–34. [CrossRef]

6. Aghaeipour, A.; Madhkhan, M.; Ashrafian, A.; Taheri Amiri, M.J.; Masoumi, P.; Asadi-Shiadeh, M.; Yaghoubi-Chenari, M.; Mosavi, A.; Nabipour, N.; Pavan, S.; et al. *Geopolymers Structure, Processing, Properties and Industrial Applications*; Woodhead Puplishing Limited: Cambridge, UK, 2017; Volume 21, ISBN 9789400776715.

7. Elemam, W.E.; Tahwia, A.M.; Abdellatief, M.; Youssf, O.; Kandil, M.A. Durability, Microstructure, and Optimization of High-Strength Geopolymer Concrete Incorporating Construction and Demolition Waste. *Sustainability* **2023**, *15*, 15832. [CrossRef]

8. Zhao, Y.; Chen, B.; Duan, H. Effect of Rice Husk Ash on Properties of Slag Based Geopolymer Pastes. *J. Build. Eng.* **2023**, *76*, 107035. [CrossRef]

9. Youssf, O.; Mills, J.E.; Elchalakani, M.; Alanazi, F.; Yosri, A.M. Geopolymer Concrete with Lightweight Fine Aggregate: Material Performance and Structural Application. *Polymers* **2023**, *15*, 171. [CrossRef]

10. Walkley, B. Geopolymers. *Encycl. Earth Sci. Ser.* **2020**, *37*, 1633–1656. [CrossRef]

11. Youssf, O.; Elchalakani, M.; Hassanli, R.; Roychand, R.; Zhuge, Y.; Gravina, R.J.; Mills, J.E. Mechanical Performance and Durability of Geopolymer Lightweight Rubber Concrete. *J. Build. Eng.* **2022**, *45*, 103608. [CrossRef]

12. Zhong, H.; Zhang, M. Engineered Geopolymer Composites: A State-of-the-Art Review. *Cem. Concr. Compos.* **2023**, *135*, 104850. [CrossRef]

13. Wang, F.; Ding, Y.; Nishiwaki, T.; Zhang, Z.; Yu, J.; Yu, K. Fully Recycled Engineered Geopolymer Composite: Mechanical Properties and Sustainability Assessment. *J. Clean. Prod.* **2024**, *471*, 143382. [CrossRef]

14. Behzad, N.; Jay, S.; Uddin, A.S.F. Tensile Strain Hardening Behavior of PVA Fiber-Reinforced Engineered Geopolymer Composite. *J. Mater. Civ. Eng.* **2015**, *27*, 4015001. [CrossRef]

15. Nematollahi, B.; Sanjayan, J.; Shaikh, F.U.A. Matrix Design of Strain Hardening Fiber Reinforced Engineered Geopolymer Composite. *Compos. Part B Eng.* **2016**, *89*, 253–265. [CrossRef]

16. Nematollahi, B.; Sanjayan, J. Effect of Different Superplasticizers and Activator Combinations on Workability and Strength of Fly Ash Based Geopolymer. *Mater. Des.* **2014**, *57*, 667–672. [CrossRef]

17. Nematollahi, B.; Sanjayan, J.; Shaikh, F.U.A. Synthesis of Heat and Ambient Cured One-Part Geopolymer Mixes with Different Grades of Sodium Silicate. *Ceram. Int.* **2015**, *41*, 5696–5704. [CrossRef]

18. Alrefaei, Y.; Dai, J.G. Tensile Behavior and Microstructure of Hybrid Fiber Ambient Cured One-Part Engineered Geopolymer Composites. *Constr. Build. Mater.* **2018**, *184*, 419–431. [CrossRef]

19. Ling, Y.; Wang, K.; Li, W.; Shi, G.; Lu, P. Effect of Slag on the Mechanical Properties and Bond Strength of Fly Ash-Based Engineered Geopolymer Composites. *Compos. Part B Eng.* **2019**, *164*, 747–757. [CrossRef]

20. Deb, P.S.; Nath, P.; Sarker, P.K. The Effects of Ground Granulated Blast-Furnace Slag Blending with Fly Ash and Activator Content on the Workability and Strength Properties of Geopolymer Concrete Cured at Ambient Temperature. *Mater. Des.* **2014**, *62*, 32–39. [CrossRef]

21. Ahmed, J.K.; Atmaca, N.; Khoshnaw, G.J. Exploring Flexural Performance and Abrasion Resistance in Recycled Brick Powder-Based Engineered Geopolymer Composites. *Beni-Suef Univ. J. Basic Appl. Sci.* **2024**, *13*, 68. [CrossRef]

22. Wu, H.; Liu, X.; Wang, C.; Zhang, Y.; Ma, Z. Micro-Properties and Mechanical Behavior of High-Ductility Engineered Geopolymer Composites (EGC) with Recycled Concrete and Paste Powder as Green Precursor. *Cem. Concr. Compos.* **2024**, *152*, 105672. [CrossRef]

23. Nair, D.G.; Jagadish, K.S.; Fraaij, A. Reactive Pozzolanas from Rice Husk Ash: An Alternative to Cement for Rural Housing. *Cem. Concr. Res.* **2006**, *36*, 1062–1071. [CrossRef]

24. Zabihi, S.M.; Tavakoli, H.R. Evaluation of Monomer Ratio on Performance of GGBFS-RHA Alkali-Activated Concretes. *Constr. Build. Mater.* **2019**, *208*, 326–332. [CrossRef]

25. Elyamany, H.E.; Abd Elmoaty, A.E.M.; Mohamed, B. Effect of Filler Types on Physical, Mechanical and Microstructure of Self Compacting Concrete and Flow-Able Concrete. *Alex. Eng. J.* **2014**, *53*, 295–307. [CrossRef]

26. Gautam, L.; Jain, J.K.; Kalla, P.; Danish, M. Engineered Cementitious Composites (ECC). *Mater. Today Proc.* **2020**, *44*, 4196–4203. [CrossRef]

27. Shilar, F.A.; Ganachari, S.V.; Patil, V.B.; Nisar, K.S.; Abdel-Aty, A.H.; Yahia, I.S. Evaluation of the Effect of Granite Waste Powder by Varying the Molarity of Activator on the Mechanical Properties of Ground Granulated Blast-Furnace Slag-Based Geopolymer Concrete. *Polymers* **2022**, *14*, 306. [CrossRef]

28. Zeyad, A.M.; Magbool, H.M.; Tayeh, B.A.; Garcez de Azevedo, A.R.; Abutaleb, A.; Hussain, Q. Production of Geopolymer Concrete by Utilizing Volcanic Pumice Dust. *Case Stud. Constr. Mater.* **2022**, *16*, e00802. [CrossRef]

29. Kabay, N.; Mert, M.; Miyan, N.; Omur, T. Pumice as Precursor in Geopolymer Paste and Mortar. *J. Civ. Eng. Constr.* **2021**, *10*, 225–236. [CrossRef]

30. *ASTM C 494/C 494M–17*; Standard Specification for Chemical Admixtures for Concrete. ASTM: West Conshohocken, PA, USA, 2000; Volume 8, pp. 3–4. [CrossRef]

31. Dong, M.; Feng, W.; Elchalakani, M.; Li, G.; Karrech, A.; May, E.F. Development of a High Strength Geopolymer by Novel Solar Curing. *Ceram. Int.* **2017**, *43*, 11233–11243. [CrossRef]

32. *AS 1012.3.5:2015*; Methods of Testing Concrete Method 3.5: Determination of Properties Related to the Consistency of Concrete—Slump Flow, T 500 and J-Ring Test. Australian Standard: Sydney, Australia, 2015; pp. 1–13.

33. *ASTM C 109*; Standard Test Method for Compressive Strength of Hydraulic Cement Mortars. ASTM International: West Conshohocken, PA, USA, 2020; Volume 4, p. 9.

34. Rokugo, K.; Yokota, H.; Sakata, N.; Kanda, T. "Recommendations for Design and Construction of High Performance Fiber Reinforced Cement Composite with Multiple Fine Cracks" Published by JSCE. *Concr. J.* **2007**, *45*, 3–9. [CrossRef]

35. *ASTM: C 1609/C 1609M–07*; Standard Test Method for Flexural Performance of Fiber-Reinforced Concrete. ASTM: West Conshohocken, PA, USA, 2007; pp. 1–9.

36. *ASTM C948-81*; Standard Test Method for Dry and Wet Bulk Density, Water Absorption, and Apparent Porosity. ASTM International: West Conshohocken, PA, USA, 1981; Volume 81, p. 2.

37. Givi, A.N.; Rashid, S.A.; Aziz, F.N.A.; Salleh, M.A.M. Assessment of the Effects of Rice Husk Ash Particle Size on Strength, Water Permeability and Workability of Binary Blended Concrete. *Constr. Build. Mater.* **2010**, *24*, 2145–2150. [CrossRef]

38. Das, S.K.; Mishra, J.; Singh, S.K.; Mustakim, S.M.; Patel, A.; Das, S.K.; Behera, U. Characterization and Utilization of Rice Husk Ash (RHA) in Fly Ash—Blast Furnace Slag Based Geopolymer Concrete for Sustainable Future. *Mater. Today Proc.* **2020**, *33*, 5162–5167. [CrossRef]

39. Patel, Y.J.; Shah, N. Enhancement of the Properties of Ground Granulated Blast Furnace Slag Based Self Compacting Geopolymer Concrete by Incorporating Rice Husk Ash. *Constr. Build. Mater.* **2018**, *171*, 654–662. [CrossRef]

40. Safari, Z.; Kurda, R.; Al-Hadad, B.; Mahmood, F.; Tapan, M. Mechanical Characteristics of Pumice-Based Geopolymer Paste. *Resour. Conserv. Recycl.* **2020**, *162*, 105055. [CrossRef]

41. Kotop, M.A.; El-Feky, M.S.; Alharbi, Y.R.; Abadel, A.A.; Binyahya, A.S. Engineering Properties of Geopolymer Concrete Incorporating Hybrid Nano-Materials. *Ain Shams Eng. J.* **2021**, *12*, 3641–3647. [CrossRef]
42. Abbas, I.S.; Abed, M.H.; Canakci, H. Development and Characterization of Eco- and User-Friendly Grout Production via Mechanochemical Activation of Slag/Rice Husk Ash Geopolymer. *J. Build. Eng.* **2023**, *63*, 105336. [CrossRef]
43. Zeyad, A.M.; Bayraktar, O.Y.; Tayeh, B.A.; Öz, A.; Özkan, İ.G.M.; Kaplan, G. Impact of Rice Husk Ash on Physico-Mechanical, Durability and Microstructural Features of Rubberized Lightweight Geopolymer Composite. *Constr. Build. Mater.* **2024**, *427*, 136265. [CrossRef]
44. Kusbiantoro, A.; Nuruddin, M.F.; Shafiq, N.; Qazi, S.A. The Effect of Microwave Incinerated Rice Husk Ash on the Compressive and Bond Strength of Fly Ash Based Geopolymer Concrete. *Constr. Build. Mater.* **2012**, *36*, 695–703. [CrossRef]
45. Khater, H.M.; El Nagar, A.M.; Ezzat, M.; Lottfy, M. Fabrication of Sustainable Geopolymer Mortar Incorporating Granite Waste. *Compos. Mater. Eng.* **2020**, *2*, 1–12.
46. Talaat Mohammed, D.; Yaltay, N. Strength and Elevated Temperature Resistance Properties of the Geopolymer Paste Produced with Ground Granulated Blast Furnace Slag and Pumice Powder. *Ain Shams Eng. J.* **2024**, *15*, 102483. [CrossRef]
47. Mehta, A.; Siddique, R. Sustainable Geopolymer Concrete Using Ground Granulated Blast Furnace Slag and Rice Husk Ash: Strength and Permeability Properties. *J. Clean. Prod.* **2018**, *205*, 49–57. [CrossRef]
48. Zeyad, A.M.; Tayeh, B.A.; Yusuf, M.O. Strength and Transport Characteristics of Volcanic Pumice Powder Based High Strength Concrete. *Constr. Build. Mater.* **2019**, *216*, 314–324. [CrossRef]
49. Luo, Y.; Bao, S.; Liu, S.; Zhang, Y.; Li, S.; Ping, Y. Enhancing Reactivity of Granite Waste Powder toward Geopolymer Preparation by Mechanical Activation. *Constr. Build. Mater.* **2024**, *414*, 134981. [CrossRef]

 *infrastructures*

Review

Crumb Rubber in Concrete—The Barriers for Application in the Construction Industry

Patricia Kara De Maeijer, Bart Craeye, Johan Blom and Lieven Bervoets

1   EMIB, Faculty of Applied Engineering, University of Antwerp, Groenenborgerlaan 171,
    2020 Antwerp, Belgium; bart.craeye@uantwerpen.be (B.C.); johan.blom@uantwerpen.be (J.B.)
2   DuBiT, Department of Industrial Sciences & Technology, Odisee University College, 9320 Aalst, Belgium
3   SPHERE, Department of Biology, University of Antwerp, Groenenborgerlaan 171, 2020 Antwerp, Belgium;
    lieven.bervoets@uantwerpen.be
*   Correspondence: patricija.karademaeijer@uantwerpen.be

Abstract: This state-of-the-art review was aimed to conduct a comprehensive literature survey to summarize experiences of crumb rubber (CR) application in concrete within the last 30 years. It shows that certain gaps prevent obtaining a coherent overview of both mechanical behaviour and environmental impact of crumb rubber concrete (CRC) to object to the stereotypes which prevent to use of CR in concrete in the construction industry. Currently, four major barriers can be distinguished for a successful CR application in the concrete industry: (1) the cost of CR recycling, (2) mechanical properties reduction, (3) insufficient research about leaching criteria and ecotoxicological risks and (4) recyclability of CRC. The application of CR in concrete has certainly its advantages and in general cannot be ignored by the construction industry. CR can be applied, for example, as an alternative material to replace natural aggregates and CRC can be used as recycled concrete aggregates (RCA) in the future. A certain diversity for the CR application can be introduced in a more efficient way when surface treatment and concrete mix design optimization are properly developed for each type of CR application in concrete for possible field applications. The role of CRC should not be limited to structures that are less dependent on strength.

Keywords: crumb rubber (CR); crumb rubber concrete (CRC); mechanical properties; durability; leaching; ecotoxicology; recyclability

Citation: Patricia Kara De Maeijer, Bart Craeye, Johan Blom and Lieven Bervoets Crumb Rubber in Concrete—The Barriers for Application in the Construction Industry. *Infrastructures* **2021**, 6, 116. https://doi.org/10.3390/infrastructures6080116

Academic Editor: Kamil Elias Kaloush

Received: 23 July 2021
Accepted: 17 August 2021
Published: 20 August 2021

Publisher's Note: MDPI stays neutral with regard to jurisdictional claims in published maps and institutional affiliations.

1. Introduction

One billion end-of-life tires (ELTs) are generated globally each year, 355 million of ELTs out of this production are related to the EU [1,2]. The majority of ELTs are landfilled and only about 5% is used for civil engineering applications. However, considering the massive requirement for construction (approximately 32 billion tons each year [3]), the application of crumb rubber (CR) as a recycled organic component in Portland cement concrete can effectively resolve the environmental pressures [4] and has the potential to abate the consumption of virgin materials while simultaneously reducing landfill dumping [5].

The use of ELTs as CR aggregates (coarse and fine granulates) in the construction industry has been intensively studied in several countries [6–9]. Several researchers referred to the decrease of the mechanical properties of rubberized concrete, some researchers reported the enhancing of the properties of the CR particles by means of pre-treatment of CR surface to be compatible with cement matrix, but less information was provided regarding CR hazardous impact to the environment and how to mitigate it.

The outcome of experimental programs showed that there are certain gaps that prevent to obtain a coherent overview of both mechanical behavior and environmental impact of crumb rubber concrete (CRC) to object the stereotypes which prevent to use CR in concrete in the construction industry.

The current definition of the gaps in research and market indicates certain barriers or stereotypes which limit the application of CR in the construction industry. Those could be listed as follows:

- The current construction materials market is rather conservative and there is a need to foresee the alternative solutions since natural aggregates resources are depleting [10];
- Constant flow of available materials and their storage, efficient time frame for the delivery of the materials within the country/countries (for example, secondary streams are very important). Meanwhile, the cost for natural aggregate keeps increasing due to the limited source and long transport distance [5];
- CRC cost analysis and comparison to the traditional concrete vary from country to country due to the limited industrial-scale production of CR [11]. The price of CR ranges from 40 to 320 EUR per ton based on its origin and fineness;
- The end-of-waste (EoW) management is crucial, and recycling is mandatory; however, the profit of recycling is limited by the manufacturing costs; also, it depends on the recycling developments within the countries (which may differ quite a lot from country to country);
- Society reaction to waste usage in construction materials due to lack of sufficient technical information and comprehension may develop certain barriers in application of different recycled materials;
- The recycled rubber products have poor mechanical performances-interfacial bond strength as the main factor which limits the application of CR in construction elements;
- Insufficient investigation of leaching behavior and ecotoxicological impact to the environment;
- Insufficient information on the recyclability of concrete containing CR;
- For the moment, there are no restrictions defined regarding CR application in concrete nor rubberized concrete aggregates.

In 2016, yearly global production of natural and synthetic rubber reached 27.3 million tons (54% synthetic), with ~70% used in the manufacture of vehicle tires. It has been recorded that around 1 billion tires are discarded every year and this number may increase up to 1.2 billion by the end of 2030 [12,13]. Tire manufacturing is a global business: there are over 160 tire manufacturers located in more than 45 countries worldwide. Tires are made to international standards and are freely traded across the world. Not each country might have natural resources for construction, but it has tires disposal. Stockpiling of used tires is very common in several countries that leads, in general, to a serious ecological threat contributing to the reduction of biodiversity since the tires also contain toxic and soluble components [14]. Tire materials are complex mixtures; as various chemicals are used during the production of tires. Tires contain a total of approximately 1.5% by weight of potentially hazardous waste compounds: copper (Cu) (0.02%), zinc (Zn) (1%, but can be up to 2%), cadmium (Cd) (max 0.001%), lead (Pb) and Pb compounds (max 0.005%), acidic solutions (0.3%). Despite consistent usage of the general ingredients, the composition of a specific tire depends on its application. For example, a common-sized all-season passenger commercial tire contains approximately 30 types of synthetic and 8 natural rubbers, 8 kinds of carbon black, steel cord for belts, polyester and nylon fiber, steel bead wire and 40 different chemicals, waxes, oils, pigments, silica and clays [15–17]. One of the suggested compositions can be taken as a reference [18]: rubber/elastomers (48%), carbon black (22%), metal (15%), additives (8%), textile (5%), zinc oxide (1%-passenger tires or 2% truck and off-road tires), sulphur (1%).

The ELTs are generally disposed of in different ways such as burning, landfilling and use as fuel. The burning of tires causes serious fire hazards, emission of potentially harmful compounds, landfilling by ELTs results in depletion of the available useful sites. The use of ELTs as fuel is economical but not attractive when compared to other products used for the same purpose.

According to European Community Directives [19,20], ELTs are banned from landfills due to the risk of pollutant release. The production of crumb rubber granulates (CRG) from ELTs has been considered as an acceptable way of utilizing this recycled waste material. However, according to Directive 2008/98/EC [20], the specific EoW criteria for CR are not defined. In general, there is a priority order for the EoW management which includes: prevention, preparation for re-use, recycling, recovery, and disposal.

In the EU the European Committee for Standardization (CEN) has classified products obtained from grinding of waste tires according to their size [21]. Particularly, the following five classes are defined:

- Cut tires, with size bigger than 300 mm,
- Shreds, with size ranging between 20 and 400 mm,
- Chips, with size ranging between 10 and 50 mm,
- Rubber granulates, with size ranging between 0.8 and 20 mm,
- Rubber dust, smaller than 0.8 mm (*limited application, but can be successfully used in concrete*).

In the last years, the suitability and efficiency of using different by-products (mainly steel fibers-used as a partial or total replacement of industrial steel fibers to produce Fiber Reinforced Concrete (FRC) [22–30] and rubber particles-used as a partial replacement of natural aggregates for producing rubberized concrete [31–36]) obtained from the recycling of waste tires for improving some concrete properties have been investigated, however, the related research is limited and leaves a wide space for further investigations [37].

With regards to the terminology, the most common frequently used terms for this type of concrete are rubberized concrete [38,39] and crumb rubber concrete (CRC) [40,41].

Mechanical shredding (particle size of 13–76 mm) and mechanical grinding (particle size 0.075–4 mm) at ambient temperature, cryogenic grinding (particle size 0.18–1.4 mm) particle size at a temperature below the glass transition temperature (-200 °C), high-pressure waterjet grinding (particle size 0.2–1 mm) are the technologies used to obtain rubber aggregates. The Scanning Electron Microscope (SEM) images of CR particles are shown in Figure 1a–c. Mechanical shredding tends to produce rough-edged particles with many hair-like appendages, along with generated heat rubber particles get aged and rubber properties are altered, the fibers are removed with an air separator and the metal with magnets. Cryogenic processing has the advantage of not generating heat throughout the process and the CR produced has little fiber and metal content. Waterjet processing under the impact of the water stream leads to micro-blasting of the rubber and directly separates the rubber from the steel belt, further wet slurry undergoes separation in a centrifuge [42].

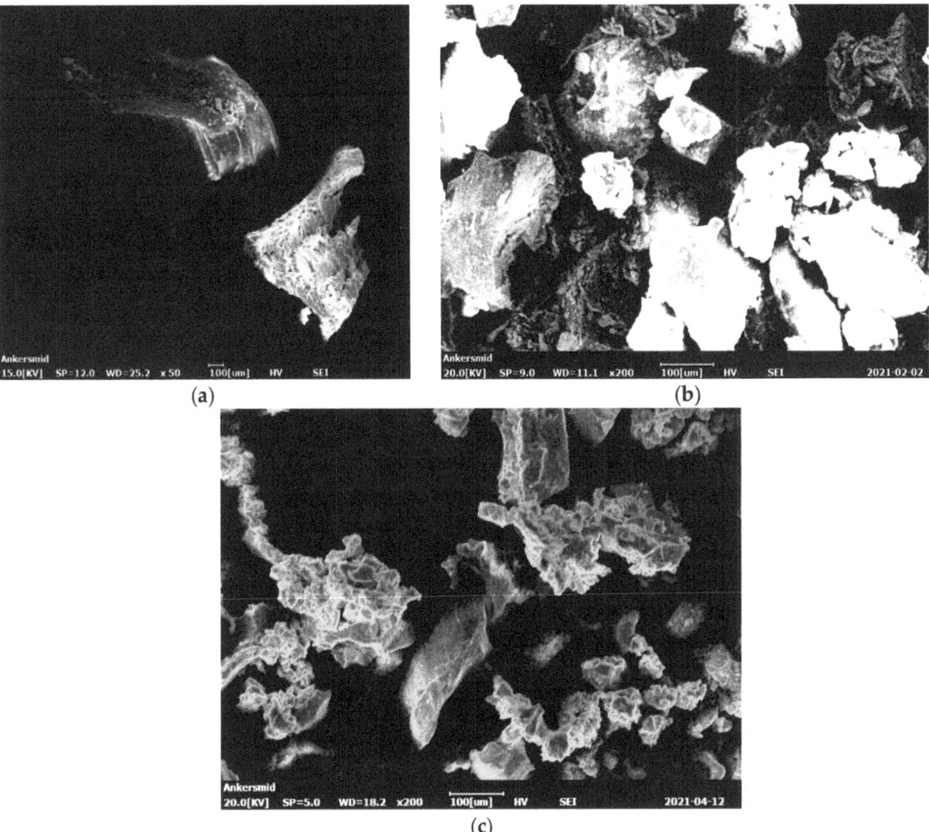

Figure 1. SEM images of the CR particles at 100 µm scale: (**a**) mechanical, (**b**) cryogenic and (**c**) waterjet.

2. Review Significance

2.1. CR application in Concrete in Large Research Projects

The tire recycling sector is constituted by large communities of small and mid-size enterprises (SMEs), estimated as more than 20,000 connected to the recycling trade in Europe. The sector is in a compelling need of looking for an innovative and decisive solution to increase the competitiveness of tires recycling. The rubberized concrete has the advantages of low density [43], good sound absorption [33], acid resistance [13,14], freeze-thaw resistance [44], chloride permeability resistance [45,46], increased damping capacity, bending impact strength and toughness [31,47–49]. These advantages make the rubberized concrete attractive for applications such as lightweight concrete [50–52], non-bearing concrete walls [53], noise screen [33,45,51], improved thermal insulation for flooring in buildings [54,55], reinforced concrete jersey barriers [38,56,57], pavement [38,58,59], railway track beds [60], reinforced column for earthquake-resistant structures [45,61,62], rubberized concrete beams with high impact resistance [63,64], expansion joints in the concrete floor [65], steel tubes filled with rubberized concrete [66,67]. These positive characteristics can be useful in supporting the development of sustainable CRC to achieve greater environmental and economic benefits.

Several international projects were carried out to utilize CR in concrete within the last decade. Among those the most successful are: SMART (7-FP, 2012–2015) [6], ANA-GENNISI (7-FP, 2014–2017) [7], RISEN (H2020, 2016–2020) [8] and the Australian Research Council (ARC) Linkage project (2016–2020) [9]. These projects involved the following countries' contributions: Australia, Bosnia and Herzegovina, Croatia, Cyprus, Czech Republic, Finland, France, Germany, Italy, Norway, Poland, Portugal, Romania, Slovenia, Spain, Sweden, The Netherlands, and the UK. It was mentioned in the SMART project [6] that the added value of a recycled product derived from the tire-shredding process is very low due to the following factors: (1) the recycled rubber products have poor mechanical performances, (2) the profit of recycling is limited by the manufacturing costs. Within the ANAGENNISI project, several demonstration projects were undertaken in five European countries to convince contractors and infrastructure owners of the benefits of the examined tire by-products. Some projects included slabs on grade, tunnel linings, precast concrete elements (rubberized poles and railway sleepers [8], safety road barriers and a repair screed application [7]. Within the ARC Linkage project [9] reinforced concrete constructed with CR from used tires as a partial replacement for sand was a viable option for the construction of residential footing slabs that utilize approximately 40% of reinforced concrete poured in Australia each year.

2.2. Mechanical Properties—The Main Barrier for Application in the Construction Industry

The first patent invention related to rubber crumb reinforced cement concrete was filed by Frankowski [68] in 1992. There were several studies related to the application of CR in concrete performed in the 90 s [38,69–74]. At that time, it was stated that rubberized concrete did not demonstrate brittle failure and had abilities to absorb a large amount of plastic energy under compressive and tensile loads, to improve shock wave absorption, to provide resistance to cracking, to be fire-resistant, to lower heat conductivity and to improve the acoustical environment (e.g., noise reduction barriers) which is advantageous in applications in structures subjected to dynamic and impact loading and non-structural elements [12]. According to statistics, about 70% of the CR powder was used for construction engineering in the United States [65]. CRC was not widely adopted in the industry since then mainly due to its low compressive strength, however, its structural properties are still relatively unexplored [36,75,76].

Topçu [77] assessed the brittleness index of rubberized concrete and concluded that concretes containing 15% rubber chips by weight gave the highest brittleness index values with low compressive strength and toughness values. Bayomy and Khatib [38] examined the behaviour of rubberized concrete and concluded that it could be suitable for non-structural purposes such as light-weight concrete walls, building facades, and architectural units. During the last two decades, there has been a significant growing interest in the use of recycled tire rubber in concrete and a significant number of scientific publications about CR in concrete were elaborated. In early 2000, a wave of pioneering effort to build CRC test sites was undertaken in the US on urban development-related projects [40]. Concrete designs of 11.9 kg up to 35.67 kg of CR per cubic meter were used in the construction of different types of concrete slabs. The building of these test slabs has provided very useful experience and the means to evaluate first-hand knowledge about CRC mixing, hauling, pumping, placing, finishing, and curing. Laboratory evaluation tests included compressive strength, thermal coefficient of expansion, fracture, shrinkage cracking and microscopic matrix analysis. The increased interest to recycle CR in different ways was observed in the late years of the last decade. There is an interest to find a solution for the application of CR in the concrete and asphalt industry looking for ways to increase the reactivity of this technogenic waste material to overcome the barriers which prevent the use of it in massive recycling in construction.

The use of tire rubber in concrete faces a challenge of lack of heat treatment, which is crucial to enable good adhesion between the tire rubber particles and other materials. Another challenge for this use lies in the potential mismatch in stiffness between the relatively soft viscoelastic rubber and the relatively rigid elastic cement-based matrix [78]. CR provides a corresponding improvement in concrete performance due to the flexibility of tire rubber aggregate, but its strength significantly decreases. Therefore, CRC has been primarily applied for concrete with low CR volume content (less than 20% of the total aggregate volume) and non-structural elements with lower bearing demand already in the 90 s [38,69]. The experimental study was carried out by Hassanli et al. [79] to understand the behavior of CRC at the structural application level. It was concluded that increasing the CR content from 0% to 18% followed by the compressive strength reduction by 31% in concrete beams.

The affecting factors of strength loss due to the replacement of fine aggregates with CR have been reported in some studies. Eldin and Senouci [69] first discovered significant reductions in CRC strength through tests. When the rubber aggregates completely replaced the fine aggregates, the compressive and tensile strength of CRC decreased by 65% and 50%, respectively. In addition, with the same rubber particle content of fine aggregate, the reduction rate of high-strength concrete was found to be greater than that of low-strength concrete [36,39,80]. The rubber type only has a marginal effect on the mechanical properties of CRC when the rubber aggregates replaced the fine aggregates. These results indicated that concrete containing rubber with fine grading had lower compressive strength than that containing rubber with coarse grading [35,38,71,72,81–85]. Because of its low modulus of elasticity, a piece of rubber aggregate acts as a large pore and does not have a significant role in the resistance to externally applied loads. Thus, the compressive and tensile strengths of concrete that contains rubber particles depend on the volume of rubber aggregate (fine and large pores) in concrete mass [81]. However, Bignozzi et al. [54] and Su et al. [4] reported that CRC with smaller CR particle size has higher compressive strength than that with larger CR particle size. In general, the replacement ratio of aggregates in rubber concrete should not exceed 20% of the total aggregates by volume [38].

Hassanli et al. [79] when studied the behavior of rubberized concrete at the structural application level, concluded that by increasing the rubber content from 0% to 18% followed by the compressive strength reduction by 31% in concrete beams, the strength reduction was only 6% and 12% in the corresponding tested beam and beam-column members, respectively. Valadares and de Brito [86] concluded that to avoid severe worsening of the mechanical properties of structural concrete the use of rubber coarse aggregates should be limited to percentage levels around 5% of the overall volume of aggregates. To minimize these negative impacts, however, one should opt for the coarser rubber aggregates since they unequivocally lead to more favourable performances, with the additional benefit of being cheaper to produce.

As mentioned above, compressive strength usually decreases with an increase of rubber content in rubberized concrete due to an increase in air voids and poor adhesion [36]. Flexural strength commonly decreases with the rubber percentage increase in concrete. Flexural strength of concrete was reduced up to 18% with 10% replacement of sand by CR and reached 32% with replacement of 50% [87]. It was decreased by 9% and 19% when 10% and 15% CR was added as partial replacement of coarse aggregate [88]. The effect of rubber in concrete on split tensile strength (STS) is analogous to flexural strength behaviour, STS generally decreases for rubberized concrete with an increase of rubber percentage due to an increase in voids and poor adhesion [36]. STS was reduced from 16.5% to 31% when 5% to 40% fine aggregate replacement with CR was used [89]. As the rubber content increased, the indirect tensile strength (ITS) decreased but the strain at failure increased. A higher tensile strain at failure is indicative of a more ductile and a more energy-absorbent mix [40]. Elastic modulus decreases for rubberized concrete with an increase of rubber percentage when compared with the reference concrete due to low elastic modulus and poor bonding of CR [36]. The addition of CR from 5% to 25% as partial replacement of fine aggregate, reduced the concrete elastic modulus by 2.44% to 31.74% [90]. For compressed CRC specimens, an increase in elastic modulus was noticed for specimens with a CR replacement ratio of up to 15% [11,91]. CRC and reference concrete have a small difference in the key mechanical properties that affect bearing strength [66], such as properties under multi-axial compression, the ratio of tensile strength to compressive strength. This ultimately leads to the similar bearing strength of CRC and reference concrete when their compressive strengths are the same.

2.3. The Fresh Properties—Improvement of Workability

The workability can be improved by the application of CR with particle fraction up to 1 mm, the mix is getting nice consistency and easier for placing (with CR content up to 15% [92]). When CR is applied above 15% then fly ash [93,94], slag [95], metakaolin [96] are used to help to reduce viscosity. The study by Su et al. [4] demonstrated that with a 20% replacement ratio, the slump of the rubberized concrete can decrease from 13.7% to 25.2% based on the CR size. Rubber aggregates have the property to keep moisture away while attracting air on their rough surface which may results in more air voids when rubber content is increased in concrete [97]. The influence on air void for CR within 15% replacement can be insignificant. CR can enhance workability by adjusting the aggregate gradation [98]. Adjusting the workability with a superplasticizer can also increase the air content [96].

The unit weight of the CRC mix decreased approximately 96 kg/m^3 for every 22 kg of CR added [40]. The density of the rubber is usually smaller than that of the water, the average density of the crumb rubber obtained is 870 kg/m^3 [99]. With the densities of the crumb rubber and the fine aggregate, the replacement of fine aggregates with crumb rubber for a certain replacement ratio by volume can be easily implemented using the density ratio between the fine aggregates and rubber, which is approximately 3 (i.e., 2580/870 = 2.97). CR can be added as high as 40–50% of the aggregate volume without major workability problems [38]. CRC mixes can give a viable alternative to the normal weight concrete when the strength is not a major factor in the design, for example, with lightweight concrete walls, building façades or other building architectural units.

2.4. Interfacial Bonding between the CR Particles and Cement Matrix

Khaloo et al. [33] noted that the reduced stiffness of rubber in relation to the other materials makes a minor contribution to global strength. Consequently, the rubber particles generate high stresses in their periphery which leads to cracking which, when widespread, causes premature rupture of the specimens. Topçu [71] explained this phenomenon based on the high tensile stresses in the direction perpendicular to loading. Therefore, the compressive strength reduction of CRC is mostly contributed by the difference of the elastic modulus between the CR and the cement matrix and the inadequate bonding between them. The elastic modulus of standard concrete is about 2.0–3.0 × 10^4 MPa. This is about 30,000 times the elastic modulus of CRC which is about 0.5–2.0 MPa [100]. Under external load, the deformation of the rubber aggregates is much larger than that of the cement mix. The bonding between them is so weak that the rubber aggregates debond from the cement. The NaOH treatment, for example, can remove the dust on the rubber surface and improve the surface hydrophilicity leading to a denser Interfacial Transition Zone (ITZ) compared to the concrete with as-received aggregate [101,102]. If the bond is improved at ITZ by any suitable and economical means then the negative effects of CR on strength properties of normal concrete may be reduced, consequently, it would be possible to effectively utilize the rubberized concrete in numerous concrete structures by the construction industry.

2.5. CR Pre-Treatment

Physical and chemical pre-treatment of CR can enhance interfacial bond strength in rubberized concrete. Although research studies have been done in this regard in the recent past, further research work is still required for the development of standard guidelines regarding the improvement of the bond between cement matrix and rubber aggregates. Literature review indicates that already investigated methods of bond improvement in rubberized concrete have resulted in improvement of compressive strength from 7% to 59% when compared with untreated CR [36]. Recently, Strukar et al. [103] recommended carrying research for structural applications of rubberized concrete concerning the limitation of rubber contents and methodology for pre-treatment of rubber aggregates. Li et al. [36] recommended the scope of further research to find the cost-effective and most efficient method of bond improvement, which is important concerning its field application. It was also suggested to do further research to investigate the durability of rubberized concrete with modified rubber surfaces.

Researchers have explored approaches such as chemical pre-treatment of scrap tires as a means of reducing the mechanical strength loss in rubberized cement composites. Li et al. [104] pre-coated the rubber particles with cement paste before being applied in concrete production. The energy-absorbing capability of the samples was found to increase with the added CR particles. Balaha et al. [105] reported that the use of *polyvinyl alcohol (PVA)* and *sodium hydroxide (NaOH)* treated CR reduced the compressive and tensile strength loss observed in these concrete mixes compared to mixes containing untreated CR. The NaOH treatment enhances the adhesion of tire rubber particles to cement paste, abrasion resistance and water absorption experiments [106]. CR mostly heavily aged during the service life and the ageing effect can attach functional groups (e.g., the carboxylic acids group) to the CR particle. After the treatment with NaOH, the acid carboxylic group will react with the alkaline content and during cement hydration, the chemical structure can provide a weak basic condition near the rubber aggregate-cement interface which will enhance ITZ properties. But in general, this method complicates the technology of concrete production, indeed it affects the hardening chemistry of the binder but does not give significant results.

Dong et al. [107] verify the feasibility of the approach [108] to improve the performance of CRC through developing a cementitious coating around rubber particles with a *silane coupling agent*. The results showed that the compressive and split tensile strengths of concrete incorporating *coated rubber* were 10–20% higher than the concrete incorporating uncoated rubber. Albano et al. [109] also observed an insignificant improvement in the compressive and splitting tensile strengths of rubberized concrete containing silane pre-treated scrap tire waste as a fine aggregate replacement material.

Rubber particles have been added to cement to form a self-healing cement system, where the rubber particles can expand on exposure to a particular fluid to close cracks, improving the sealing performances of the cement [110,111].

Improvement in the compressive, tensile and flexural strengths of concrete samples containing waste tire pre-treated with *sulphur compounds* was reported by Chou et al. [112]. Using a UV based pre-treatment on the CR has been shown to reduce strength losses [113]. For the pre-coating CR, water and 3 μm *limestone powder* were mixed in a Hobart mixer at a low speed [46]. He et al. [114] in their paper excavated the engineering application potential of rubberized concrete by using urea solution which could be beneficial for application in CRC to improve the mechanical strength and impact resistance. However, it is well-known that urea ($CO(NH_2)_2$), a chemical compound that provides reagents for calcium carbonate formation, is already among the most used precursors being added to the biological culture medium for some decades [115–117]. Biel and Lee [74] experimented with a special cement *Magnesium Oxychloride type* to enhance the bonding strength between rubber particles and cement. Regarding the different grinding processes of the CR (mechanical or cryogenic), even though there is a small difference in compressive strength in favour of cryogenic CR, there is no significant benefit to favour one or the other (furthermore, cryogenic CR is more expensive) [86].

It has recently been suggested that tire rubber ash can also be used to enhance concrete microstructure [56]. Combining surface treatment and particle size reduction might produce Portland cement concrete that can be used in varieties of applications. Reducing the particle size down to the same order of magnitude of cement powders will both reduce the "flaw" induced by the recycling and increase the stiffness of the particles. Enhanced pre-treatment of the rubber surface and mixing minerals can improve the strength of CRC [46]. However, surface treatment does have the potential to help improve the bond, it will not change the fundamental fact of the stiffness incompatibility between the CR and other constituents of Portland cement concrete.

2.6. Usage of Supplementary Cementitious Materials (SCMs) along with CR in Concrete

The use of supplementary cementitious materials (SCMs) such as silica fume (10% [118], 15% [101]), fly ash and metakaolin can slightly enhance the mechanical properties of CRC. Guneyisi et al. [119] believe that silica fume helps mostly the adherence between rubber aggregates and the binder because of the small size of the CR particles. Azevedo et al. [120] reported that a CRC mix containing 30% SCMs (15% fly ash + 15% metakaolin) exhibited a 23% reduction in compressive strength in comparison to the reference samples. Onuaguluchi and Panesar [46] indicated a clear correlation in the minimization of concrete strength loss and resistance to chloride penetration when silica fume was used as a bonding agent in conjunction with CR content. Such concrete mixes would be highly suitable for marine environment structures, which are more susceptible to accelerated deterioration due to corrosion-causing chlorides. Furthermore, [121] observed an increase in the abrasion resistance and compressive strength of rubberized concrete with the addition of silica fume.

2.7. CR Application in High-Performance Concrete

The brittle failure of the control high-performance concrete and non-brittle failure of CRC highlights an increase in the ductile nature when fine aggregates are partly replaced with CR [122]. CR application in high-performance concrete may be limited by the effects of increased replacement levels. The CRC containing 0–12.5% CR can be an optimal replacement level which could indicate a successful usage within high strength concrete mixes [12]. The addition of silica fume as a bonding agent to CRC with CR replacement level up to 12.5% can be beneficial to exhibit minimal losses in compressive, flexural tensile, and pull-off strengths, as well as an increase in abrasion resistance and water penetration, and a decrease in water absorption [14]. The addition of super-absorbent polymers (SAPs) to CRC can improve the pore size distribution and pore connectivity to have a positive impact on the internal self-drying effect during hydration and its volume stability for concrete with low w/c ratios [123].

2.8. Durability

Freeze-thaw action is the major reason for the cracking of concrete and ageing of concrete structures. The addition of CR can effectively enhance the concrete freeze-thaw resistance [124–126]. Richardson et al. [127] indicated that a C40 (40 MPa) concrete mixture containing 0.6% CR by weight of concrete displayed optimal durability factors when CR models exhibited minimal internal and surface damage throughout 56 freeze-thaw cycles; contrasting with reference concrete models that demonstrated failure after 28 cycles. Si et al. [128] demonstrated that the added CR can lead to a lower mass loss and dynamic modulus decrease after 500 freeze-thaw cycles. The enhancement of the freeze-thaw resistance is caused by the fact that the ductile CR can provide extra space for ice expansion and increase the air void content [129]. The provided extra space by CR can be also favourable for the expansion of the alkali-silica reaction (ASR) gel due to swelling. Afshinnia and Poursaee [130] examined the ASR expansion of mortar samples with 16% and 24% replaced CR, ASR expansion of the samples decreased by 43% and 39%, respectively.

A higher percentage of CR aggregates in concrete causes an increase in water absorption. Several researchers [43,131,132] observed an increase in the immersed water absorption of concrete with the incorporation of CR. In contrast, Segre et al. [133] and Oikonomou and Mavridou [134] reported reductions in the absorbed water of rubberized cement composites as the CR content of mixes increased. Depth of water penetration was observed to increase by 0% to 225% compared to reference concrete when CR contents were increased by 2.5% to 20% [135]. The presence of CR aggregate in concrete shows encouraging behaviour for abrasion resistance too [135]. Appropriate content and size of CR, and pre-treatment of CR results in enhanced abrasion resistance of rubberized concrete.

Drying shrinkage in concrete commonly increases with the addition of CR aggregates [136]. It has been observed that the free shrinkage increases because of strain capacity enhancement when <4 mm crumb rubber aggregate replacement is used. It has been suggested that this incremental change in shrinkage could be due to the incorporation of the low stiffness aggregate (i.e., rubber particles) leading to a reduction in 'internal restraint' and a consequent increase in the length change resulting from shrinkage [137]. However, it was decreased when rubber aggregates were pre-treated with sodium hydro-oxide solution [128]. It was because of improved adhesion between cement matrix and rubber aggregates.

It was observed that prolonged exposure to chloride ions and higher CR contents causes to increase in its penetration depth [138]. For small rubber contents (about 5% to 12.5%), chloride ion penetration in rubberized concrete is normally reduced. However, it has reverse behavior when rubber content is more than 12.5% [135]. The CRC sample with NaOH treated CR have higher chloride resistance [139].

Incorporating CR can alter the fine pore pressure of the concrete and the degree of corrosion of the inner steel of the concrete can be reduced [140]. Thomas et al. [135] indicated the carbonation depth will not increase within the 12.5% rubber replacement ratio. However, Bravo and de Brito [132] demonstrated the carbonation depth will increase with the CR replacement ratio. The influence of added CR on concrete carbonation resistance is controversial.

Concrete is deteriorated by acid attack and may disintegrate due to reaction with alkaline ingredients during the hydration process in concrete. However, rubberized concrete exhibits high acid resistance. Reference concrete exhibited greater loss of weight at 90 and 180 days in 3% H_2SO_4 solution. However, it was decreased with a rise in rubber aggregate contents [13]. The durability of CRC is significantly affected by sulphate attack [135]. Loss of mass is generally taken as an evaluation factor to check the resistance of rubberized concrete against sulphate attack. Carbonation depth was observed to be decreased by 38% to 15% upon adding 10% to 25% of CR powder in samples of mortar having a 0.56 w/c ratio. In opposite to that, carbonation depth was increased by 44% to 200% with the addition of CR powder by 5% to 25% in mortar samples having a 0.51 w/c ratio [141].

2.9. Leaching and Ecotoxicology

Most studies mentioned above were mostly focused on the mechanical and durability properties but less on ecotoxicological and leaching issues for CRC. The potential ecotoxicological impact of CR remains largely unknown but has been recognized as an information gap that needs to be addressed.

The application of CR in concrete production can lead to environmental concerns due to the presence of trace metals and polycyclic aromatic hydrocarbons (PAHs) [142]. However, the surrounding cementitious materials can well confine the trace metals or volatile organics exited in rubber particles based on the Toxicity Characteristic Leaching Procedure (TCLP) test [143].

Zn is often cited as the candidate most likely responsible for observed CRG/tire wear particles (TWP) leachate toxicity [1,15]. Zn is used in tires as a vulcanization activator by manufacturers, and zinc leachate is believed to contribute predominantly to the leachate toxicity within the surrounding environment. For example, zinc toxicity was shown to be related to a disruption of calcium ion uptake in *Daphnia magna* [144]. The presence of Zn in the highest concentrations doesn't necessarily indicate the highest risk of toxicity since there are other chemicals present in CR which can contribute to toxicity at the lower concentrations [1].

Several publications originated from the US tried to bring attention to the toxicological side of the CRG application problem about the reuse of disposal tires. The crumb rubber material (CRM) has been widely used as in fill for artificial turf installations in the US since 1960. Concerns have been expressed that those metals, including manganese (Mn), iron (Fe) and Zn derived from tire rubber could be transferred to the environment and affect organisms having direct contact with recycled products, tire wear particles or illegal dumps. For example, chemical additives such as Zn and PAHs were widely detected in the leachate from tire rubber [145–148]. Zhang et al. [147] reported that the levels of PAHs and

Zn in CR used as infill for artificial turf were above health-based soil standards, and lead (Pb) in the CRM was highly bioaccessible in synthetic gastric fluid at relatively low levels. The leaching from CRG is assumed to occur more rapidly due to the increased surface area when compared to whole tires. The toxic leaching occurs from an equivalent source and poses an increasing threat to the environment, indicating that CRG particulates and the leaching chemicals will continue to pose a threat to wildlife long after their disposal [1,149]. The leaching test based on the TCLP standard demonstrated that the concentration of the leached Zn can exceed the limit and can lead to potential pollution [150]. Kardos and Durham [143] conducted the TCLP test to examine the leaching potential of the concrete prepared with CR particles and the results demonstrated the concentration of both leached trace metals and volatile organics were below the limit. Compared to dense cement concrete materials, the confinement in asphalt concrete material is relatively limited, for example. Field tests of the asphalt pavement with CRM demonstrated the leached concentration of Benzothiazole, 2(3H)-benzothiazolone, mercury (Hg) and aluminium (Al) reached potential harmful levels [151]. Benzothiazole (BT) and its derivatives, derived primarily from tire wear particles, were detected ubiquitously in environmental samples [152,153].

The pH level can significantly influence the leaching potential of both trace metals and hydrocarbons. The study [154] demonstrated that organics and metals can be more easily leached out under basic and acid conditions, respectively. Similarly, the study by Selbes et al. [155] also demonstrated that trace metals can be more easily extracted from CR under acidic conditions and the amount of dissolved organic carbon (DOC) was significantly increased under basic conditions. Another factor that can influence the leaching potential is the ageing effect of the waste organic component; however, the influence is still not very clear.

Besides the environmental impact caused by the production of raw materials and handling of demolition waste, it must be considered that potentially harmful substances can be leached from the materials during their service life because of their exposure to water in the form of rain, surface water or groundwater. Trace metals can also be leached out which can originate from natural or secondary raw materials or additives as well as from primary and secondary fuels used in the production process of cement [156]. Their release is dependent on different factors, e.g., the chemical composition of the leachate [157] and the binding mechanisms of the substances [158].

To date, several European countries including Belgium, Germany and The Netherlands carry out investigations concerning the release of potentially harmful substances from hardened concrete. The first standard—the tank test, NEN 7345 [159]—was developed in 1995 in the Netherlands and was revised in 2004 as NEN7375 [160]. The corresponding German method was published by the German Committee for Reinforced Concrete (DAf-Stb) in 2005 as a guideline [161]. A harmonized European technical specification, the DSLT, was published in 2014 [21]. Currently, leaching is tested according to the Flemish Standard CMA 2/II/A.9.1 [162] in Belgium. According to the test procedures (commonly used: column and diffusion tests) the usual set of determining compounds are: sulphate (SO4), chloride (Cl), bromide (Br), fluoride (F), arsenic (As), barium (Ba), cadmium (Cd), cobalt (Co), chromium (Cr), Cu, molybdenum (Mo), nickel (Ni), Pb, antimony (Sb), selenium (Se), tin (Sn), vanadium (V) and Zn). However, there is a certain concern that crushed powder can accelerate the leaching and does not reflect reality. The comparison of leaching criteria is given in [159,162] and for some trace metals it is shown in Table 1. In the USA, the EPA method 1315 [163] is used to evaluate the leaching behavior of monolithic materials and compacted granular materials. In these tests, pre-stored hardened concrete samples are exposed to deionized water with periodic renewal of the water. The renewal times and the liquid-to-surface area ratio differ between the test methods. For the evaluation of the DSLT results, limits for the cumulative release are defined in Germany [164] and the Netherlands [165]. All the standardized tests are intended to be performed on hardened materials only. However, for some applications, concrete may be exposed to water already in a fresh state during the construction phase. At the beginning of hydration, there is no phase boundary between the water and the fresh cement paste, and soluble substances are not yet fixed in the hydration products. Therefore, it can be assumed that cementitious materials show a different leaching behavior in the fresh state and the leaching of the fresh material should be investigated separately [156].

Table 1. Comparison of leaching criteria.

| | Belgium (Flanders) | | The Netherlands | |
	Shaped Applications NEN 7345 (mg/m^2)	Non-Shaped Applications CMA 2/II/A.9.1 (mg/kg d.s.)	Shaped Applications NEN 7345 (mg/m^2)	Non-Shaped Applications NEN 7345 (mg/m^2)
Cd	1.1	0.03	3.8	0.04
Cu	25	0.8	98	0.9
Pb	60	1.3	400	2.3
Zn	90	2.8	800	4.5

2.10. Recycling of CRC

All over the globe the construction industry takes 50% of raw materials from nature, consumes 40% of total energy and produces 50% of total waste [166]. The approach 'concrete to concrete' in general preserves the natural aggregates and solves the problem of reducing the space and energy required for the landfill disposal of concrete waste. Recycled concrete aggregate (RCA) as such, holds original aggregate with hydrated cement paste. RCA are available in size of 20 mm to 50 mm; finer fractions also are available. The density of RCA is normally lower than the density of natural aggregate. The main reason for the lower density is that the adhered mortar is lighter in weight compared to the same volume of natural aggregate, which decreases the density. The density of RCA is not constant for all the crushed concrete aggregate. It depends upon the specific aggregate. The shape and gradation of RCA are dependent on the production method and type of crusher used for the manufacturing of RCA. Recycled concrete as aggregates is relatively weaker in strength as compared to natural aggregates. The crushing and abrasion value of RCA is also lower than the natural aggregates. If several SCMs are used in concrete mix design which normally are meant for cement replacement, improvement of certain properties, there is a possibility that RCA has a positive effect on the concrete properties [167]. It can be mentioned that the recycling of CRC aggregates should not be a problem when CR is a part of the hardened cement matrix.

3. Discussion

The outcome of the current state-of-the-art review and performed research so far shows certain gaps that prevent a complete overview of mechanical behavior and environmental impact (leaching behavior, ecotoxicology, and recyclability) of CRC to object to the stereotypes which prevent to use CR in concrete in the construction industry. The four major barriers can be distinguished for a successful CR application in the concrete industry: (1) the cost of CR recycling, (2) mechanical properties reduction, (3) insufficient research about leaching criteria and ecotoxicological risks and (4) recyclability of CRC.

Recycling a large number of organic aggregates into concrete can be developed for reduced environmental impacts and natural resources. There is a certain deficiency of river sand in Belgium, which lately was partly replaced by crushed limestone. During the past years, a shift has been observed within the concrete sector from the use of river sand from The Netherlands and/or Germany to Belgian marine sand. At present, about 3 million tons or 2 million m^3 of Belgian marine sands is extracted each year, of which 80% is used in the construction industry [168]. The mining and quarry industry should guarantee an adequate and continuous supply of raw materials as a producer of construction minerals to the construction sector to sustain the economic development of the country [169]. The effect of a shortage of these construction materials results in increased construction cost and then transferring the burden to the end-users, which in turn affects the national development. Aggressive consumption reduces the non-renewable aggregate resources; therefore, proper planning and prevention are essential to avoid the impact of shortage and address the issues that may affect the supply for future development. All stakeholders involved in the construction industry need to start shifting towards production methods adopted by a sustainability model to maintain security and preserve the long-term availability of construction mineral supplies. One of the best ways suggested is the practice of the mandate for a sustainable concept through recycling activities in the construction and/or concrete industries.

Several actions should take place to bring CRC to the market. CR surface treatment (that leads to interface bonding between CR and cement paste), and concrete mix design optimization need to be properly developed for each type of CR application in concrete for possible field applications (see Section 2.1). The production of high-performance concrete containing CR needs to be added as the normal practice [5]. The cost analysis among various surface treatment methods must be thoroughly studied and the cost-effectiveness of different types of CRC and CRC construction elements production must be provided.

It is usually considered that the price of processed CR is higher than natural aggregates and might not be price-wise reasonable for application in concrete. However, if to consider the landfilling cost of waste rubber tires [170] and processed CR (mechanically ground) cost [171] for the application in concrete, there might be a contradiction in opinions (see Tables 2 and 3).

Table 2. Sand prices for application in concrete.

Volume (m^3)	Price of Sand (€) [172]	Sand in Certain Concrete Mix [92]		
		Sand Mass (kg)	Price of Sand (€)	Price of Sand (15%) (€)
1	139	770	71.35	**10.70**
10	265	7700	136.03	**20.41**
20	405	15,400	207.90	**31.19**
50	825	38,500	423.50	**63.53**
100	1525	77,000	782.83	**117.43**

Table 3. Processed CR (15%) price vs. sand (15%) in CRC.

Volume (m^3)	Price of CR (15%) (€) [171]	Landfill Tax (15%) (€)		Price of CR (15%) Incl. Possible Landfill Tax (€)		Difference in Price CR vs. Sand (€)	
		Flanders	Wallonia	Flanders	Wallonia	Flanders	Wallonia
1	4.86	4.08	4.54	**0.79**	**0.32**	−9.92	−10.38
10	48.64	40.78	45.44	**7.86**	**3.20**	−12.55	−17.21
20	97.28	81.56	90.89	**15.72**	**6.39**	−15.47	−24.79
50	243.20	203.91	227.22	**39.29**	**15.98**	−24.23	−47.55
100	486.4	407.82	454.44	**78.58**	**31.96**	−38.84	−85.47

It has been already pointed out by Kazmi et al. [11] that incorporated landfill tax of waste tires in CRC cost calculations is assumed to be part of a processed CR cost, and hence, the cost of CR is considered zero.

In Table 3, the possible difference in cost when 15% of CR (mechanically ground) is used in concrete to replace 15% of sand (in Belgium, incl. Flanders and Wallonia regions) is shown. The approximate price for sand is given considering that 1 m^3 is about 1500 kg, however the final sand price can be higher depending on the region of delivery [172]. If to consider that the final price of CR consists of actual processed CR price minus possible landfill tax, then CR (mechanically ground) application in concrete benefits with a sand replacement for 15%.

CRC still need to be tested and modelled in detail before it can be widely used in construction [75]. However, the UniSA (Australia), RMIT University (Australia) and University of Salerno (Italy) researchers consider that the role of CRC may not need to be limited to structures that are less dependent on strength. In several studies [24,26,27,37,61,173,174], they have shown that a wide variety of techniques in the CRC production processes can be used to improve its structural properties.

4. Conclusions

The application of CR in concrete has certainly its advantages and in general, cannot be ignored by the construction industry. CR can be applied, for example, as an alternative material to replace natural aggregates and CRC can be used as RCA in the future. A certain diversity for the CR application can be introduced in a more efficient way when CR surface treatment and concrete mix design optimization are properly developed for each type of CR application in concrete for possible field applications. The role of CRC should not be limited to structures that are less dependent on strength:

- The workability of rubberized concrete decreases with the increase in CR content and particle size. However, it can be improved with the inclusion of admixtures such as superplasticizers, silica fume, SCMs like fly ash, slagand metakaolin.
- The density of rubberized concrete decreases substantially with the increase in CR content or fineness, due to the lower specific gravity and air-entraining capability of CR. This makes rubberized concrete useful for lightweight structures.
- The pre-treatment of CR influences positively on ITZ. If the bond is improved at ITZ by any suitable and economical means then the negative effects of CR on strength properties of normal concrete may be reduced, consequently, it would be possible to effectively use the rubberized concrete in numerous concrete structures by the construction industry.
- The optimal CR replacement range based on the mechanical, physical and durability properties could vary between 10–20% for fine aggregates replacement and 5% for coarse aggregates replacement.
- Within the optimal range and pre-treatment of CR: (1) freeze-thaw resistance, chloride ion penetration resistance, acid resistance and abrasion resistance are enhanced; (2) ASR, drying shrinkage and carbonation will not increase. However, CRC is significantly affected by sulphate attacks.
- The incorporation of CR significantly improves the dampness ratio of concrete. The vibration absorption capacity and increased dampness absorption make CRC an ideal construction material in structures under dynamic load including railways sleepers, seismic prone structures, concrete columns, bridges etc.
- It has been confirmed the surrounding cementitious materials can well confine the trace metals or volatile organics exited in rubber particles based on the TCLP test.
- The cost analysis among various surface treatment methods must be thoroughly studied and the cost-effectiveness of different types of CRC and CRC construction elements production must be provided.

Author Contributions: Writing—original draft preparation, P.K.D.M.; writing—review, J.B., L.B. and B.C. All authors have read and agreed to the published version of the manuscript.

Funding: This research received no external funding.

Institutional Review Board Statement: This study does not involve any studies on humans or animals.

Informed Consent Statement: Not applicable.

Conflicts of Interest: The authors declare no conflict of interest.

References

1. Halsband, C.; Sørensen, L.; Booth, A.M.; Herzke, D. Car tire crumb rubber: Does leaching produce a toxic chemical cocktail in coastal marine systems? *Front. Environ. Sci.* **2020**, *8*, 125. [CrossRef]
2. Song, W.-J.; Qiao, W.-G.; Yang, X.-X.; Lin, D.-G.; Li, Y.-Z. Mechanical properties and constitutive equations of crumb rubber mortars. *Constr. Build. Mater.* **2018**, *172*, 660–669. [CrossRef]
3. Seddik Meddah, M. Recycled aggregates in concrete production: Engineering properties and environmental impact. *MATEC Web Conf.* **2017**, *101*, 05021. [CrossRef]

4. Su, H.; Yang, J.; Ling, T.-C.; Ghataora, G.S.; Dirar, S. Properties of concrete prepared with waste tyre rubber particles of uniform and varying sizes. *J. Clean. Prod.* **2015**, *91*, 288–296. [CrossRef]

5. Guo, S.; Hu, J.; Dai, Q. A critical review on the performance of portland cement concrete with recycled organic components. *J. Clean. Prod.* **2018**, *188*, 92–112. [CrossRef]

6. SMART. Final Report Summary—SMART (Sustainable Moulding of Articles from Recycled Tyres). Available online: https://cordis.europa.eu/project/id/286465/reporting (accessed on 29 April 2021).

7. ANAGENNISI. Final Report Summary—ANAGENNISI (Innovative Reuse of All Tyre Components in Concrete). Available online: https://cordis.europa.eu/project/id/603722 (accessed on 29 April 2021).

8. RISEN. Periodic Reporting for Period 1—RISEN (Rail Infrastructure systems Engineering Network). Available online: https://cordis.europa.eu/project/id/691135/reporting (accessed on 29 April 2021).

9. ARC Linkage Project. Reinforced crumbed Rubber Concrete for Residential Construction. Available online: https://www.unisa.edu.au/research/scarce-resources-and-circular-economy/research-projects/reinforced-crumbed-rubber-concrete-for-residential-construction/ (accessed on 29 April 2021).

10. Leal Filho, W.; Hunt, J.; Lingos, A.; Platje, J.; Vieira, L.W.; Will, M.; Gavriletea, M.D. The unsustainable use of sand: Reporting on a global problem. *Sustainability* **2021**, *13*, 3356. [CrossRef]

11. Kazmi, S.M.S.; Munir, M.J.; Wu, Y.-F. Application of waste tire rubber and recycled aggregates in concrete products: A new compression casting approach. *Resour. Conserv. Recy.* **2021**, *167*, 105353. [CrossRef]

12. Sofi, A. Effect of waste tyre rubber on mechanical and durability properties of concrete—A review. *Ain Shams Eng. J.* **2018**, *9*, 2691–2700. [CrossRef]

13. Bisht, K.; Ramana, P.V. Waste to resource conversion of crumb rubber for production of sulphuric acid resistant concrete. *Constr. Build. Mater.* **2019**, *194*, 276–286. [CrossRef]

14. Thomas, B.S.; Gupta, R.C.; Panicker, V.J. Recycling of waste tire rubber as aggregate in concrete: Durability-related performance. *J. Clean. Prod.* **2016**, *112*, 504–513. [CrossRef]

15. Wagner, S.; Hüffer, T.; Klöckner, P.; Wehrhahn, M.; Hofmann, T.; Reemtsma, T. Tire wear particles in the aquatic environment—A review on generation, analysis, occurrence, fate and effects. *Water Res.* **2018**, *139*, 83–100. [CrossRef]

16. Sienkiewicz, M.; Kucinska-Lipka, J.; Janik, H.; Balas, A. Progress in used tyres management in the European Union: A review. *Waste Manag.* **2012**, *32*, 1742–1751. [CrossRef]

17. Lapkovskis, V.; Mironovs, V.; Kasperovich, A.; Myadelets, V.; Goljandin, D. Crumb rubber as a secondary raw material from waste rubber: A short review of end-of-life mechanical processing methods. *Recycling* **2020**, *5*, 32. [CrossRef]

18. ETRMA. *European Tyre and Rubber Industry—Statistics*; ETRMA: Brussels, Belgium, 2014.

19. Directive 1999/31/EC of 26 April 1999 on the Landfill of Waste, EUR-Lex, Publications Office of the European Union, Luxembourg, Luxembourg. Available online: https://eur-lex.europa.eu/legal-content/EN/TXT/?uri=CELEX%3A01999L0031-20180704&qid=1626971916106 (accessed on 22 July 2021).

20. Directive 2008/98/EC of the European Parliament and of the Council of 19 November 2008 on Waste and Repealing Certain Directives, EUR-Lex, Publications Office of the European Union, Luxembourg, Luxembourg. Available online: https://eur-lex.europa.eu/legal-content/EN/TXT/?uri=CELEX%3A32008L0098&qid=1626973255571 (accessed on 22 July 2021).

21. PD CEN/TS16637-2:2014. *Construction Products—Assessment of Release of Dangerous Substances—Part 2: Horizontal Dynamic Surface Leaching Test*; CEN-CENELEC Management Centre: Brussels, Belgium, 2014.

22. Pilakoutas, K.; Neocleous, K.; Tlemat, H. Reuse of tyre steel fibres as concrete reinforcement. *Proc. Inst. Civ. Eng.* **2004**, *157*, 131–138. [CrossRef]

23. Neocleous, K.; Tlemat, H.; Pilakoutas, K. Design issues for concrete reinforced with steel fibers, including fibers recovered from used tires. *J. Mater. Civil. Eng.* **2006**, *18*, 677–685. [CrossRef]

24. Aiello, M.A.; Leuzzi, F.; Centonze, G.; Maffezzoli, A. Use of steel fibres recovered from waste tyres as reinforcement in concrete: Pull-out behaviour, compressive and flexural strength. *Waste Manag.* **2009**, *29*, 1960–1970. [CrossRef]

25. Barros, J.A.; Zamanzadeh, Z.; Mendes, P.J.; Lourenço, L. Assessment of the potentialities of recycled steel fibres for the reinforcement of cement based materials. In Proceedings of the 3rd Workshop: The New Boundaries of Structural Concrete, Session C—New Scenarios for Concrete, ACI Italy Chapter, Bergamo, Italy, 3–4 October 2013; pp. 1–11.

26. Caggiano, A.; Xargay, H.; Folino, P.; Martinelli, E. Experimental and numerical characterization of the bond behavior of steel fibers recovered from waste tires embedded in cementitious matrices. *Cem. Concr. Comp.* **2015**, *62*, 146–155. [CrossRef]

27. Martinelli, E.; Caggiano, A.; Xargay, H. An experimental study on the post-cracking behaviour of Hybrid Industrial/Recycled Steel Fibre-Reinforced Concrete. *Constr. Build. Mater.* **2015**, *94*, 290–298. [CrossRef]

28. Onuaguluchi, O.; Banthia, N. Durability performance of polymeric scrap tire fibers and its reinforced cement mortar. *Mater. Struct.* **2017**, *50*, 158. [CrossRef]

29. Medina, N.F.; Medina, D.F.; Hernández-Olivares, F.; Navacerrada, M.A. Mechanical and thermal properties of concrete incorporating rubber and fibres from tyre recycling. *Constr. Build. Mater.* **2017**, *144*, 563–573. [CrossRef]

30. Chen, M.; Chen, W.; Zhong, H.; Chi, D.; Wang, Y.; Zhang, M. Experimental study on dynamic compressive behaviour of recycled tyre polymer fibre reinforced concrete. *Cem. Concr. Comp.* **2019**, *98*, 95–112. [CrossRef]

31. Siddique, R.; Naik, T.R. Properties of concrete containing scrap-tire rubber—An overview. *Waste Manag.* **2004**, *24*, 563–569. [CrossRef]

32. Papakonstantinou, C.G.; Tobolski, M.J. Use of waste tire steel beads in Portland cement concrete. *Cem. Concr. Res.* **2006**, *36*, 1686–1691. [CrossRef]
33. Khaloo, A.R.; Dehestani, M.; Rahmatabadi, P. Mechanical properties of concrete containing a high volume of tire–rubber particles. *Waste Manag.* **2008**, *28*, 2472–2482. [CrossRef] [PubMed]
34. Ganjian, E.; Khorami, M.; Maghsoudi, A.A. Scrap-tyre-rubber replacement for aggregate and filler in concrete. *Constr. Build. Mater.* **2009**, *23*, 1828–1836. [CrossRef]
35. Aiello, M.A.; Leuzzi, F. Waste tyre rubberized concrete: Properties at fresh and hardened state. *Waste Manag.* **2010**, *30*, 1696–1704. [CrossRef]
36. Li, Y.; Zhang, S.; Wang, R.; Dang, F. Potential use of waste tire rubber as aggregate in cement concrete—A comprehensive review. *Constr. Build. Mater.* **2019**, *225*, 1183–1201. [CrossRef]
37. Barros, J.A.O.; Frazão, C.; Caggiano, A.; Folino, P.; Martinelli, E.; Xargay, H.; Zamanzadeh, Z.; Lourenço, L. Cementitious composites reinforced with recycled fibres. In *Recent Advances on Green Concrete for Structural Purposes: The Contribution of the EU-FP7 Project EnCoRe*; Barros, J.A.O., Ferrara, L., Martinelli, E., Eds.; Springer: Cham, Switzerland, 2017; pp. 141–195. [CrossRef]
38. Khatib, Z.K.; Bayomy, F.M. Rubberized Portland cement concrete. *J. Mater. Civil. Eng.* **1999**, *11*, 206–213. [CrossRef]
39. Aslani, F.; Khan, M. Properties of high-performance self-compacting rubberized concrete exposed to high temperatures. *J. Mater. Civil. Eng.* **2019**, *31*, 04019040. [CrossRef]
40. Kaloush, K.E.; Way, G.B.; Zhu, H. Properties of crumb rubber concrete. *Transp. Res. Rec.* **2005**, *1914*, 8–14. [CrossRef]
41. Yang, L.-H.; Han, Z.; Li, C.-F. Strengths and flexural strain of CRC specimens at low temperature. *Constr. Build. Mater.* **2011**, *25*, 906–910. [CrossRef]
42. Loderer, C.; Partl, M.N.; Poulikakos, L.D. Effect of crumb rubber production technology on performance of modified bitumen. *Constr. Build. Mater.* **2018**, *191*, 1159–1171. [CrossRef]
43. Mohammed, B.S.; Anwar Hossain, K.M.; Eng Swee, J.T.; Wong, G.; Abdullahi, M. Properties of crumb rubber hollow concrete block. *J. Clean. Prod.* **2012**, *23*, 57–67. [CrossRef]
44. Yung, W.H.; Yung, L.C.; Hua, L.H. A study of the durability properties of waste tire rubber applied to self-compacting concrete. *Constr. Build. Mater.* **2013**, *41*, 665–672. [CrossRef]
45. Rashad, A.M. A comprehensive overview about recycling rubber as fine aggregate replacement in traditional cementitious materials. *Int. J. Built Environ. Sustain.* **2016**, *5*, 46–82. [CrossRef]
46. Onuaguluchi, O.; Panesar, D.K. Hardened properties of concrete mixtures containing pre-coated crumb rubber and silica fume. *J. Clean. Prod.* **2014**, *82*, 125–131. [CrossRef]
47. Najim, K.B.; Hall, M.R. Crumb rubber aggregate coatings/pre-treatments and their effects on interfacial bonding, air entrapment and fracture toughness in self-compacting rubberised concrete (SCRC). *Mater. Struct.* **2013**, *46*, 2029–2043. [CrossRef]
48. Hamdi, A.; Abdelaziz, G.; Farhan, K.Z. Scope of reusing waste shredded tires in concrete and cementitious composite materials: A review. *J. Build. Eng.* **2021**, *35*, 102014. [CrossRef]
49. Zheng, L.; Sharon Huo, X.; Yuan, Y. Experimental investigation on dynamic properties of rubberized concrete. *Constr. Build. Mater.* **2008**, *22*, 939–947. [CrossRef]
50. Youssf, O.; Hassanli, R.; Mills, J.E.; Abd Elrahman, M. An experimental investigation of the mechanical performance and structural application of LECA-Rubcrete. *Constr. Build. Mater.* **2018**, *175*, 239–253. [CrossRef]
51. Zhang, B.; Poon, C.S. Sound insulation properties of rubberized lightweight aggregate concrete. *J. Clean. Prod.* **2018**, *172*, 3176–3185. [CrossRef]
52. Kashani, A.; Ngo, T.D.; Mendis, P.; Black, J.R.; Hajimohammadi, A. A sustainable application of recycled tyre crumbs as insulator in lightweight cellular concrete. *J. Clean. Prod.* **2017**, *149*, 925–935. [CrossRef]
53. El-Gammal, A.; Abdel-Gawad, A.K.; El-Sherbini, Y.; Shalaby, A. Compressive strength of concrete utilizing waste tire rubber. *J. Eng. Trends Eng. Appl. Sci.* **2010**, *1*, 96–99. [CrossRef]
54. Bignozzi, M.C.; Sandrolini, F. Tyre rubber waste recycling in self-compacting concrete. *Cem. Concr. Res.* **2006**, *36*, 735–739. [CrossRef]
55. Najim, K.B.; Hall, M.R. A review of the fresh/hardened properties and applications for plain-(PRC) and self-compacting rubberised concrete (SCRC). *Constr. Build. Mater.* **2010**, *24*, 2043–2051. [CrossRef]
56. Reda Taha, M.M.; El-Dieb, A.S.; Abd El-Wahab, M.A.; Abdel-Hameed, M.E. Mechanical, fracture, and microstructural investigations of rubber concrete. *J. Mater. Civil. Eng.* **2008**, *20*, 640–649. [CrossRef]
57. Uygunoğlu, T.; Topçu, İ.B. The role of scrap rubber particles on the drying shrinkage and mechanical properties of self-consolidating mortars. *Constr. Build. Mater.* **2010**, *24*, 1141–1150. [CrossRef]
58. Mohammadi, I.; Khabbaz, H. Shrinkage performance of crumb rubber concrete (CRC) prepared by water-soaking treatment method for rigid pavements. *Cem. Concr. Comp.* **2015**, *62*, 106–116. [CrossRef]
59. Meddah, A.; Beddar, M.; Bali, A. Use of shredded rubber tire aggregates for roller compacted concrete pavement. *J. Clean. Prod.* **2014**, *72*, 187–192. [CrossRef]
60. Rahman, M.M.; Usman, M.; Al-Ghalib, A.A. Fundamental properties of rubber modified self-compacting concrete (RMSCC). *Constr. Build. Mater.* **2012**, *36*, 630–637. [CrossRef]
61. Youssf, O.; ElGawady, M.A.; Mills, J.E. Static cyclic behaviour of FRP-confined crumb rubber concrete columns. *Eng. Struct.* **2016**, *113*, 371–387. [CrossRef]

62. Medina, N.F.; Garcia, R.; Hajirasouliha, I.; Pilakoutas, K.; Guadagnini, M.; Raffoul, S. Composites with recycled rubber aggregates: Properties and opportunities in construction. *Constr. Build. Mater.* **2018**, *188*, 884–897. [CrossRef]
63. Al-Tayeb, M.M.; Abu Bakar, B.H.; Akil, H.M.; Ismail, H. Performance of rubberized and hybrid rubberized concrete structures under static and impact load conditions. *Exp. Mech.* **2013**, *53*, 377–384. [CrossRef]
64. Al-Tayeb, M.M.; Abu Bakar, B.H.; Ismail, H.; Akil, H.M. Effect of partial replacement of sand by recycled fine crumb rubber on the performance of hybrid rubberized-normal concrete under impact load: Experiment and simulation. *J. Clean. Prod.* **2013**, *59*, 284–289. [CrossRef]
65. Li, W.; Huang, Z.; Wang, X.C.; Wang, J.W. Review of crumb rubber concrete. *Appl. Mech. Mater.* **2014**, *672–674*, 1833–1837. [CrossRef]
66. Xu, X.; Zhang, Z.; Hu, Y.; Wang, X. Bearing strength of crumb rubber concrete under partial area loading. *Materials* **2020**, *13*, 2446. [CrossRef]
67. Duarte, A.P.C.; Silva, B.A.; Silvestre, N.; de Brito, J.; Júlio, E.; Castro, J.M. Finite element modelling of short steel tubes filled with rubberized concrete. *Compos. Struct.* **2016**, *150*, 28–40. [CrossRef]
68. Frankowski, R. Patent US 5290356 Rubber Crumb-Reinforced Cement Concrete. U.S. Patent 5290356, 23 April 1992.
69. Eldin, N.N.; Senouci, A.B. Observations on rubberized concrete behavior. *Cem. Concr. Comp.* **1994**, *16*, 287–298. [CrossRef]
70. Zandi, I.; Lepore, J.; Rostami, H. Particulate rubber included concrete compositions. U.S. Patent 5456751, 14 November 1994.
71. Topçu, İ.B. The properties of rubberized concretes. *Cem. Concr. Res.* **1995**, *25*, 304–310. [CrossRef]
72. Fattuhi, N.I.; Clark, L.A. Cement-based materials containing shredded scrap truck tyre rubber. *Constr. Build. Mater.* **1996**, *10*, 229–236. [CrossRef]
73. Toutanji, H.A. The use of rubber tire particles in concrete to replace mineral aggregates. *Cem. Concr. Comp.* **1996**, *18*, 135–139. [CrossRef]
74. Biel, T.D.; Lee, H. Magnesium oxychloride cement concrete with recycled tire rubber. *Transp. Res. Rec.* **1996**, *1561*, 6–12. [CrossRef]
75. Mills, J.; Gravina, R.; Zhuge, Y.; Ma, X.; Skinner, B. Crumbed Rubber Concrete: A Promising Material for Sustainable Construction. 2018. Available online: https://www.scientia.global/wp-content/uploads/Mills-Zhuge-Skinner-Ma-Gravina/Mills-Zhuge-Skinner-Ma-Gravina.pdf (accessed on 1 February 2021).
76. Li, Y.; Zhang, X.; Wang, R.; Lei, Y. Performance enhancement of rubberised concrete via surface modification of rubber: A review. *Constr. Build. Mater.* **2019**, *227*, 116691. [CrossRef]
77. Topçu, İ.B. Assessment of the brittleness index of rubberized concretes. *Cem. Concr. Res.* **1997**, *27*, 177–183. [CrossRef]
78. Li, C.-F.; Xie, J.; Liu, C.; Guo, S.; Deng, J. Fire performance of high-strength concrete reinforced with recycled rubber particles. *Mag. Concr. Res.* **2011**, *63*, 187–195. [CrossRef]
79. Hassanli, R.; Youssf, O.; Mills, J.E. Experimental investigations of reinforced rubberized concrete structural members. *J. Build. Eng.* **2017**, *10*, 149–165. [CrossRef]
80. Ling, T.-C. Prediction of density and compressive strength for rubberized concrete blocks. *Constr. Build. Mater.* **2011**, *25*, 4303–4306. [CrossRef]
81. Eldin, N.N.; Senouci, A.B. Rubber-tire particles as concrete aggregate. *J. Mater. Civil. Eng.* **1993**, *5*, 478–496. [CrossRef]
82. Issa, C.A.; Salem, G. Utilization of recycled crumb rubber as fine aggregates in concrete mix design. *Constr. Build. Mater.* **2013**, *42*, 48–52. [CrossRef]
83. Liu, F.; Zheng, W.; Li, L.; Feng, W.; Ning, G. Mechanical and fatigue performance of rubber concrete. *Constr. Build. Mater.* **2013**, *47*, 711–719. [CrossRef]
84. Yang, G.; Chen, X.; Xuan, W.; Chen, Y. Dynamic compressive and splitting tensile properties of concrete containing recycled tyre rubber under high strain rates. *Sādhanā* **2018**, *43*, 178. [CrossRef]
85. Chan, C.W.; Yu, T.; Zhang, S.S.; Xu, Q.F. Compressive behaviour of FRP-confined rubber concrete. *Constr. Build. Mater.* **2019**, *211*, 416–426. [CrossRef]
86. Valadares, F.; Bravo, M.; de Brito, J. Concrete with used tire rubber aggregates: Mechanical performance. *ACI Mater. J.* **2012**, *109*, 283–292.
87. da Silva, F.M.; Gachet Barbosa, L.A.; Lintz, R.C.C.; Jacintho, A.E.P.G.A. Investigation on the properties of concrete tactile paving blocks made with recycled tire rubber. *Constr. Build. Mater.* **2015**, *91*, 71–79. [CrossRef]
88. Jokar, F.; Khorram, M.; Karimi, G.; Hataf, N. Experimental investigation of mechanical properties of crumbed rubber concrete containing natural zeolite. *Constr. Build. Mater.* **2019**, *208*, 651–658. [CrossRef]
89. Ismail, M.K.; Hassan, A.A.A. Impact resistance and mechanical properties of self-consolidating rubberized concrete reinforced with steel fibers. *J. Mater. Civil. Eng.* **2017**, *29*, 04016193. [CrossRef]
90. Bharathi Murugan, R.; Natarajan, C. Investigation of the behaviour of concrete containing waste tire crumb rubber. In *Advances in Structural Engineering*; Matsagar, V., Ed.; Springer: New Delhi, India, 2015; pp. 1795–1802. [CrossRef]
91. Wu, Y.-F.; Kazmi, S.M.S.; Munir, M.J.; Zhou, Y.; Xing, F. Effect of compression casting method on the compressive strength, elastic modulus and microstructure of rubber concrete. *J. Clean. Prod.* **2020**, *264*, 121746. [CrossRef]
92. Van Tigchelt, S.; Roelandt, D.; Kara De Maeijer, P.; Craeye, B. Use of Crumb Rubber as Inert Material for Sustainable Concrete Applications. Master's Thesis, UAntwerp, Antwerpen, Belgium, 2021.
93. Ganesan, N.; Bharati Raj, J.; Shashikala, A.P. Flexural fatigue behavior of self compacting rubberized concrete. *Constr. Build. Mater.* **2013**, *44*, 7–14. [CrossRef]

94. Güneyisi, E. Fresh properties of self-compacting rubberized concrete incorporated with fly ash. *Mater. Struct.* **2010**, *43*, 1037–1048. [CrossRef]
95. Lotfy, A.; Hossain, K.M.A.; Lachemi, M. Mix design and properties of lightweight self-consolidating concretes developed with furnace slag, expanded clay and expanded shale aggregates. *J. Sustain. Cem. Based Mater.* **2016**, *5*, 297–323. [CrossRef]
96. Ismail, M.K.; Hassan, A.A.A. Use of metakaolin on enhancing the mechanical properties of self-consolidating concrete containing high percentages of crumb rubber. *J. Clean. Prod.* **2016**, *125*, 282–295. [CrossRef]
97. Alsaif, A.; Bernal, S.A.; Guadagnini, M.; Pilakoutas, K. Freeze-thaw resistance of steel fibre reinforced rubberised concrete. *Constr. Build. Mater.* **2019**, *195*, 450–458. [CrossRef]
98. Youssf, O.; Mills, J.E.; Hassanli, R. Assessment of the mechanical performance of crumb rubber concrete. *Constr. Build. Mater.* **2016**, *125*, 175–183. [CrossRef]
99. Bekhiti, M.; Trouzine, H.; Asroun, A. Properties of waste tire rubber powder. *Eng. Technol. Appl. Sci. Res.* **2014**, *4*, 669–672. [CrossRef]
100. Gregori, A.; Castoro, C.; Marano, G.C.; Greco, R. Strength reduction factor of concrete with recycled rubber aggregates from tires. *J. Mater. Civil. Eng.* **2019**, *31*, 04019146. [CrossRef]
101. Pelisser, F.; Zavarise, N.; Longo, T.A.; Bernardin, A.M. Concrete made with recycled tire rubber: Effect of alkaline activation and silica fume addition. *J. Clean. Prod.* **2011**, *19*, 757–763. [CrossRef]
102. Mohammadi, I.; Khabbaz, H.; Vessalas, K. Enhancing mechanical performance of rubberised concrete pavements with sodium hydroxide treatment. *Mater. Struct.* **2016**, *49*, 813–827. [CrossRef]
103. Strukar, K.; Kalman Šipoš, T.; Miličević, I.; Bušić, R. Potential use of rubber as aggregate in structural reinforced concrete element—A review. *Eng. Struct.* **2019**, *188*, 452–468. [CrossRef]
104. Li, Z.; Li, F.; Li, J.S.L. Properties of concrete incorporating rubber tyre particles. *Mag. Concr. Res.* **1998**, *50*, 297–304. [CrossRef]
105. Balaha, M.M.; Badawy, A.A.M.; Hashish, M. Effect of using ground waste tire rubber as fine aggregate on the behaviour of concrete mixes. *Indian J. Eng. Mater. Sci.* **2007**, *14*, 427–435. Available online: http://nopr.niscair.res.in/bitstream/123456789/22 5/1/IJEMS%2014%286%29%20%282007%29%20427-435.pdf (accessed on 1 February 2021).
106. Segre, N.; Joekes, I. Use of tire rubber particles as addition to cement paste. *Cem. Concr. Res.* **2000**, *30*, 1421–1425. [CrossRef]
107. Dong, Q.; Huang, B.; Shu, X. Rubber modified concrete improved by chemically active coating and silane coupling agent. *Constr. Build. Mater.* **2013**, *48*, 116–123. [CrossRef]
108. Huang, B.; Shu, X.; Cao, J. A two-staged surface treatment to improve properties of rubber modified cement composites. *Constr. Build. Mater.* **2013**, *40*, 270–274. [CrossRef]
109. Albano, C.; Camacho, N.; Reyes, J.; Feliu, J.L.; Hernández, M. Influence of scrap rubber addition to Portland I concrete composites: Destructive and non-destructive testing. *Compos. Struct.* **2005**, *71*, 439–446. [CrossRef]
110. Cavanagh, P.H.; Johnson, C.R.; Roy-Delage, L.; Sylvaine, D.; Gerard, G.; Cooper, I.; Guillot, D.J.; Bulte, H.; Dargaud, B. Self-healing cement-novel technology to achieve leak-free wells. In Proceedings of the SPE/IADC Drilling Conference, Amsterdam, The Netherlands, 20–22 February 2007. [CrossRef]
111. Le Roy-Delage, S.; Comet, A.; Garnier, A.; Presles, J.L.; Bulté-Loyer, H.; Drecq, P.; Rodriguez, I.U. Self-healing cement system—A step forward in reducing long-term environmental impact. In Proceedings of the IADC/SPE Drilling Conference and Exhibition, New Orleans, LA, USA, 2–4 February 2010. [CrossRef]
112. Chou, L.-H.; Lin, C.-N.; Lu, C.-K.; Lee, C.-H.; Lee, M.-T. Improving rubber concrete by waste organic sulfur compounds. *Waste Manag. Res.* **2010**, *28*, 29–35. [CrossRef] [PubMed]
113. Ossola, G.; Wojcik, A. UV modification of tire rubber for use in cementitious composites. *Cem. Concr. Comp.* **2014**, *52*, 34–41. [CrossRef]
114. He, L.; Cai, H.; Huang, Y.; Ma, Y.; Van den bergh, W.; Gaspar, L.; Valentin, J.; Vasiliev, Y.E.; Kowalski, K.J.; Zhang, J. Research on the properties of rubber concrete containing surface-modified rubber powders. *J. Build. Eng.* **2021**, *35*, 101991. [CrossRef]
115. Mwaluwinga, S.; Ayano, T.; Sakata, K. Influence of urea in concrete. *Cem. Concr. Res.* **1997**, *27*, 733–745. [CrossRef]
116. Stocks Fischer, S.; Galinat, J.K.; Bang, S.S. Microbiological precipitation of CaCO$_3$. *Soil Biol. Biochem.* **1999**, *31*, 1563–1571. [CrossRef]
117. De Muynck, W.; Debrouwer, D.; De Belie, N.; Verstraete, W. Bacterial carbonate precipitation improves the durability of cementitious materials. *Cem. Concr. Res.* **2008**, *38*, 1005–1014. [CrossRef]
118. Gesoğlu, M.; Güneyisi, E. Strength development and chloride penetration in rubberized concretes with and without silica fume. *Mater. Struct.* **2007**, *40*, 953–964. [CrossRef]
119. Güneyisi, E.; Gesoğlu, M.; Özturan, T. Properties of rubberized concretes containing silica fume. *Cem. Concr. Res.* **2004**, *34*, 2309–2317. [CrossRef]
120. Azevedo, F.; Pacheco-Torgal, F.; Jesus, C.; Barroso de Aguiar, J.L.; Camões, A.F. Properties and durability of HPC with tyre rubber wastes. *Constr. Build. Mater.* **2012**, *34*, 186–191. [CrossRef]
121. Kang, J.; Zhang, B.; Li, G. The abrasion-resistance investigation of rubberized concrete. *J. Wuhan Univ. Technol. Mat. Sci. Edit.* **2012**, *27*, 1144–1148. [CrossRef]
122. Mohajerani, A.; Burnett, L.; Smith, J.V.; Markovski, S.; Rodwell, G.; Rahman, M.T.; Kurmus, H.; Mirzababaei, M.; Arulrajah, A.; Horpibulsuk, S.; et al. Recycling waste rubber tyres in construction materials and associated environmental considerations: A review. *Resour. Conserv. Recy.* **2020**, *155*, 104679. [CrossRef]

123. Zhu, H.; Wang, Z.; Xu, J.; Han, Q. Microporous structures and compressive strength of high-performance rubber concrete with internal curing agent. *Constr. Build. Mater.* **2019**, *215*, 128–134. [CrossRef]
124. Gonen, T. Freezing-thawing and impact resistance of concretes containing waste crumb rubbers. *Constr. Build. Mater.* **2018**, *177*, 436–442. [CrossRef]
125. Shu, X.; Huang, B. Recycling of waste tire rubber in asphalt and portland cement concrete: An overview. *Constr. Build. Mater.* **2014**, *67*, 217–224. [CrossRef]
126. Girskas, G.; Nagrockienė, D. Crushed rubber waste impact of concrete basic properties. *Constr. Build. Mater.* **2017**, *140*, 36–42. [CrossRef]
127. Richardson, A.E.; Coventry, K.A.; Ward, G. Freeze/thaw protection of concrete with optimum rubber crumb content. *J. Clean. Prod.* **2012**, *23*, 96–103. [CrossRef]
128. Si, R.; Guo, S.; Dai, Q. Durability performance of rubberized mortar and concrete with NaOH-Solution treated rubber particles. *Constr. Build. Mater.* **2017**, *153*, 496–505. [CrossRef]
129. Thomas, B.S.; Chandra Gupta, R. Properties of high strength concrete containing scrap tire rubber. *J. Clean. Prod.* **2016**, *113*, 86–92. [CrossRef]
130. Afshinnia, K.; Poursaee, A. The influence of waste crumb rubber in reducing the alkali–silica reaction in mortar bars. *J. Build. Eng.* **2015**, *4*, 231–236. [CrossRef]
131. Gesoğlu, M.; Güneyisi, E. Permeability properties of self-compacting rubberized concretes. *Constr. Build. Mater.* **2011**, *25*, 3319–3326. [CrossRef]
132. Bravo, M.; de Brito, J. Concrete made with used tyre aggregate: Durability-related performance. *J. Clean. Prod.* **2012**, *25*, 42–50. [CrossRef]
133. Segre, N.; Joekes, I.; Galves, A.D.; Rodrigues, J.A. Rubber-mortar composites: Effect of composition on properties. *J. Mater. Sci.* **2004**, *39*, 3319–3327. [CrossRef]
134. Oikonomou, N.; Mavridou, S. Improvement of chloride ion penetration resistance in cement mortars modified with rubber from worn automobile tires. *Cem. Concr. Comp.* **2009**, *31*, 403–407. [CrossRef]
135. Thomas, B.S.; Gupta, R.C. Long term behaviour of cement concrete containing discarded tire rubber. *J. Clean. Prod.* **2015**, *102*, 78–87. [CrossRef]
136. Alsaif, A.; Koutas, L.; Bernal, S.A.; Guadagnini, M.; Pilakoutas, K. Mechanical performance of steel fibre reinforced rubberised concrete for flexible concrete pavements. *Constr. Build. Mater.* **2018**, *172*, 533–543. [CrossRef]
137. Turatsinze, A.; Bonnet, S.; Granju, J.L. Potential of rubber aggregates to modify properties of cement based-mortars: Improvement in cracking shrinkage resistance. *Constr. Build. Mater.* **2007**, *21*, 176–181. [CrossRef]
138. Alsaif, A.; Bernal, S.A.; Guadagnini, M.; Pilakoutas, K. Durability of steel fibre reinforced rubberised concrete exposed to chlorides. *Constr. Build. Mater.* **2018**, *188*, 130–142. [CrossRef]
139. Guo, S.; Dai, Q.; Si, R.; Sun, X.; Lu, C. Evaluation of properties and performance of rubber-modified concrete for recycling of waste scrap tire. *J. Clean. Prod.* **2017**, *148*, 681–689. [CrossRef]
140. Zhu, H.; Liang, J.; Xu, J.; Bo, M.; Li, J.; Tang, B. Research on anti-chloride ion penetration property of crumb rubber concrete at different ambient temperatures. *Constr. Build. Mater.* **2018**, *189*, 42–53. [CrossRef]
141. Gheni, A.A.; Alghazali, H.H.; ElGawady, M.A.; Myers, J.J.; Feys, D. Durability properties of cleaner cement mortar with by-products of tire recycling. *J. Clean. Prod.* **2019**, *213*, 1135–1146. [CrossRef]
142. Li, X.; Berger, W.; Musante, C.; Mattina, M.I. Characterization of substances released from crumb rubber material used on artificial turf fields. *Chemosphere* **2010**, *80*, 279–285. [CrossRef]
143. Kardos, A.J.; Durham, S.A. Strength, durability, and environmental properties of concrete utilizing recycled tire particles for pavement applications. *Constr. Build. Mater.* **2015**, *98*, 832–845. [CrossRef]
144. Muyssen, B.T.A.; De Schamphelaere, K.A.C.; Janssen, C.R. Mechanisms of chronic waterborne Zn toxicity in Daphnia magna. *Aquat. Toxicol.* **2006**, *77*, 393–401. [CrossRef] [PubMed]
145. Stephensen, E.; Adolfsson-Erici, M.; Celander, M.; Hulander, M.; Parkkonen, J.; Hegelund, T.; Sturve, J.; Hasselberg, L.; Bengtsson, M.; Förlin, L. Biomarker responses and chemical analyses in fish indicate leakage of polycyclic aromatic hydrocarbons and other compounds from car tire rubber. *Environ. Toxicol. Chem.* **2003**, *22*, 2926–2931. [CrossRef]
146. Wik, A.; Dave, G. Environmental labeling of car tires—toxicity to Daphnia magna can be used as a screening method. *Chemosphere* **2005**, *58*, 645–651. [CrossRef]
147. Zhang, J.; Han, I.-K.; Zhang, L.; Crain, W. Hazardous chemicals in synthetic turf materials and their bioaccessibility in digestive fluids. *J. Expo. Sci. Environ. Epidemiol.* **2008**, *18*, 600–607. [CrossRef]
148. Kanematsu, M.; Hayashi, A.; Denison, M.S.; Young, T.M. Characterization and potential environmental risks of leachate from shredded rubber mulches. *Chemosphere* **2009**, *76*, 952–958. [CrossRef]
149. Turner, A.; Rice, L. Toxicity of tire wear particle leachate to the marine macroalga, Ulva lactuca. *Environ. Pollut.* **2010**, *158*, 3650–3654. [CrossRef]
150. Bocca, B.; Forte, G.; Petrucci, F.; Costantini, S.; Izzo, P. Metals contained and leached from rubber granulates used in synthetic turf areas. *Sci. Total Environ.* **2009**, *407*, 2183–2190. [CrossRef]
151. Azizian, M.F.; Nelson, P.O.; Thayumanavan, P.; Williamson, K.J. Environmental impact of highway construction and repair materials on surface and ground waters: Case study: Crumb rubber asphalt concrete. *Waste Manag.* **2003**, *23*, 719–728. [CrossRef]

152. Reddy, C.M.; Quinn, J.G. Environmental chemistry of benzothiazoles derived from rubber. *Environ. Sci. Technol.* **1997**, *31*, 2847–2853. [CrossRef]
153. Kloepfer, A.; Jekel, M.; Reemtsma, T. Determination of benzothiazoles from complex aqueous samples by liquid chromatography-mass spectrometry following solid-phase extraction. *J. Chromatogr. A* **2004**, *1058*, 81–88. [CrossRef]
154. Downs, L.A.; Humphrey, D.N.; Katz, L.E.; Rock, C.A. Water Quality Effects of Using Tire Chips below the Groundwater Table. University of Maine, Orono, ME, USA. 1996. Available online: https://www.ustires.org/sites/default/files/LEA_004_USTMA.pdf (accessed on 18 July 2021).
155. Selbes, M.; Yilmaz, O.; Khan, A.A.; Karanfil, T. Leaching of DOC, DN, and inorganic constituents from scrap tires. *Chemosphere* **2015**, *139*, 617–623. [CrossRef] [PubMed]
156. Overmann, S.; Lin, X.; Vollpracht, A. Investigations on the leaching behavior of fresh concrete—A review. *Constr. Build. Mater.* **2021**, *272*, 121390. [CrossRef]
157. Hartwich, P.; Vollpracht, A. Influence of leachate composition on the leaching behaviour of concrete. *Cem. Concr. Res.* **2017**, *100*, 423–434. [CrossRef]
158. Vollpracht, A.; Brameshuber, W. Binding and leaching of trace elements in Portland cement pastes. *Cem. Concr. Res.* **2016**, *79*, 76–92. [CrossRef]
159. NEN7345. *Uitloogkarakteristieken—Bepaling van de Uitloging van Anorganische Componenten uit Vormgegeven en Monolitische Materialen met een Diffusieproef—Vaste Grond- en Steenachtige Materialen*; The Royal Netherlands Standardization Institute: Delft, The Netherlands, 1995. (In Dutch)
160. NEN7375. *Leaching Characteristics—Determination of the Leaching of Inorganic Components from Moulded or Monolitic Materials with a Diffusion Test—Solid Earthy and Stony Materials*; The Royal Netherlands Standardization Institute: Delft, The Netherlands, 2004.
161. DAfStb. *Determination of the Release of Inorganic Substances Byleaching from Cement-Bound Building Materials*; Deutscher Ausschuss für Stahlbeton: Berlin, Germany, 2005.
162. CMA2/II/A.9.1. VLAREMA—The Flemish Standard. 2016. Available online: https://navigator.emis.vito.be/mijn-navigator?woId=43993 (accessed on 1 March 2021).
163. SW-846. *Test Method 1315. Mass Transfer Rates of Constituents in Monolithic or Compacted Granular Materials Using a Semi-Dynamic Tank Leaching Procedure*; United States Environmental Protection Agency (EPA): Washington, DC, USA, 2017.
164. Deutsches Institut für Bautechnik (DIBt). *Muster-Verwaltungsvorschrift Technische Baubestimmungen*; Model Administrative Regulation Fortechnical Building Regulations; Deutsches Institut für Bautechnik: Berlin, Germany, 2019. (In German)
165. Dutch Soil Quality Decree: Regulation of 13 December 2007, No. DJZ2007124397. Available online: https://wetten.overheid.nl/BWBR0023085/2018-11-30 (accessed on 18 July 2021). (In Dutch)
166. Oikonomou, N.D. Recycled concrete aggregates. *Cem. Concr. Comp.* **2005**, *27*, 315–318. [CrossRef]
167. Kara, P. The next generation ecological self compacting concrete with glass waste powder as a cement component in concrete and recycled concrete aggregates. In Proceedings of the 3rd Workshop on The New Boundaries of Structural Concrete, University of Bergamo—ACI Italy Chapter, Bergamo, Italy, 3–4 October 2013; pp. 21–30, ISBN 9788890429279.
168. FPSE, Sand and Gravel Extractionin the Belgian Part of the North Sea. Federal Public Service Economy, SMEs, Self-Employed and Energy. 2014. Available online: www.vliz.be/imisdocs/publications/265503.pdf (accessed on 30 April 2021).
169. Ismail, S.; Hoe, K.W.; Ramli, M. Sustainable aggregates: The potential and challenge for natural resources conservation. *Procedia Soc. Behav. Sci.* **2013**, *101*, 100–109. [CrossRef]
170. CEWEP, Landfill Taxes and Bans Overview. Available online: https://www.cewep.eu/wp-content/uploads/2017/12/Landfill-taxes-and-bans-overview.pdf (accessed on 30 April 2021).
171. RECYTYRE, Recytyre in Sprekende Cijfers. Available online: https://www.recytyre.be/nl/recytyre-sprekende-cijfers (accessed on 30 April 2021). (In Dutch)
172. Wat Kost een Kuub Zand. Available online: https://grondverzet.nu/wat-kost-een-kuub-zand (accessed on 30 April 2021). (In Dutch)
173. Youssf, O.; ElGawady, M.A.; Mills, J.E.; Ma, X. An experimental investigation of crumb rubber concrete confined by fibre reinforced polymer tubes. *Constr. Build. Mater.* **2014**, *53*, 522–532. [CrossRef]
174. Youssf, O.; Hassanli, R.; Mills, J.E. Mechanical performance of FRP-confined and unconfined crumb rubber concrete containing high rubber content. *J. Build. Eng.* **2017**, *11*, 115–126. [CrossRef]

Review

Sustainable Assessment of Concrete Repairs through Life Cycle Assessment (LCA) and Life Cycle Cost Analysis (LCCA)

Neel Renne, Patricia Kara De Maeijer, Bart Craeye, Matthias Buyle and Amaryllis Audenaert

[1] Built Environment Assessing Sustainability (BEASt), Energy and Materials in Infrastructure and Buildings (EMIB), University of Antwerp, 2020 Antwerp, Belgium

[2] Durable Building in Team (DuBiT), Department of Industrial Sciences & Technology, Odisee University College, 9320 Aalst, Belgium

[3] Unit Sustainable Materials Management, Flemish Institute for Technological Research (VITO), 2400 Mol, Belgium

* Correspondence: neel.renne@uantwerpen.be

Abstract: Nowadays, a vast number of concrete structures are approaching the end of their expected service life. The need for maintenance and repair is high due to the continued deterioration of the existing building inventory and infrastructure, resulting in a large need for concrete repair in the near future. Reinforcement corrosion is the most important deterioration mechanism, causing (i) severe concrete damage (cracking along reinforcement and the spalling of the cover concrete) and (ii) loss in steel section. Therefore, appropriate repair techniques for corrosion damage are the main focus of this review paper. With the European transition towards a circular economy and with sustainable development goals in mind, it is also important to consider the environmental impact along with the technical requirements and life cycle cost. In order to improve the sustainability of concrete structures and repairs over their life cycle, life cycle assessment (LCA) and life cycle cost analysis (LCCA) should be applied. However, more research efforts are needed in this field for further development and refinement. This literature review tries to adress this need by compiling existing knowledge and gaps in the state-of-the-art. A comprehensive literature survey about concrete repair assessment through LCA and LCCA is performed and showed a high potential for further investigation. Additionally, it was noticed that many differences are present between the studies considering LCA and/or LCCA, namely, the considered (i) structures, (ii) damage causes, (iii) repair techniques, (iv) estimated and expected life spans, (v) LCCA methods, (vi) life cycle impact assessment (LCIA) methods, etc. Therefore, due to the case specificity, mutual comparison is challenging.

Keywords: corrosion concrete damage; repair; rehabilitation; life cycle assessment (LCA); life cycle cost analysis (LCCA)

Citation: Neel Renne, Patricia Kara De Maeijer, Bart Craeye, Matthias Buyle and Amaryllis Audenaert Sustainable Assessment of Concrete Repairs through Life Cycle Assessment (LCA) and Life Cycle Cost Analysis (LCCA). *Infrastructures* **2022**, *7*, 128. https://doi.org/10.3390/infrastructures7100128

Academic Editor: Chris Goodier

Received: 13 August 2022
Accepted: 21 September 2022
Published: 26 September 2022

Publisher's Note: MDPI stays neutral with regard to jurisdictional claims in published maps and institutional affiliations.

1. Introduction

The European construction industry reached a peak in the manufacturing of reinforced concrete (RC) structures in the 1960s and 1970s. To date, the majority of these structures is either approaching or has already reached the end of their expected service life. Consequently, the need for maintenance and repair is high. Due to the continued deterioration of the existing building inventory and infrastructure, a large volume of concrete repair is expected [1]. To make it more tangible, it has been estimated that approximately 50% of Europe's annual construction budget is spent on refurbishment and repair, which confirms the importance of sustainable concrete repair [2]. Moreover, the construction sector uses about 50% of the Earth's raw materials and produces 50% of its waste [3,4]. Besides, the carbon emissions and energy demand associated with concrete use are mostly attributable to cement production and represents 5 to 8% of the total CO_2 emissions from human activities and approximately 12 to 15% of the total industrial energy demand [3,5,6]. Lastly, it is also known that the construction sector of the EU is the largest consumer of natural resources and the largest producer of waste at the same time. These facts ask for more circular and environmentally friendly approaches in the construction sector to achieve the transition towards a circular economy (CE). According to the definition of the European Commission, a circular economy aims to maintain the value of products, materials, and resources for as long as possible by returning them into the product cycle at the end of their use, while minimizing the generation of waste [7]. Hence, conforming to the vision on CE, service life extension and the reuse of elements is crucial to reduce the environmental impact, as fewer products are discarded and fewer new materials are extracted. Concrete can be a durable material with a satisfactory performance over an acceptably long service life period. Nevertheless, numerous deterioration processes can affect all structures and materials, especially if preventive maintenance is not applied. This fact confirms the need for a through-life maintenance/repair management approach for RC structures in order to maximize the service life and delay the need for the demolishing of damaged RC structures. The latter is not desirable in light of the principles of the sustainable and circular economy and confirms the high value of service life extension.

Damage to RC structures can be related to defects in concrete or to reinforcement corrosion. The former can have many causes like mechanical, chemical, physical, or accidental (e.g., fire related, blasting, impact, etc.) [8]. Yet 50 to 80% of the damage to RC structures is induced by reinforcement corrosion initiated by the carbonation of the surrounding concrete and/or chloride ingress [9,10]. Therefore, corrosion control is one of the most important considerations that impact the durability of concrete [1]. Corrosion affects the durability of RC structures and is usually manifested in the form of concrete damage (i) cracking, the spalling of the concrete cover caused by the expansion of corrosion products around the reinforcement (Figure 1), and (ii) a reduction of the cross-section of the rebars with a reduced bearing capacity of the element as a consequence. The concrete cover is the main protective mechanism against weather and other aggressive effects. When the concrete cover is damaged, the reinforcement steel diameter reduction increases, eventually resulting in a decrease or loss of structural safety. This phenomenon is more critical in the case of pitting corrosion, as one of the most destructive localized forms of corrosion, initiated by chlorides, compared to uniform corrosion due to carbonation [11].

Figure 1. Examples of concrete spalling due to reinforcement corrosion.

In light of the principles of the circular economy, demolishing damaged RC structures is not desirable. This has been confirmed by several studies rating rehabilitation more environmentally friendly than rebuilding [12,13]. In addition, in our recently published paper about balconies, we have shown that demolishing and rebuilding had the highest

LCCA/LCA score [14]. Therefore, repair and maintenance to extend the technical service life should be the first priority. In this context, concrete repair and the main deterioration caused by reinforcement corrosion indicate the relevance for the determination of an optimal repair method regarding corrosion-caused concrete damage. The application of an appropriate repair method for the occurred damage is important to ensure the intended service life extension. Concrete repair could have a limited service life extension in case of poor design or execution or due to the lack of inspection prior to the repair. This can lead to insufficient repair and the fast reappearance of damage (e.g., due to the halo-effect) after a relatively short time, further increasing their life cycle cost [15]. Qu et al. [16] highlighted the importance to detect and determine the cause of corrosion before any solution is selected to repair or reduce corrosion. Defining the repair strategy prior to the diagnosis and condition assessment of the existing concrete structure is something that should be avoided at all times. Regarding corrosion damage, there are several concrete repair methods available to prevent or stop the corrosion process: preserving or restoring passivity, increasing resistivity, cathodic control/protection (CP), and the control of anodic areas [8]. It must be noticed that CP has been applied to concrete structures worldwide for more than 25 years [15,17,18].

Currently, the selection of a repair method is mostly done based on technical requirements and initial cost, but without a life cycle perspective on costs nor on environmental impact. However, besides this classic approach, the induced impact on the environment should be also considered along with the life cycle of the structure [19]. Wang [20] studied the repair of concrete tunnels and also highlighted the importance of further studying the integration of the economic and environmental effects within the maintenance and/or repair strategy. In order to determine an optimal repair method taking into account both criteria, life cycle assessment (LCA) and life cycle cost analysis (LCCA) should be applied. LCA is a holistic method to determine the environmental impact of a product or process with a systematic set of procedures for compiling and examining the inputs and outputs of materials and energy during the entire life cycle [21]. A life cycle is the interlinked stages of a product or service system, from the extraction of natural resources to final disposal (cradle-to-grave). LCCA is a systematic or analytical method to determine the economic performance of a product or process during the entire life cycle, when the initial cost is taken into account, along with future cash flows incurred throughout the lifespan over a predefined period of analysis [22]. The future cash flows are often taken into account by discounting, which compares costs and revenues at different stages in time and emphasizes the importance of present cash flows rather than future ones due to inflation and the earning power of money [22]. Therefore, present and future costs or revenues cannot be compared without considering the opportunity value of time. The latter can be defined as the economic return that could be earned on investments (e.g., funds) as the best alternative [23]. The discount rate, which takes this opportunity value of time into account, is used to adjust future cash flows into the present. Hence, choosing the most appropriate discount rate is a critical step as it is dependent on the cost of required investments, the anticipated level of risk, and the opportunity cost (benefits that are missed out when choosing one alternative over another) of the investment.

A life cycle approach based on LCCA and LCA has a wide range of benefits compared to short-term decisions, e.g., considering the effect of choices, avoiding problem-shifting to another life cycle phase, maximizing the potential of RC structures instead of just patching up, etc. For concrete structures, Matos et al. [24] illustrated the applicability and advantages of using LCCA. Namely, the best repair strategy of several alternatives can be chosen due to a comparison of costs over the entire life cycle. Therefore, the integration of LCCA is necessary to achieve a comprehensive and long-term analysis, which positively affects the returns on investment. Hájek et al. [25] and Vieira et al. [26] emphasized the importance of subjects like performing a detailed LCA using data sets with local relevance and the need for more attention for the lack of a holistic assessment of environmental impacts, the lack of applications that consider regional and technological variations, and the neglection of life cycle phases. However, a rigorous life cycle analysis is not always feasible, particularly in regions where the exact economic and environmental data are not available [27]. Besides, several other studies have indicated the high value and holisticness of these methods, which shows the valuable application of supporting the decision-making process [28–31].

The service life has a major impact on the results of a LCA and LCCA and could lead to a wide range of results [32,33]. In order to take the appropriate service life (extension) into account, it is important to determine it in an accurate way. In most cases, an approximate service life based on other research, manufacturers' data, or empirical analysis of in situ performance is used for service life estimations [34]. However, this will often result in inaccurate results due to varying in situ life spans compared with the considered ones for the analysis. The service life of a concrete repair is dependent on the materials' properties, material and system composition, quality of design and installation, damage mechanisms, expected maintenance regimes, and climate and exposure [32,34]. In order to determine the (extended) service life of RC structures susceptible to corrosion and their repair, predictive models could be used to describe corrosion initiation, propagation, and the corresponding deterioration by a probability of failure [35].

Studies considering both economic and environmental criteria at concrete repairs for service life extension are rather limited [12–14,27,36–41]. In general, there is a lack of LCA results of service life-extending concrete repair techniques. This is confirmed by Palacios-Munoz et al. [42], who mentioned that most of the literature on LCA focuses on new constructions, while refurbishment is dealt with to a lower extent. Additionally, according to Vilches et al. [43], most LCAs focus on energy refurbishment, while there are almost no LCA studies that consider the environmental impact of building system repair, rehabilitation, or retrofitting. Overall, the available studies (i) include mostly only one of the two previously mentioned assessment methods (LCA or LCCA), (ii) consider a limited number of repair techniques, and/or (iii) are mostly about particular structures/case studies (no generality). Moreover, there are also uncertainties about the long-term effect of some interventions (e.g., galvanic sacrificial anodes) on the end-of-life (EoL). These facts indicate that there is a strong need for more research in the field of the sustainability assessment of concrete repair and maintenance for further development and refinement. This literature review tries to address this need by compiling current knowledge and gaps in the state-of-the-art. For this reason, this research compiles the relevant published papers, related to concrete repair for corrosion-damaged concrete elements and structures, that have evaluated environmental and economic impacts using LCA and LCCA. This is in accordance with the advice of Scope et al. [36], who stated that, for future meta-analyses about the subject, the scope should be narrowed to specific structures or structural components (such as bridges, road pavements, building frames, water mains, etc.) and the underlying causes for each 'maintenance' measure. Therefore, the current paper attempts to focus on repair methods for concrete damage caused by the corrosion of the reinforcement and to make a clear distinction between the different structures. To conclude, a comprehensive literature survey is conducted to summarize existing knowledge and define the state-of-the-art. In this manner, further research recommendations can be formulated.

2. Methods

The main objective of this review paper is to critically review the current state-of-the-art regarding the assessment through LCA and LCCA of concrete repair techniques for corrosion-damaged RC structures. Therefore, the literature review has been focused on studies that have evaluated any type of repair interventions using these methods, to propose recommendations for further research by answering the following research questions:

- Which concrete repair principles are available?
- To what extent are LCA and LCCA incorporated in the selection process of concrete repairs?
- What are the benefits and drawbacks of assessment through LCA and LCCA?

- What are the knowledge gaps for the accurate sustainability assessment of concrete repairs?

Relevant publications were collected and identified by an extensive bibliographic search using the bibliographic databases Web of Science and Google Scholar. In order to select relevant publications about concrete repair assessment through LCCA and LCA, an initial scope was performed based on several search sources that are shown in Table 1. Based on this, a first distinction of articles, theses, book chapters, and conference proceedings could be made. The searches with Google Scholar resulted in a very extensive list of references sorted by relevance. Therefore, the first 200 references were screened here. Secondly, with Web of Science, a more restricted list of search results was obtained, for which the numbers of records are shown in Table 1. Lastly, based on the bibliography of relevant papers, 50 more records were selected for further evaluation.

Table 1. Methodology: used search strings within bibliographic databases; n = number of records.

Database	Search Term 1	Search Term 2	n
Google Scholar	*With all words:* Life cycle cost analysis concrete repair		100
	With all words: Life cycle assessment analysis concrete repair		100
Web of Science	*All Fields:* Life cycle cost analysis concrete repair		153
	All Fields: Life cycle assessment analysis concrete repair		98
	All Fields: Concrete repair methods	*All Fields:* Life cycle	181
	Title: Concrete repair	*All Fields:* Life cycle	42
	Title: Corrosion	*All Fields:* Concrete AND Life cycle	257
	Articles' sources		50

The majority of documents were excluded after a first screening of titles, abstracts, and keywords. These irrelevant papers had no or small relevance to the search terms of Table 1. Therefore, the second screening of sources was applied to the full-text papers that discussed in any way the sustainability assessment of repairs for corrosion-damaged concrete at different kinds of structures (e.g., buildings, pavements, bridges, tunnels, etc.). For screening 2, the most important selection criterion was the need for the incorporation of the different repair principles of EN 1504-9 or the evaluation of demolishing versus repair. Therefore, studies about structural design strategies and concrete compositions were excluded. In addition, several studies with important insights about the service life prediction of reinforced concrete structures were also selected, as it is an important aspect of LCA and LCCA. A process chart of the methodology with the number of records before and after screening 1 and 2 can be seen in Figure 2.

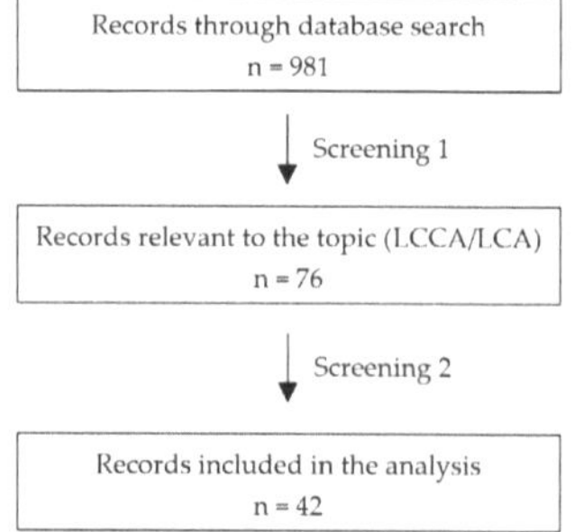

Figure 2. Process chart for paper selection: concrete repair assessment through LCCA and LCA.

For the first research question, the available repair principles according to EN 1504-9 were discussed in detail [8]. In this manner, a good understanding of concrete repair itself can be achieved first. Subsequently, the second research question was answered by mapping the different available studies of which an overview was made. It was noticed that only 10 studies (limited) included both LCA and LCCA at concrete repair. The other ones only included one of both methods or did not include repair comprehensively. In addition, for the second and third research questions, the studies were deeply analyzed and discussed. By this approach, knowledge gaps and recommendations for further research were formulated to overcome the indicated shortcomings. The obtained insights can be used in/for further research but also by companies in the industry who perform concrete repair.

3. Protection and Repair Methods Related to Reinforcement Corrosion

Many concrete repair principles are available. However, the appropriate methods to restore reinforced concrete structures are formulated in the European Standard EN 1504-9 [8]. An overview of the ones (principle 7 until principle 11) that can be used for damage caused by reinforcement corrosion are shown in Figure 3. The other principles (principles 1-6) of EN 1504-9 are related to defects in the concrete itself and are therefore not the scope of this study. In order to obtain a good understanding of the available methods and their underlying principles, they will be discussed in the following section.

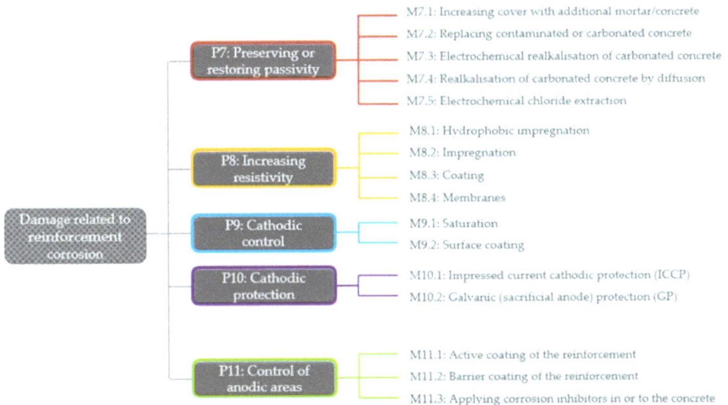

Figure 3. Overview repair techniques for deterioration related to reinforcement corrosion, based on EN 1504-9; P = principle, M = method [8].

3.1. Preserving or Restoring Passivity

Passivation is the process by which a material becomes self-protective against corrosion by means of a protective film formation on the surface. For example, iron is passivated whenever it oxidizes to produce a solid product and corrodes whenever the product is ionic and soluble. This behavior can be depicted on the color-coded Pourbaix diagram (see Figure 4). This potential-pH diagram represents the stability of iron as a function of potential and pH. The red and green regions represent conditions under which the oxidation of iron produces soluble and insoluble products, respectively [44].

The first method with the objective of preserving or restoring the passivity (principle P7) is an increase in the cover with an additional layer of mortar or concrete on places where the reinforcement is still passivated. Lee et al. [45] covered a corrosion-inhibiting mortar in their research and showed its relevance as a repair method. In contrast, a replacement of the (chloride) contaminated or carbonated concrete is also possible. With this technique, the concrete is removed entirely and replaced by new mortar or concrete, wherein the reinforcement is situated in contaminated concrete due to chloride ingress or carbonation. With this repairing technique, there is a risk for continued corrosion due to the incipient anode (or halo) effect. Mechanisms that may cause incipient anode activity include repair/parent material interface effects, residual chloride contamination within the parent concrete, and/or vibration damage to the steel/parent concrete interface during repair area preparation [46]. Krishnan et al. [47] indicated that patch repair (i.e., only replacing damaged loose concrete with repair mortar) without galvanic anodes can lead to another major repair within five years due to the continued corrosion caused by the halo effect and the residual chloride effect. Thirdly, electrochemical realkalization of the carbonated concrete can be used to re-passivate the concrete as additional corrosion protection. On places where the reinforcement is active or passive, it increases the alkalinity, and so the passivity of the carbonated concrete is being restored. However, according to NBN EN 1504-9 [8], electrochemical methods may cause the embrittlement of susceptible prestressing steel and induce an alkali-aggregate reaction with potential susceptible aggregates, a decrease in frost resistance due to an increase in moisture contents, or corrosion in adjacent structures if submerged under water. Moreover, with the realkalization of carbonated concrete by diffusion, the alkalinity of the carbonated concrete is restored through diffusion from the surface, where a highly alkaline cementitious concrete or mortar is applied. However, this method is still limited in application, and not much experience has been gained according to the Standard EN 1504-9 [8]. Besides, the performed works have a variable success rate. Finally, electrochemical chloride extraction can also be applied to restore the concrete's passivity. Due to chloride ingress, the reinforcement can become passive and active on different positions. Electrochemical chloride extraction reduces the chloride ion content in concrete around the rebars and provides passivity and additional corrosion protection. Although the possible negative side-effects of the H_2 embrittlement of the prestressing steel of electrochemical methods also need to be considered here.

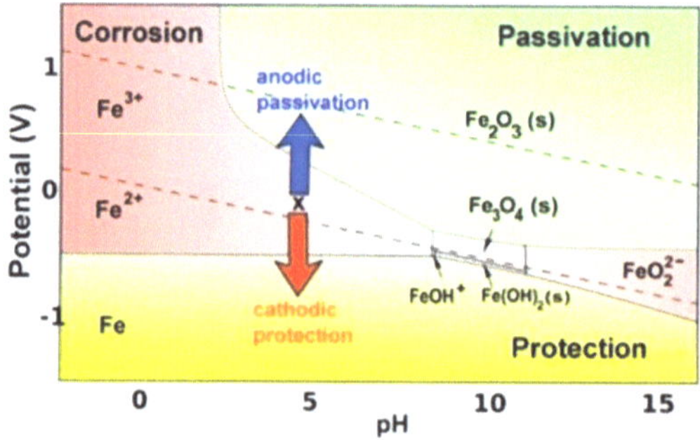

Figure 4. Pourbaix diagram [44].

3.2. Increasing Resistivity

Another principle for repairing deteriorated concrete structures due to the corrosion of the reinforcement is increasing resistivity (principle P8). This can be done by hydrophobic impregnation, impregnation, coatings, or membranes (see Figure 5). The hydrophobic impregnation technique (Figure 5a) provides a water-repellent surface that is created by the internal coating of pores and capillarities without filling them. There is no film applied on the concrete surface, and there is little or no change in its appearance [48]. According to EN 1504-9 [8], hydrophobic impregnation reduces the moisture content of concrete but could cause an increase in the carbonation rate as a potential negative side effect. Besides, the surface porosity is reduced with the impregnation by the treatment of the concrete, and the surface strength is increased. With the impregnation technique, the pores and capillaries are partially or totally filled (Figure 5b) [48]. In addition, a similar method is the application of a coating through which a continuous protective layer is applied to the concrete surface (Figure 5c) [48]. However, a side effect is that the surface coating could enclose moisture in the concrete and can break down the adhesion or reduce frost resistance [8]. An alternative repair scenario to increase the resistivity is the application of a membrane that is a preformed sheet or a liquid-applied membrane. It is part of a waterproofing membrane system that prevents water ingress and, if needed, even the ingress of potential pollutants [49–51].

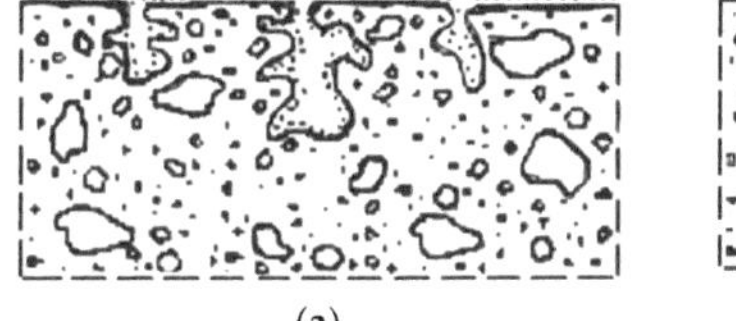 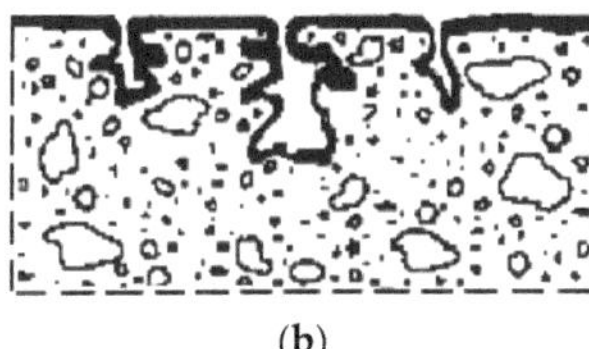 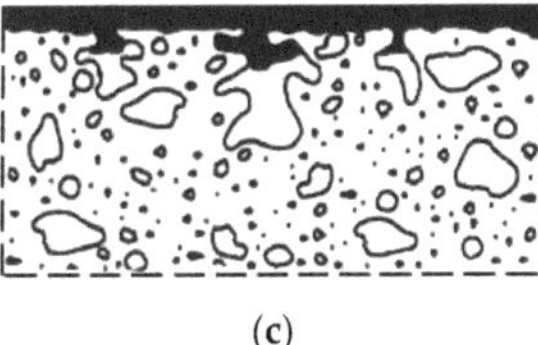

(a) (b) (c)

Figure 5. Schematic drawing of: (**a**) hydrophobic impregnation; (**b**) impregnation; (**c**) coating [48].

3.3. Cathodic Control

The third general principle for corrosion deterioration repairment is cathodic control (principle P9), which has the purpose of limiting oxygen content to all possible cathodic areas so that corrosion cells are stifled by the inactivity of cathodes [8]. In other words, the potentially cathodic areas are unable to drive an anodic reaction that can be accomplished by saturation or surface coating [50]. The first option is to use coatings on the steel surface (saturation treatment), which limits the available oxygen content. Another possibility is the application of an inhibitor on the concrete surface, which forms a film on the rebars' surface and protects it from oxygen.

3.4. Cathodic Protection

Cathodic protection (CP) is a technique (principle P10) that applies an electrical potential on the reinforcement and is especially appropriate in the case of significant chloride contamination or extensive carbonation depth [8]. The working principle is visualized in Figure 4, in the yellow part of the diagram, where the iron can be protected by keeping the potential below the oxidation potential with the use of a more active metal or an impressed current. The former principle is called galvanic sacrificial anode protection (GP), wherein a less noble (more active) metal is used. Besides, the use of an impressed current is known as impressed current cathodic protection (ICCP). Within this technique, several anode systems are present: e.g., conductive coating, titanium (Ti) mesh, Ti-probes, Ti-strips, etc. Therefore, for the same protection principle, variations could be possible regarding the exact configuration.

In the 1970s, CP by Stratfull showed the effectiveness of the principle [52]. The two main influencing factors of the corrosion rate of steel in atmospherically exposed RC structures are the water content and the pore structure. Therefore, the cathodic control of the corrosion rate due to a limited availability of O_2 is relevant only under long-term immersion, i.e., when all gaseous and dissolved O_2 is depleted in concrete [53]. An overview of CP systems with a conductive coating over a period of 25 years is presented by van den Hondel and van den Hondel [54]. They concluded that the lifetime extension of concrete structures of at least 15 to 20 years may be well achievable. Similarly, Polder et al. [55] conducted a survey of CP systems based on data from 150 structures. In practice, the service lives of CP systems without major intervention of 10 to 25 years have occurred. Wilson et al. [56] formulated the advantages and disadvantages of the two CP systems, ICCP and GP. GP should be the most suitable for small and targeted repairs, repairs wherein budgets are limited, and repairs wherein the service life extension has to be around 10 years. This is also confirmed by Krishnan et al. [47], who concluded that galvanic anodes are successful in controlling chloride-induced corrosion for about 10 to 14 years. Furthermore, ICCP is generally used to treat substantial corrosion problems at large structures and surface areas, where the service life extension should be more than 25 years or where access and traffic management are challenging and very costly. Regardless of the level of chloride contamination, ICCP is always a possible repair technique. Besides, it limits the amount of concrete removal to the physically damaged parts, and the continuous monitoring of the effectiveness of the system is present. However, ICCP needs a yearly check-up (depolarization) and the electronic components need to be checked and maintained. If executed properly, a long-term corrosion control can be achieved, and it even counteracts the effect of concrete contamination and the incipient anode problem (halo effect) [8]. Secondly, galvanic sacrificial anodes (mostly of zinc), which are connected with the rebars, consist of a more active or less noble metal (more negative reduction potential or more positive electrochemical potential) compared to the reinforcement by which the sacrificial anodes will corrode instead of the rebars connected with it. When the anodes are placed in the concrete structure, they are embedded in mortar, which intercepts the reaction products of the corrosion reaction. Once the GCP anodes are sacrificed, the protection of the steel reinforcement stops. As an example, an investigation at the historic KBC Tower in Antwerp, Belgium can be mentioned. It was indicated that only an ICCP system could resolve the underlying corrosion problem of the structural steel frame by protecting it with a low-voltage protective electrical current. Traditional masonry repair would not solve the underlying corrosion problem with the structural steel, because it would not succeed in mitigating ongoing corrosion damage. Therefore, ICCP was installed through joints in the exterior façade to protect about 25% of the structure [57].

Kamde et al. [58] did research about the long-term performance of galvanic anodes for the protection of steel-reinforced concrete structures. They reported that alkali-activated galvanic anodes can protect steel rebars from corrosion for at least 12 years. After this period, the pores in the encapsulating mortar will be partially filled with zinc corrosion products, resulting in substantial pore blockage around the zinc metal. As a result, a reduction in the pH buffer in the vicinity of the zinc metal is achieved as a natural consequence of anode dissolution (and OH^- reduction).

However, the continuous and long-term corrosion of zinc can be achieved by using adequate encapsulating mortar with (i) activators and (ii) humectants. Activators increase the dissolution kinetics of anodes and maintain a highly corrosive environment around the zinc metal. Humectants are hygroscopic materials, which maintain adequate humidity around the anode metal for continuous corrosion. Two types of activators were applied:
Two types of activators applied:

- *Halide activators*: such as fluoride, chloride, bromide, and iodide act as catalysts to maintain a continuous corrosive environment around the anode metal. The mitigation of the soluble corrosion products through encapsulating mortar aids the continuous corrosion of the metal.
- *Alkali activators*: such as lithium hydroxide, sodium hydroxide, and potassium hydroxide, which help in maintaining the pH of the encapsulating mortar to more than 14, thereby, keeping the zinc active. A reduction of pH at the galvanic metal-encapsulating mortar occurs due to their consumption.

Frequently used humectants are lithium bromide, lithium nitrate, and calcium chloride, etc.

Lastly, hybrid cathodic protection (HCP) has also recently been introduced in the market. It combines the high-level performance of an impressed current system with the long-term maintenance-free capabilities of a galvanic cathodic prevention system. In a first phase, a high charge density is applied on the steel, which passivates active corrosion. Consequently, in a second phase, the passivity of the steel is maintained by galvanic anodes [59,60]. To date, hybrid systems for reinforced concrete comprise discrete zinc-based anodes installed in predrilled cavities. Such a system has a design life of 25 to 30 years, according to Brueckner et al. [60]. Further, they mention that discrete anodes are generally suited in the case of targeted protection.

3.5. Control of Anodic Areas

By means of anodic area control (principle P11), potentially anodic reactions of reinforcement are avoided to take part in the corrosion reaction by protecting the repaired areas from the future ingress of aggressive agents (carbonation, chlorides) [50]. It can be accomplished by four methods, namely the active coating or barrier coating of the reinforcement or by applying corrosion inhibitors in or to the concrete. Active coatings contain active pigments, which may function as an anodic inhibitor or by a sacrificial galvanic action. When it is not possible to remove all contaminated concrete, it can control incipient anode formation by treating the surface of the reinforcement in the patch repair. Secondly, barrier coatings form barriers on the surface of the corrosion-free reinforcement. It is vital that the coating is defect-free and that it completely encapsulates the entire circumference. Besides, it is important to consider the effect of the coating on the bond between the reinforcement and the concrete according to EN 1504-9 [8]. Thirdly, applying corrosion inhibitors in or to the concrete changes the steel rebar's surface or form a passive film over it. They can be used by addition to the concrete repair product or system or by application to the concrete surface, which is followed by migration to the position of the reinforcement. Regarding the latter, the penetration to the depth of the reinforcement is obviously crucial. It is important to notice that some inhibitors work by the control of both anodic and cathodic areas.

3.6. Concrete Damage Repair

Besides the repair of the steel reinforcement in concrete structures according to principles P7 to P11, the damaged concrete itself also needs proper repair (principles P1 to P6). Hence, after the rebars, the concrete is consecutively restored with respect to principles 1 to 6: (P1) protection against ingress, (P2) moisture control, (P3) concrete restoration, (P4) structural strengthening, (P5) increasing physical resistance, and (P6) resistance to chemicals.

4. Sustainability Assessment

The amount of research work done with regard to the concrete repair/maintenance decision-making process through LCCA and particularly LCA is rather limited. Namely, almost no LCA studies exist that consider the environmental impact of system repair, while, due to the high number of structures, renovations will be a key factor in the future of the European building sector [42,43]. However, when studies exist, they mostly only include one of the two assessment methods and consider a limited number of repair methods. The current review gathers a comprehensive list of selected references that are relevant to this subject. An overview of them can be seen in Table 2. For each reference a variety of information is indicated: the (1) covered assessment method(s), (2) subject, (3) reference type, (4) the year of publication, (5) the county of the main author, (6) the potential incorporation of a case study, (7) the potential consideration of corrosion damage, (8) covered repair techniques, (9) the potential prediction of service life extension, (10) the LCCA method, and (11) the LCIA method. Lastly, in order to indicate the significance of the references being relevant to the subject of this research, those are rated from 1 to 5 ("1"- almost completely out of scope, "2"- less relevant, "3"- partly relevant, "4"- relevant, "5"- highly relevant). The relevance is assessed based on the extent to which different repair principles related to EN 1504-9 are compared through LCCA and/or LCA. In addition, papers with more general insights about concrete repair combined with LCCA and/or LCA were also indicated as valuable. Lastly, several studies about service life prediction could also be marked as (partly) relevant. Ranking 1 was given to papers with very specific repair methods, not really considering the principles of EN 1504-9. However, their results were valuable to include them. Based on the clear ranking, the reader can immediately see which references are also worth checking out related to this paper.

Table 2. Overview of state-of-the-art sustainability assessment concrete repair through LCCA and LCA.

LCCA	LCA	Ref.	Subject	Source Type	Year	Country [61]	Case Study	Corrosion	Repair Technique													SL Pre-Diction	LCCA Method	LCIA Method	Relevance
									CC	CR	PR	cSC	rSC	IM	CP	CE	ER	RD	CI	HA				
General																									
✓	✓	[36]	Types of sustainability assessments applied to 'maintenance' interventions using concrete- or cement-based composite materials.	RP	2021	DE	+−	+	+	+	+	+	+	+	+	−	−	−	−	−	Overview	Overview	Overview	5
✓ (limited)	✓ (limited)	[27]	An alternative to a LCCA named the repair index method (RIM), which enables the possibility of including other non-technical requirements that would be difficult to quantify in a LCCA.	JP	2003	ES	−	++	−	−	+	−	−	−	+	+	+	−	+	+	SLA	RIM	RIM	5
✓	−	[62]	Investigation of the time-dependent capacity of a corroded circular RC column by using nonlinear finite element analysis.	JP	2019	IR	+	++	+	−	−	+	−	−	−	−	−	−	−	−	++	n.m.	−	2
✓	−	[24]	Overview of the ongoing works for a state-of-the-art report (bulletin) regarding LCCA analyses of concrete assets.	CP	2018	DK	+−	−	−	−	−	−	−	−	−	−	−	−	−	−	−	−	−	3
✓	−	[55]	Assessment of the performance of CP systems in practice with information on 150 reinforced concrete structures (RCS).	JP	2014	NL	−	++	−	−	−	−	−	−	++	−	−	−	−	−	+− (SLA)	NPV	−	3
✓	−	[47]	Market study on the performance and LCCA of CP using galvanic anodes in RCS in India and worldwide.	JP	2021	IN	+	++	−	+−	+	−	+−	−	++	−	−	−	−	−	SLA	FV	−	3
✓	−	[45]	Study on the concrete carbonation in the presence of repair materials using the maintenance periods and repair cost according to the coefficient of variation (CV) of the carbonation depth.	JP	2020	USA	−	++	−	−	−	+	−	−	−	−	−	+	−	−	++	Repair cost only	−	4
✓	−	[63]	Determination of important life cycle variables: expected time lost in repairs, reliability of the system, and the cost of operation and failure.	JP	2014	USA	+	+	−	−	−	−	−	−	−	−	−	−	−	−	++	−	−	3
−	✓	[25]	General discussion on LCA application to concrete structures + case study floors.	JP	2011	CZ	+	−	−	−	+−	−	−	−	−	−	−	−	−	−	SLA	−	m.i.s.	2

Table 2. *Cont.*

LCCA	LCA	Ref.	Subject	Source Type	Year	Country [61]	Case Study	Corrosion	Repair Technique	SL Pre-Diction	LCCA Method	LCIA Method	Relevance
–	✓	[26]	Literature review conducted to present the state-of-the-art of LCA methodological practices in the manufacturing of common concrete and concrete with aggregates derived from recycled waste.	RP	2016	BR	+−	–	– – – – – – – – – –	Overview	–	Overview	2
–	✓	[64]	Probabilistic sustainability framework for the design of concrete repairs and rehabilitation to achieve targeted improvements in quantitative sustainability indicators.	JP	2014	USA	+	+	+ – – – – – – –	++	–	ReCiPe TRACI EI 99	5
–	✓	[65]	Framework and methodology for quantifying the ecological effects and impacts from various methods and systems for the repairs and maintenance of concrete structures (CS).	CP	2001	NO	+	+	– – + + + – – – – +	SLA	–	m.i.s.	5
Buildings													
CCO	✓	[66]	Influence of design strategies on the economic and environmental performance of 30-story residential RC building.	JF	2018	BR	++	–	– – – – – – – – –	–	n.m.	CML	2
✓	limited	[13]	Assessment of the cost/benefit ratio for total demolition vs. refurbishment on a 40-year-old detached single house.	JP	2013	PT	++	–	– – – – – – – – –	SLA	CBA	MF EE	2
✓	✓	[12]	Literature review: compares different LCA works for refurbished and new buildings + real LCA and LCCA case study for a classified ancient building.	RP	2015	PT	+	–	– – – – – – – – –	RP: Overview Case: SLA	RP: Overview Case: Sum	RP: Overview Case: m.i.s.	3
✓	–	[67]	Probabilistic assessment method of service life and life cycle maintenance strategies + reliability function of structural safety performance based on hazard rate/function of a deterioration RC building during a rare earthquake.	JP	2010	JP	+	++	– +− + + – – – + – – – –	++	NPV	–	3

Table 2. *Cont.*

LCCA	LCA	Ref.	Subject	Source Type	Year	Country [61]	Case Study	Corrosion	Repair Technique											SL Pre-Diction	LCCA Method	LCIA Method	Relevance
−	✓	[43]	Summary of the recent contributions related to the environmental evaluations of building refurbishment and renovation using LCA.	RP	2017	ES	+−	+−	−	−	−	+	−	−	−	−	−	−	−	+−	−	Overview	4
−	✓	[42]	New approach to estimate building lifespans based on their structures durability (degradation models of reinforced concrete structures) + refurbishment versus demolition and new building evaluated from an environmental point of view.	JP	2019	ES	+	+	−	−	−	−	−	−	−	−	−	−	−	+	−	GWP	5
			Civil infrastructure (Bridges, tunnels, …)																				
✓	✓	[37]	A framework for the maintenance-scheme optimization of existing bridges based on the genetic algorithm (GA).	JP	2018	CN	+	+−	−	−	+−	+−	−	−	−	−	−	−	−	+	NPV	EI 99	2
✓	✓	[38]	Evaluation of (the economic and environmental impacts of) 18 different design alternatives for an existing concrete bridge deck exposed to chlorides.	JP	2019	ES	++	++	+	−	−	+	+	−	+	−	−	+	+	+	NPV	ReCiPe 2008	3
✓	−	[20]	Methods and technology for concrete repair, waterproofing work, tunnel rehabilitation, and eco-efficient repair + tunnel performance evaluation.	JP	2018	JP	−	+	−	−	+	+	−	−	−	−	−	−	−	+	NPV	−	2
✓	−	[68]	Probabilistic and deterministic LCCAs for an entirely FRP-reinforced concrete bridge and a conventional RC prestressed concrete (PC).	JP	2021	USA	+−	++	−	−	+	−	−	−	+	−	−	−	−	SLA	NPV	−	2
✓	−	[69]	Probabilistic framework to estimate the LCCA associated with bridge decks constructed with different reinforcement alternatives.	JP	2021	USA	+	++	+	+−	+−	−	+	−	−	−	−	−	−	+	NPV	−	2
✓	−	[70]	Describes an approach for agencies to enhance bridge investment decisions.	JP	2015	SE	+	−	−	+−	+−	−	−	+−	−	−	−	−	−	SLA	NPV EAC	−	3
✓	−	[71]	Development of a rational method for the most cost-effective intervention schedule for bridges, where the structural safety is maintained with the minimum possible LCCA.	JP	2018	CA	+	+	−	+	−	−	−	+	−	−	−	−	−	+	EAC	−	3
✓	−	[72]	LCCA for various options to prevent or remediate corrosion damage in an example bridge exposed to de-icing salts, locally aggravated by the leakage of expansion joints.	JP	2016	NL	++	++	−	+	−	+	−	−	+	−	−	+	+	SLA	NPV	−	3
✓	−	[73]	Framework for the prediction of deterioration.	JP	2010	JP	+	++	−	−	+	−	−	−	+	−	−	−	−	+	Sum	−	3

Table 2. *Cont.*

LCCA	LCA	Ref.	Subject	Source Type	Year	Country [61]	Case Study	Corrosion	Repair Technique												SL Pre-Diction	LCCA Method	LCIA Method	Relevance
✓	–	[74]	Overview of recent research about life cycle engineering for civil and marine structural systems and future research directions.	JP	2016	USA	–	+	–	–	–	–	–	–	–	–	–	–	–	–	+	CBA	–	4
–	✓	[75]	Potential for using a self-healing engineered cementitious composite (SH-ECC) for the rehabilitation of bridges.	CP	2018	BE	–	±	–	–	–	–	–	–	–	–	–	–	–	–	SLA	–	GWP	1
–	✓	[76]	Comparison of the different solutions for bridge rehabilitation from an environmental point of view.	JP	2013	FR	+	–	–	+	–	+	–	–	–	–	–	–	–	–	SLA	–	GWP (CML)	1
–	✓	[77]	Comprehensive LCA to study the environmental impact of interventions on an existing bridge using PE-UHPFRC.	JP	2019	CH	+	+	–	–	–	+	–	–	–	–	–	–	–	–	SLA	–	m.i.s.	1
–	✓	[78]	Analysis of the environmental implications of several prevention strategies through a LCA using a prestressed bridge deck as a case study.	JP	2018	ES	++	++	+	+	+	+	+	+	–	–	–	–	–	+	+	–	EI 99 EPS ReCiPe	3
–	✓	[79]	Probabilistic service life prediction models for determining the time to repair + probabilistic LCA models for measuring the impact of a repair	JP	2020	USA	+	+	–	+	–	–	–	–	–	–	–	–	–	–	++	–	TRACI ReCiPe	4
–	✓	[80]	Service life prediction models combining deterioration mechanisms with limit states + LCA models for the impact of a given repair, rehabilitation, or strengthening.	CP	2011	USA	–	±	–	–	–	–	–	–	–	–	–	–	–	–	+	–	n.m. (GWP)	4
Pavements																								
✓	✓	[39]	Investigate the environmental, economic, and social impacts of the three most widely adopted rigid pavement choices through LCA.	JP	2016	USA	+	±	–	–	–	–	–	–	–	–	–	–	–	–	SLA	NPV	m.i.s.	1
✓	✓	[40]	Literature review repair of concrete pavements.	RP	2018	USA	±	±	–	–	+	–	–	–	–	–	–	–	–	–	Overview	Overview	m.i.s.	2
✓	–	[22]	Review of existing methodologies in the wider field of LCCA for road projects with a highlight on critical processes and the identification of hotspots in order to increase the robustness of LCCA. frameworks.	JP	2020	BE	–	–	–	–	±	–	–	–	–	–	–	–	–	–	Overview	Overview	–	3

Table 2. *Cont.*

LCCA	LCA	Ref.	Subject	Source Type	Year	Country [61]	Case Study	Corrosion	Repair Technique												SL Pre-Diction	LCCA Method	LCIA Method	Relevance
Others: more specific (floorings, columns)																								
✓	✓	[81]	Environmental and economic LCA of three different floor systems.	JP	2018	UK	++	–	–	–	–	–	–	–	–	–	–	–	–	–	SLA	NPV	TRACI	1
✓	✓	[41]	Evaluation of environmental impacts and costs of a structural element (slab) with varying of concrete cover thickness using LCA and LCCA.	JP	2021	BR	+–	+	–	–	–	–	–	–	–	–	–	–	–	–	+	Sum	CML4.4	2
✓	✓	[14]	LCA and LCCA of life-extending repair methods for RC balconies.	JP	2022	BE	++	++	–	+	+	+	+	+–	+	+–	–	–	–	+	SLA	NPV	ReCiPe v1.13	5
✓	–	[82]	Repair strategies are examined for their economical relevance to LCCA.	CP	2013	AT	+	++	+	+	+	+	–	+	+	–	–	–	–	–	+	NPV	–	4
–	✓	[83]	Simplified methodology for the size strengthening of beams and to provide the application of LCA to the selected techniques.	JP	2018	ES	+	+	–	–	–	–	–	–	–	–	–	–	–	–	+–	–	CED GWP	1

Legend of symbols and abbreviations: ✓ = assessment method included; – = not included; +– = mentioned but not included/not expressly included; + = included; ++ = focused; RP = review paper; CCO = construction-cost-only; JP = journal paper; CP = conference paper; CC = concrete cover; CR = conventional repair; PR = patch repair; cSC = concrete surface coating; rSC = reinforcement surface coating; IM = impregnation; CP = cathodic protection; CE = electrochemical chloride extraction; ER = electrochemical realkanization carbonated concrete; RD = realkalization carbonated concrete by diffusion; CI = corrosion inhibitors; HA = hydrophobic agents; SL = service life; SLA = service life assumption; n.m. = method not mentioned; NPV = net present value; FV = future value; CBA = cost-benefit analysis; Sum = sum up without discounting; EAC = equivalent annual cost; m.i.s. = manual indicator selection; EE = embodied energy; MF = material flow; EI = ecoindicator; GWP = global warming potential; CED = cumulative energy demand; ReCiPe, EI, Traci, CML, and EPS are standard LCIA methods.

4.1. Assessment through LCA and LCCA

Overall, LCA and LCCA studies mainly focus on the material level, e.g., "green concrete" [84,85], or on material choices during the design phase [66,81]. However, in order to get a better idea about (1) the incorporation of LCA and LCCA during concrete repair selection and (2) the drawbacks and benefits of these methods, the selected references are discussed.

To start with, based on Table 2, the publication evolution over the past decades and where the authors are situated is visualized in Figure 6. The gathered data represent results from papers conducted in over 22 countries. Most research reported in international publications is done in Europe, followed by North America, but results from countries like China, India, Iran, Japan, Korea, and Brazil are present as well. The first publication originates from 2001, but the investigations on the subject remained limited for several years. However, in the past decade, a high increase in the number of papers can be seen. This indicates the popularity of the subject and, consequently, the relevance and need of/for concrete repair.

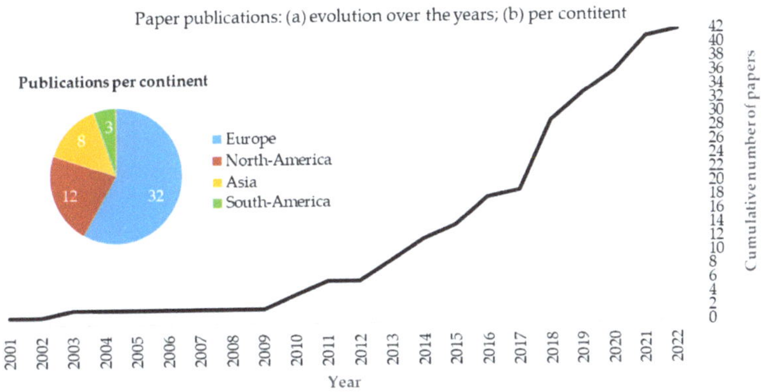

Figure 6. Situating papers about concrete repair in time and geographically.

Subsequently, the 10 references that consider (in limited ways) the two methods for concrete repair in their research are reviewed [12–14,27,36–41]. Scope et al. [36] explored and synthesized the sustainability potential of maintenance and repair methods using concrete- and cement-based composite materials in a review article. However, there is no sufficient information provided on repair techniques specific for corrosion-caused concrete damage. Therefore, the paper cannot recommend any particular repair technique in general based on their literature review. Nevertheless, according to Scope et al. [36], there is a trend towards more holistic types of assessment; environmental and economic sustainability dominates, with global warming and energy consumed being the most often reported. In addition, the importance of long-term orientation and life cycle thinking for sustainable maintenance strategies is highlighted.

Ferreira et al. [12] and Gaspar and Santos [13] investigated whether refurbishing an existing structure is environmentally and/or economically profitable compared to new construction based on a case study. The former indicated that refurbishment is environmentally beneficial compared to a new equivalent construction [12]. However, the gains were not as high as commonly suggested, mainly due to the massive use of structural steel and shotcrete. In this case study, an earthquake-safe building needs to be achieved, causing the need for a high amount of extra material. In contrast, as far as cost is concerned, refurbishment was found to be less competitive due to the high construction cost for seismic and structural strengthening, which are very intrusive works. Therefore, in this case, additional solutions should be developed, and financing facilities should be studied that could support refurbishment works. In addition, whether the highest impact contribution is due to labor or materials is not clarified by the paper. Moreover, Gaspar and Santos [13] reported a similar building cost but a lower environmental impact for refurbishment compared to new construction. This can be explained by a lower cost in terms of materials for refurbishment, but the larger building period and better-skilled workforce due to the complexity of the building process. The environmental advantage can be explained by less matter, embodied energy consumption, and demolition waste.

Moreover, Xie et al. [37] and Navarro et al. [38] emphasized the need for preventive maintenance in order to reduce the cost and environmental impact over the life cycle of bridges. Maintenance optimization should result in significant reductions of life cycle impacts if compared to maintenance undertaken at the end of the service life. A well-considered initial time and time interval of periodic maintenance would effectively decrease the bridge's life cycle environmental impact. However, the reduction of the life cycle cost of the bridge caused by maintenance-scheme optimization is not significant due to a discounting effect [37]. Navarro et al. [38] also evaluated different preventive bridge-design alternatives over 100 years, for which the following order was obtained for environmental impact (less beneficial first): stainless steel-galvanized steel-organic inhibitor-migratory inhibitor-ICCP-sealant product-hydrophobic treatment. For the economic cost, the order has some changes: galvanized steel-stainless steel-ICCP-organic inhibitor-hydrophobic treatment-sealant product-migratory inhibitor. The effect of the addition of silica fume, fly ash, and polymers and the effect of w/c ratio and concrete cover was also investigated for different scenarios. As these principles give very varying results according to their quantity, they are not further discussed. The difference between the impacts can be explained by the considered maintenance interval, used materials and processes, transport, etc. The optimization of the maintenance intervals reduces the economic and environmental life cycle impacts up to 13 and 19%, respectively, compared to essential maintenance.

Furthermore, Andrade and Izquierdo [27] developed a method to select repairs based on safety, serviceability, environmental impact, durability, and economy requirements, which they called the repair index method (RIM). A rating is given for each section based on four levels and is added to a total value after multiplying with a ranking of importance. Therefore, it is not a very transparent comparison but could be valuable when there is a good balance between the evaluation requirements. Another advantage is the possibility of incorporating non-technical requirements, as of the social and legal types. In the research, RIM is used to rate five repair options for corrosion-damaged RCS, which gave the following sequence from less to most beneficial: electrochemical treatment-inhibitors-cathodic protection-hydrophobic agents-patching. However, this is only a general rating and is not case specific. Lastly, the biggest drawback is that, by the incorporation of the environmental and economic impact in the same rating, objectivity is lost.

Choi et al. [39] investigated the economic, environmental, and social impacts of three major rigid-pavement rehabilitation alternatives. They indicated the high value of the assumption of a life cycle perspective due to the fact that the initial cost can be recouped by long-term sustained benefits. Their study indicated that a continuously reinforced concrete pavement is the most sustainable choice and is much preferable to jointed reinforced concrete pavement or jointed plain concrete pavement, which, respectively, have, on average, 1.6- and 1.2-times higher environmental, economic, and social impacts. This is the result of the former requiring fewer resources while providing more durability and longer performance. In addition, Wang [40] discussed the LCA of the repair of concrete pavements and mentioned that future LCA studies should consider the time value of environmental impacts, as discounting frequently occurs in LCCA. Besides, it has been stated that routine and minor maintenance need to be considered better, certainly for small projects. Wang indicates joint resealing, slab stabilization, partial-depth repairs, full-depth repairs, load transfer restoration, and diamond grinding and grooving as common preservation and routine maintenance techniques of concrete pavements. Further, crack and seat, rubblization, concrete overlay, and asphalt overlay are major rehabilitation treatments usually conducted towards the end of pavement service life.

Related to specific structural elements, Wittocx et al. [14] evaluated five different frequently used repair techniques for reinforced concrete balconies (30 m^3 and 600 m^2) by means of a LCA and LCCA: (i) patch repair, (ii) conventional repair, (iii) galvanic cathodic protection, (iv) impressed current cathodic protection, and (v) the total replacement of the element. For a lifetime extension of 5 years, patch repair is indicated as the most preferable option, as the structure is restored with a minimum of intervention. However, when a service life extension of up to 40 years is requested, different options (conventional, GCP, ICCP) are found to be more sustainable. The most extensive scenario is the total replacement of the balconies and involves the highest environmental and financial impact for the described functional units. Additionally, the paper also highlighted the unknown effect of sacrificial anodes on the end-of-life when concrete containing these elements is reused. A sensitivity analysis on the effect of the service life of PR, the coating application at CP, the volume of the contaminated concrete at CR, the end-of-life characteristics of sacrificial anodes, the amount of zinc at GCP, labor and material costs, and the repair mortar composition on the LCCA/LCA ranking is also included.

Finally, Menna Barreto et al. [41] investigated the influence of varied concrete cover thicknesses in the life cycle of a reinforced concrete structure. They did not consider different repair methods; however, their study provides relevant insights about the determination of the service life at different cover thicknesses. By the prediction of the service life, it was shown that an increase in concrete cover thickness enhances the structure's durability and reduces costs and environmental impacts per year. However, the cracking potential of concrete in tensile stress zones is not taken into account. It might increase and that will lead to a negative impact on the service life of a reinforced concrete structure.

An overview of the different studies evaluating different repair/prevention techniques through LCCA and LCA can be found in Table 3. First of all, the papers assessing concrete repair through both assessment methods were found to be limited, which shows the need for further research as the incorporation at the selection process of concrete repairs can reduce the economic and environmental impact. In this manner, more clear and correct conclusions can be made. It can also be noticed that a different set of techniques is compared for each study, which makes it difficult to compare the results between each other. Besides, the assumed service life (extension) for their assessment also differs, which makes mutual comparison even more difficult. This is also stated by Marinković et al. [86], who showed the high influence of the service life. Secondly, differences between the two assessment methods are also present by which not one optimal repair can be indicated. In some cases, they have the same optimal repair; however, this is rather rare. The results of the different studies show the case specificity for the repair impacts and the impossibility of generalization. However, low labor-intensive techniques like patch repair seem to be a good choice for short service life extensions. In order to get more comparable results, standardizing the functional unit (to the extent feasible), expanding system boundaries, improving data quality, and examining a larger array of environmental indicators would be helpful, which Santero et al. also propose [87]. Lastly, the performed studies show, based on the exposure classes (Exp. class), that the repair principles of EN 1504-9 are appropriate in the most severe environments. However, the environmental aggressivity should be considered to estimate the repair's correct performance.

Table 3. Overview of repair strategies comparison through LCCA and LCA (abbreviations: Table 2).

Ref.	SL (y.)	Exp. Class	LCCA (−→+)	LCA (−→+)
[38]	100	XC4-XS1-XF2	Galvanized steel-stainless steel-ICCP-organic inhibitor-hydrophobic treatment-sealant product-migratory inhibitor	Stainless steel-galvanized steel-organic inhibitor-migratory inhibitor-ICCP-sealant product-hydrophobic treatment
[27]	Varies	Varies	Electrochemical treatment-inhibitors-CP-hydrophobic agents-PR	Electrochemical treatment-inhibitors-CP-hydrophobic agents-PR

Table 3. Cont.

Ref.	SL (y.)	Exp. Class	LCCA (−→+)	LCA (−→+)
[39]	50	XC4-XD3-XF2	Jointed reinforced concrete pavement-jointed plain concrete pavement-continuously reinforced concrete pavement	Jointed reinforced concrete pavement-jointed plain concrete pavement-continuously reinforced concrete pavement
[14]	5	XC4-XS1-XF3	New-GCP-CR-ICCP-PR	New-ICCP-CR-GCP-PR
	20		New-GCP-CR-PR-ICCP	New-PR-ICCP-CR-GCP
	40		New-GCP-CR-ICCP	New-ICCP-GCP-CR
[41]	50/100	Varies	Concrete covers mutually	Concrete covers mutually

4.2. Assessment through LCA or LCCA

Ghodoosi et al. [71] concluded that frequent minor repairs reduce the life cycle cost by reducing the number of major costly repairs. The same conclusion is drawn based on LCA by the results of a study by Navarro et al. [78]. The importance of preventive maintenance was stated by a high number of studies, which is shown in Figure 7. In fact, 12 out of 42 (26 %) papers did this based on LCCA. Similarly, for LCA, 7 out of 42 (17 %) highlighted the same conclusion based on LCA. This shows why maintenance should be included from the beginning of the structure's life span to reduce the economic and environmental impact. However, according to the research of Kumar and Gardoni [63], it may be more advantageous to have frequent repairs for a long-term service life, but for a short-term service life, it may not be advantageous or may even be disadvantageous. Therefore, the desired positive effects of an operation strategy take some time to take effect.

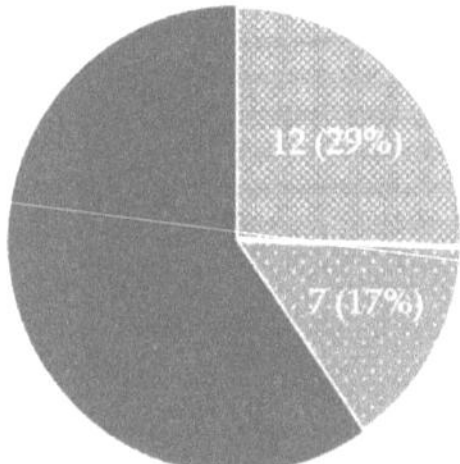

LCCA: Studies highlighting the importance of preventive maintenance
LCA: Studies highlighting the importance of preventive maintenance
Total number of studies (# 42 = whole circle)

Figure 7. Visualization of the importance of preventive maintenance.

Wittocx et al. [14] emphasized the economic, as well as the environmental, impact of the advantage of refurbishment instead of demolishing and rebuilding. Palacios-Munoz et al. [83] also showed that strengthening is more environmentally sustainable than rebuilding a new structure, even in the case of damage. The suitability of a solution is, however, strongly depending on the characteristics of the original element. In their research, four strengthening techniques of RC beams are evaluated: carbon-fiber-reinforced polymer, reinforced concrete section increasing, steel placed with mechanical anchorages, and steel placed with epoxy resin. The first and third technique are indicated as the most sustainable if the main purpose is increasing the bending capacity and if no degradation is present. This is due to a reduction of the material requirements due to the higher mechanical properties and due to the avoidance of harmful epoxy resin. However, when degradation is present, the suitability of the solution strongly depends on the geometry of the beam. Increasing the reinforced concrete section is more suitable when a large increase in the bending capacity is required, rather than for low ones due to the high workload. For the life cycle cost, no extra studies evaluated refurbishment versus new construction beside the ones discussed in Section 4.1. Nevertheless, for the environmental impact, more papers evaluated this manner. More particular, six out of six studies (of Table 2) that investigated this through LCA indicted refurbishment was preferable compared to rebuilding (Figure 8). The same conclusion was emphasized by one study for LCCA. However, also for LCCA, one study concluded that there was an equal life cycle cost for (i) refurbishment and (ii) demolition and rebuilding. Once, refurbishment was indicated as less preferable regarding the life cycle cost. Therefore, it can be concluded that, in general, refurbishment is more sustainable regarding the environmental impact, but, for the economic impact, it is case-specific. The most important factor here is the labor intensity of the work.

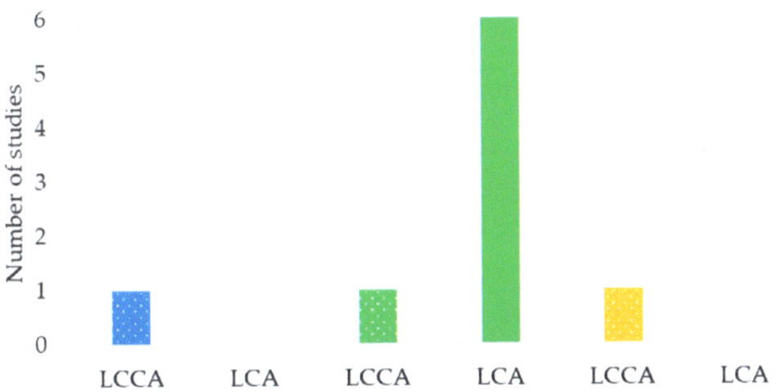

Figure 8. Visualization of sustainability refurbishment vs. demolition + rebuilding (new).

Regarding the economic impact, the paper of Polder et al. [55] showed that the life cycle cost of CP systems on concrete structures can be predicted, taking into account failure rates based on field data. They concluded that the cost of the replacement of components is relatively small compared to the cost of inspection and electrical checkups. Besides, based on the life cycle cost of 30 repair projects, Krishnan et al. [47] confirmed that the use of a CP strategy can lead to life cycle-cost savings of up to about 90% about 30 years after the first repair. Consequently, CP and cathodic prevention (CPrev) are more beneficial from an economic point of view than PR. Furthermore, CP and CPrev strategies can enhance the service life to as long as needed by the replacement of anodes at regular intervals and at minimal cost (5% of the first repair).

Moreover, the total life cycle cost with preventive measures using stainless steel reinforcement, (repeated) hydrophobic treatment, and cathodic prevention in the joint areas of an example bridge are compared with conventional concrete repair and CP by Polder et al. [72]. Stainless steel reinforcement and the hydrophobic treatment of concrete were reported as the most preferable maintenance options for a life span from 35 to 100 years. For stainless steel, this can be explained by a higher initial cost but no need for maintenance at all. Hydrophobic treatment has an average initial construction cost but also a low cost during the other life cycles. However, for a shorter life span until 35 years, cathodic protection should be more preferable. The differences between the results of Polder et al. [72] and Navarro et al. [38] can be explained by the other configuration and size of the case study. Similarly, five types of repair methods for infrastructure RC structures (e.g., bridges) were compared by Islam and Kishi [73]: CP with a conductive polymer and with a titanium mesh, patching, and two types of overlays (i.e., concrete and hot mixed asphalt with a membrane). Patching should have the highest life cycle cost, whereas concrete overlay has the lowest. The high life cycle cost can be explained by the maximum variable cost of repair. Likewise, the lowest life cycle cost is due to a low variable cost and a longer life span of repair as well.

Farahani [62] investigated, using a nonlinear finite element analysis, the time-dependent capacity of a corroded round-shaped RC column. More particularly, the influence of several scenarios on the column's performance due to chloride-induced corrosion was investigated. For the repair scenarios, five concrete surface coatings are included: acrylic-modified cementitious: type D (CPD); epoxy polyurethane (PU); aliphatic acrylic (AA); acrylic-modified cementitious: type E (CPE); and styrene acrylate (SA) The equivalent concrete cover thicknesses were calculated as: 14.4, 31.2, 38.9, 27.6, and 12.6 mm. With a cost of 85, 77, 55, 84, and 54 USD respectively, AA can therefore be indicated as the optimal concrete surface coating. In addition, four increasing concrete cover thicknesses (i.e., 10, 15, 20, and 25 mm) and using new longitudinal and horizontal reinforcements after the initial cracking of the concrete cover are also investigated as repair scenarios. Out of all repair scenarios, a 20 mm increasing concrete cover thickness adding to the initial concrete cover of 70 mm was found to be the most beneficial for a service life of 40 years.

Furthermore, Binder [82] analyzed the life cycle cost of a set of repair methods (i.e., concrete facing, patch repair, patch repair with hydrophobic impregnation, CP with titanium mesh, and CP with conductive coating) for chloride-contaminated columns. Patch repair with hydrophobic impregnation and cathodic protection with a titanium mesh turned out to be the most cost-effective strategy, taking into account the full service life extension (75 years) of the structural component. In contrast, patch repair only had a 40% higher value, resulting in the highest life cycle cost, mainly due to its low service life extension and, therefore, the high need for maintenance. Moreover, CP with a coating and a concrete overlay are the next repair options with increases in the life cycle cost of 35% and 16%, respectively, compared to the two most optimal options. Lastly, it is stated that the life cycle cost of the cathodic protection principle (CP-Mesh) could be further reduced by the optimization of the service life of the electronic components.

Cadenazzi et al. [68] compared two bridge design alternatives: reinforced bridge with traditional carbon steel (CS) vs. fiber-reinforced polymers (FRP). The life cycle cost includes (1) a direct cost that covers the initial construction cost and subsequent maintenance and repair and (2) a user cost covering losses due to traffic delay, work-zone crashes, and environmental impact. The CS alternative is found to be a high-risk design alternative with a higher cost spread and an increased life cycle cost of 30%. This can be explained by a more intensive maintenance strategy over 100 years, which overcomes the difference between the initial costs of the techniques. At CS, patch repair and cathodic protection are applied, while at FRP, only patch repair should be needed. Similarly, the life cycle cost of bridge decks constructed with different reinforcement alternatives is investigated by Shen et al. [69]. Results show that the concrete cover, chloride exposure condition, average daily traffic, and number of traffic lanes have a significant effect on the life cycle cost of reinforced concrete bridge decks, especially for those constructed with conventional reinforcement. In addition, they highlighted that conventional rebars provide the lowest direct cost for a service life up to approximately 28 years; afterwards, corrosion-resistant alternatives provide the lowest direct life cycle cost. This can be explained by the additional expenses associated with maintenance and repair actions for conventional reinforcement. Out of galvanized rebar, epoxy coated rebar, and martensitic micro-composite formable steel (MMFX) rebar applied at a case study, MMFX was found to have the lowest life cycle cost, which was approximately one-third of conventional reinforcement. Lastly, Safi et al. [70] show the high value of the implementation of LCCA in bridge procurement in order to indicate the most cost-efficient bridge design over its life cycle. Based on several case studies, the initial investment can differ by up to 50%, while the maintenance cost could generally differ by up to 15% between different designs. This highlights the advantage of considering a life cycle approach instead of only the initial construction cost. However, it is important to acknowledge that the science of LCCA is far from perfect. Its findings can be biased by the perceptions and forecasts of future costs, the reliability of the data used, the discount rates applied, the stages of the asset life cycle included in the analysis, and life cycle plans.

Regarding the environmental impact, patch repair with shotcreting and hydrophobic surface protection were compared by Årskog et al. [65]. It was pointed out that the impacts from the patch repair strongly exceed that of the hydrophobic surface protection: the use of energy (MJ/m^2) is 21.6 times higher, global warming ($kg\ CO_2\ eq/m^2$) is 46.9 times higher, acidification ($g\ SO_2\ eq/m^2$) is 125 times higher, eutrophication ($g\ SO_2\ eq/m^2$) is 126.8 times higher, and photo-oxidant formation ($g\ Ethene\ eq/m^2$) is 5.5 times higher.

Furthermore, the global warming potential of bridge rehabilitation with different types of ultra-high performance fiber-reinforced concrete (UHPFRC) and the comparison of them with more standard solutions was investigated by Habert et al. [76]. A traditional rehabilitation system using conventional concrete (C30/37) plus a waterproofing membrane, and a rehabilitation system with UHPFRC solutions are analyzed. Regarding the latter, classic UHPFRC and ECO-UHPFRC with limestone filler as cement replacement are considered. Results show, over a service life of 60 years, a higher impact for the traditional system and UHPFRC compared with ECO-UHPFRC, with, respectively, an impact 40% and 28% higher. The lower impact compared to the traditional system can be explained by lower maintenance and repair volume needs. Moreover, the study shows that the impact due to the production of materials is the major contributor to the environmental impact whatever the rehabilitation systems used. Similarly, Hajiesmaeili et al. [77] showed, respectively, 55% and 29% decreases in the environmental impact of polyethylene (PE) UHPFRC compared with the replacement with a new traditional RC bridge and the conventional UHPFRC method. The considered impact categories are global warming potential (GWP), cumulative energy demand (CED), and ecological scarcity (UBP). In addition, Van den Heede et al. [75] compared the rehabilitation with self-healing engineered cementitious composite (SH-ECC) with ordinary Portland cement (OPC) concrete and UHPFRC repair. Considering a standard error distribution, OPC concrete had the highest environmental impact, followed by SH-ECC and UHPFRC, with lower values of 55–70% and 59–74%, respectively.

Lastly, Navarro et al. [78] analyzed the environmental implications of several prevention strategies using a prestressed bridge deck exposed to chlorides as a case study. Results show that environmental impacts of the structure can be reduced substantially by considering specific preventive designs, such as adding silica fume to concrete, reducing its water to cement ratio, or applying hydrophobic or sealant treatments. In this manner, a reduction of up to 30 to 40% of the reference environmental impact can be achieved due to less intensive maintenance. Other techniques like stainless steel reinforcement, polymer addition, and concrete cover increases are less efficient in their case study. However, increasing the concrete cover can still reduce the environmental life cycle impacts of the deck by 45% if compared to the reference alternative.

An overview of the different studies evaluating different repair/prevention techniques through LCCA can be found in Table 4. Also, herein are variating results obtained by which mutual comparison is not obvious. It can be noticed that the intended service life extension is of paramount importance because repair can be the most preferable option, as well as the least one as another value is assumed. From the results, it can be noticed that patch repair seems to be less economical and that techniques like hydrophobic treatment and cathodic protection would be valuable options.

In conclusion, in Table 5, an overview of the papers comparing different repairs through LCA is presented. The first thing that stands out is the amount of research about UHPFRC repair for bridges, which was indicated as a sustainable repair. Other papers evaluating other techniques are uncommon. However, based on two studies, hydrophobic treatment seems like a good option for concrete repair, but individual evaluation is still necessary. Lastly, it can also be noticed that the amount of research about concrete repair assessment through LCA is limited. Therefore, an effort in the academic field is needed.

LCCA and LCA are extensive and time-consuming assessment methods demanding large amounts of data. Nevertheless, the benefit of reducing the economic and environmental impact of concrete repair outweighs these drawbacks.

Table 4. Overview of repair strategies comparison through LCCA (abbreviations: Table 2).

Ref.	SL (y.)	Exp. Class	LCCA (−→ +)
[55]	25	Varies	CP mutually
[47]	5–100	XC2-XS3-XF1	PR-CP-cathodic prevention
[72]	35	XC2-XS3-XF4	Cathodic prevention-CR-stainless steel reinforcement-hydrophobic treatment-CP
	35–100		CR-CP-cathodic prevention-stainless steel reinforcement-hydrophobic treatment
[73]	1–50	/	PR-CP-concrete overlay
[62]	40	XC2-XS3-XF1	Concrete surface coatings mutually vs. concrete cover thicknesses
[82]	5	Varies	Concrete facing-CP titanium mesh-PR with hydrophobic impregnation-CP coating-PR
	20		PR-CP coating-PR with hydrophobic impregnation-concrete facing-CP titanium mesh
	40		PR-concrete facing-CP coating-PR with hydrophobic impregnation-CP titanium mesh
	75		PR-CP coating-concrete facing-CP titanium mesh-PR with hydrophobic impregnation
[68]	100	XC4-XS3-XF2	Traditional carbon steel-fiber-reinforced polymers (FRP)
[69]	75	XC4-XD3/XS-XF2	Rebar alternatives mutually
[45]	100	XC	Corrosion inhibiting mortar-organic alkaline inhibitor-inhibiting surface coating-water-based paint
[67]	Varies	XC4-XD1-XF1	Combination of repair strategies

Table 5. Overview of repair strategies comparison through LCA (abbreviations: Table 2).

Ref.	SL (y.)	Exp. Class	LCA (−→+)
[83]	50	Varies	Beam-strengthening techniques mutually
[65]	10	/	PR-hydrophobic surface treatment
[76]	60	XD2/XD3	Traditional system (CR)-UHPFRC-ECO UHPFRC
[77]	100	XD3-XF4	New-conventional UHPFRC-PE UHPFRC
[78]	100	XC4-XS1-XF2	Stainless steel-galvanized steel-sealant product-hydrophobic treatment
[75]	60	XC4-XD3/XS1-XF2	OPC concrete-SHECC-UHPFRC
[64]	100	Varies	Concrete cover mutually

4.3. Service Life Prediction

According to the FIB Model Code [88], the "direct consequence of passing this limit state [of depassivation] is only that possible future protective measures for repair become more expensive". For that reason, the limit state of depassivation is often associated with relaxed target probabilities of failure (P_0), usually in the order of 1 to 12%, and in the design stage, P_0 should be chosen as a function of the cost of repair (during the intended service life of the structure) relative to the cost of construction [89].

Focusing on concrete repair strategies, a general probabilistic sustainability design framework for the design of concrete repairs and rehabilitation is presented by Lepech et al. [80]. The framework consists of two types of models: (i) service life prediction models and (ii) LCA models. In this manner, the time to the first repair combining one or several deterioration mechanisms and the environmental impact of it can be determined. The relevance of such a framework in order to improve the quantitative environmental sustainability indicators is presented, but the need for more research still remains, to allow further implementation. In a follow-up paper, it was tested for a 40 mm and 80 mm deep concrete cover repair, of which the 80 mm was found the most sustainable over the lifespan (100 years) of the structure. The thicker repair has a higher impact but a greater durability, so the cumulative impact over the life cycle is reduced. This shows the importance of taking an appropriate service life into account and choosing the right intervention for an intended life span. A great deal of research still remains in the development and validation of methods and tools [64]. Moreover, the framework is further extended, and a new mathematical approach to simplify it is presented by Zirps et al. [79]. For probabilistic service life prediction models, they used Fick's law, which is a simple method to approach the diffusion of chloride and does not capture all aspects of the complex nature of this process. The research showed that such a framework can provide an engaging tool for the sustainability-focused probabilistic design of reinforced concrete infrastructure.

Existing carbonation models predict service life based on deterministic theories, like, for example, in the study of Farahani [62]. Therefore, based on deterministic and probabilistic methods, Lee et al. [45] investigated concrete carbonation in the presence of repair materials using the maintenance periods and repair cost according to a CV of the carbonation depth. The CV value indicates the variability of the actual structure and the concrete quality. For the carbonatation depth, a carbonation probability equation is implemented using Monte Carlo simulations considering the carbonation depth distribution and the probability distribution of the cover thickness as random variables. Out of water-based paint, organic alkaline inhibitors, inhibiting surface coating, and corrosion-inhibiting mortar (CIM) as repair materials, CIM was found to be the best carbonation inhibitor. However, it has also the highest life cycle cost at the intended service life of 100 years due to a high residual life span. When, for example, a life span of 80 years is considered, CIM is by far the most beneficial option with a 2.4 to 3.1 times higher life cycle cost for the other repairs. Lastly, the difference between the deterministic and probabilistic LCCA models was highlighted. The probabilistic model will predict more efficient maintenance by adjusting the intended service life or selecting the appropriate repair material. Nevertheless, when the CV decreases, the probabilistic cost approaches the deterministic repair cost.

Thirdly, Palacios-Munoz et al. [42] evaluated the influence of the lifespan in a comparative LCA by considering three different approaches to determine the buildings' lifespans: default value, statistical, and durability-based. Due to the common practice of considering a default value for lifespans, LCA involves a high risk of programmed obsolescence in the building sector. Therefore, statistical or durability-based determined lifespans are introduced in the paper. Palacios-Munoz et al. [42] mentioned that statistical studies of buildings' lifespan provide the most realistic results. However, the results can be accurate in general terms but are not representative for the particular analyzed building. Lastly, corrosion due to carbonation is considered for the durability-based approach since it is the most frequent degradation phenomenon. The durability-based estimated value of lifespan has an uncertainty that derives from the degradation model due to simplification. So, it is important to simulate the degradation of the concrete structure accurately.

Moreover, Chiu et al. [67] developed a deterioration model to estimate the deterioration risk induced by chloride ingress resulting from failure and severe spalling or cracking during earthquakes. This method focuses on the probabilistic assessment method of service life and life cycle maintenance strategies. Regarding the former, a reliability function of structural safety performance is used, based on the hazard rate or hazard function of a deterioration RC building during a rare earthquake. For repair selection, probabilistic effect assessment models for considering the recurrence of deterioration in repaired areas and the deterioration proceeding in unrepaired areas were developed. In this manner, the system can be used to determine the optimal life cycle maintenance strategy. Furthermore, the developed system was tested in a case study for five types of repair works containing (i) finishing renewal, (ii) patch repair, (iii) chloride removal, and (iv) steel supplementation. The results revealed that maintenance strategies that include steel supplementation are effective in reducing the life cycle cost of RC buildings located in regions with a high hazard of chloride ingress and seismic activity.

Furthermore, new approaches like the renewal-theory-based life cycle analysis (RTLCA) are developed. Kumar and Gardoni [63] propose such a model and describe it as a novel probabilistic formulation for the life-cycle analysis of deteriorating systems. The formulation includes equations to obtain important life cycle variables like the expected time lost in repairs, the reliability of the system, and the cost of operation and failure. RTLCA minimizes the need for computationally expensive simulations and offers analytical equations to estimate the life cycle performance measures for a system. The model is tested for the life cycle analysis of a RC bridge where the structure is repaired whenever the instantaneous probability of failure exceeds an acceptable limit. The study shows the importance of frequent repairs in the case of a long-term service life. However, for a short service life, frequent repairs could be disadvantageous.

Lastly, Ghodoosi et al. [71] developed a method as a new procedure to predict the most proper intervention strategy for bridges where the structural reliability was maintained with the minimum life cycle intervention cost. The innovative combination of reliability analysis at the system level, nonlinear finite-element modeling, and genetic algorithm (GA)-based life cycle-cost optimization meant to assist decision-makers in planning bridge maintenance and rehabilitation in a more practical manner including safety and budget limitations criteria. The optimization results proved that the application of minor intervention activities significantly reduces the life cycle cost when compared with the conventional case in which no preventive measure is implemented. However, the entailed minimum cost of implementing only minor intervention activities might be significantly higher when compared with a case in which a combination of essential and preventive measures is applied. There exist various intervention methods in which each may entail different costs and bridge life cycles. For instance, the innovative application of FRP laminates for strengthening the reinforced concrete deck may result in higher costs and a longer bridge life cycle as compared with conventional techniques, an issue of concern for future work in this context.

To conclude, it is shown that assuming an appropriate life span is extremely important in order to achieve reliable results. Several studies are available predicting the service life through prediction methods and models. However, these approaches are often a simplification of reality and are not always reliable. Furthermore, many different approaches are present. To obtain a better overview, a more comprehensive and detailed literature review should be performed on this subject. In addition, Qu et al. [16] also stated that more research is needed about a comprehensive forecast of conveying and degradation mechanisms in both cracked and uncracked concrete. Frangopol and Soliman [74] also mentioned that methodologies for processing the large amount of data for damage diagnosis and prognosis in existing structures are still required. Lastly, Taffese and Sistonen [90] also stated that performing more research on the service life prediction of repaired concrete structures using advanced modeling techniques is necessary.

4.4. End-of-Life Characteristics
End-of-Life Galvanic Sacrificial Anodes

In order to prevent or stop the reinforcement corrosion of RCS, CP can be applied. One method that can be used is the use of galvanic sacrificial anodes (mostly of zinc) that are connected to the rebars. The anodes consist of a more active or less noble metal compared to the reinforcement by which the sacrificial anodes will corrode instead of the rebars connected with it. When the anodes are placed in the concrete structure, they are embedded in mortar that intercepts the reaction products of the corrosion reaction (zinc oxide). In the end-of-life phase of the concrete structure, they are crushed and often reused together with the concrete rubble. However, the effect of galvanic sacrificial anodes on the environmental impact is still unclear [14]. If the reclaimed concrete aggregates are used in road foundations, the zinc corrosion products could leach into the groundwater system. However, it is unclear if and to what extent this leaching will happen in reality. Some general research has been done about this subject but not specifically about the leaching behavior of reclaimed concrete with residual fractions of zinc oxide.

According to de la Fuente et al. [91], the formation of corrosion products in an atmospheric environment is a complex and continuously changing process. The degree of complexity and the rate of change depend on the type of atmosphere and the various factors involved. According to Thomas et al. [92], corrosion chiefly occurs in alkaline conditions by the formation of zinc hydroxide complexes or zinc oxides that could protect the surface depending on local pH and potential at the metal surface. Zinc forms immediately a fine film of zincite (ZnO) when it is exposed to any environment [93,94]. However, when water is present, this film is promptly transformed into zinc hydroxide ($Zn(OH)_2$). These products are found in an atmospheric environment, so if all of these or even more could be formed by a sacrificial anode (alkaline environment) is still unclear. According to Vera et al. [93], the most important insoluble zinc corrosion products, besides ZnO, in a marine environment are simonkolleite ($Zn_5(OH)_8Cl_2 \cdot H_2O$), hydrozincite ($Zn_5(CO_3)_2(OH)_6$), and zinc and sodium hydroxyl-chlorosulfate ($NaZn_4Cl(OH)_6SO_4 \cdot 6H_2O$).

These corrosion products include soluble products such as zinc chloride ($ZnCl_2$) and zinc sulfate ($ZnSO_4$), which can leach by rainfall and can be detected in subsequent runoff solutions. The research of Santana et al. [95] investigated the atmospheric corrosion of zinc samples exposed at 25 test sites with different climatic and pollution conditions during a two-year exposure program. The composition and distribution of the corrosion products of zinc were analyzed qualitatively by X-ray diffraction (XRD). They also found that simonkolleite ($Zn_5(OH)_8Cl_2$) and hydrozincite ($Zn_4CO_3(OH)_6 \cdot H_2O$) are the most frequently observed corrosion products. However, in smaller amounts are zinc oxysulfate ($Zn_3O(SO_4)_2$), zinc hydroxysulfate ($Zn_4SO_4(OH)_6$), zinc diamminehydroxynitrate ($Zn_5(OH)_8(NO_3)_2 \cdot 2NH_3$), and zinc chlorohydroxysulfate ($NaZn_4Cl(OH)_6SO_4 \cdot 6H_2O$). An example of an occurring corrosion reaction at the galvanic anode can be seen in Equations (1)–(3). Zinc reacts with both acids and bases to form salt [58]. According to Kamde et al. [58], the rate of the corrosion of zinc is high at a pH less than 6 (acidic) and greater than 12.5 (basic).

$$Zn \rightarrow Zn^{2+} + 2e^- \tag{1}$$

$$Zn^{2+} + 4OH^- \rightarrow Zn(OH)_4^{2-} \tag{2}$$

$$Zn(OH)_4^{2-} \rightarrow ZnO + 2OH^- + H_2O \tag{3}$$

An advantage of highly alkaline encapsulating mortar (pH > 14) is that the zinc corrosion products exist as soluble zincate ions ($Zn(OH)_4^{2-}$). They move into the pores of the encapsulating mortar due to their solubility where they precipitate out as zinc oxide once supersaturation occurs. On the other hand, a layer of white zinc corrosion products (zinc oxides/hydroxides) will surround the unreacted zinc metal. Dugarte and Sagüés [96] indicated that the anodes stop functioning due to encapsulating mortar failing to provide an adequate environment for continuous corrosion after about a quarter of the galvanic metal is consumed.

The study of Diotti et al. [97] investigated the leaching behavior of construction and demolition wastes and recycled aggregates. They found that the leaching of zinc is not critical, but this is obviously not related to (only) aggregates from concrete with sacrificial anodes. Besides, the influence of grain size and volumetric reduction on the release of contaminants was also investigated. Material crushing leads to higher pollutant release due to the increase of the contact surfaces between recycled concrete aggregates (RCA) and leaching agents. At the same time, sieving operations can also lead to greater fine fractions that cause high releases. However, the difference is only limited. Several studies identified the high releases of Zn at neutral or alkaline pH values [98–101]. Besides, high releases of Zn (metal cations) were also detected at lower pH values [100]. Therefore, Zn is highly released in both acidic and alkaline environments.

The study of Vera et al. [93] also investigated the precipitation runoff from zinc in a marine environment to define the pH valu, and the Cl^-, SO_4^{2-}, and Zn^{2+} ion concentrations.

The pH values for the runoff solutions are similar to those for the rainwater samples and vary between pH 6.1 and 7.1. The amount of chloride ion and sulfate concentration in the runoff is dependent on the location (atmospheric chloride and SO_2). The zinc concentrations that were measured monthly for the runoff solutions are well-correlated with the amount of rainfall, the rainfall periodicity, and the duration of the dry periods between rainfall events. So, to conclude, the different corrosion products and the amount of it leaching by rainfall is highly dependent on the environment and the rainfall characteristics. According to several studies of Kukurugya et al. [102–104] wherein the leaching behavior of furnace sludge/dust was investigated, the leaching of zinc is dependent on the concentration of the fluid (here acid), temperature, and leaching time. According to the study of Kara De Maeijer et al. [105], wherein the leaching behavior of a crumb rubber in concrete was investigated, cementitious materials can well confine trace metals such as zinc. However, some leaching is still possible.

The previous section shows nicely that the leaching behavior of galvanic anodes is an important point of attention. Based on the mentioned studies and the absence of the information about the corrosion products of sacrificial anodes and the leaching behavior of concrete aggregates containing it, it can be concluded that further research is needed. Besides the amount of leaching, the form in which it leaches out is also important, a stable non-toxic form is namely less bad than a heavy carcinogenic form. Therefore, leaching tests with concrete containing used sacrificial anode parts (alkaline environment) based on the Belgium environment and rainfall would be of high value.

5. Conclusions

In light of the principles of a sustainable and circular economy, the appropriate repair of (damaged) RC structures should be applied by which a service life extension can be obtained. A clear European standard (EN 1504-9) is present, that discusses the different repair principles to restore reinforced concrete structures. However, there is no consensus about the repair selection for when which type is the most ideal. Therefore, convenient discussion-making should be applied. In order to improve the sustainability of concrete structures and repairs over their life cycle, LCA and LCCA should be incorporated.

Based on the review, the application of LCA and LCCA for concrete repair decision-making shows certainly its advantage. Namely, a reduction of the environmental and/or financial impact during the total service life can be achieved. However, the available research about this subject is rather limited, which shows a clear research gap and the potential for further investigation. In addition, the existing studies are not complementary with each other due to the consideration of different concrete structures, assessment methods, damage causes, and repair methods. Therefore, mutual comparison is often not possible, and thus, generally applicable conclusions cannot be made. However, studies investigating refurbishment versus new construction agree that the former strategy should be environmentally beneficial, but regarding the economic cost, there are varying results. Three studies were found, of which each one state that refurbishment is more, the same, or less beneficial than rebuilding. Generally, the cost for repair should be lower, but when intensive work needs to be done (e.g., for seismic resistance), this could differ. In addition, many studies highlight the importance of preventive maintenance instead of curative repair in order to reduce the cost and environmental impact over the life cycle. Nevertheless, for a short-term service life extension, a curative approach may be advantageous. So, considering a life cycle perspective is of high value to determine when the initial cost can be recouped by long-term sustained benefits. With respect to the most beneficial repair option, no general statements can be made due to the case specificity. It was seen that particular repairs were labeled as more and less favorable in different cases. Anyway, when research is done about a specific construction and repair method, the listed findings could be very valuable. Based on the papers considering both LCA as well as LCCA, low labor-intensive techniques like patch repair should be a good choice for short service life extensions. In contrast, when only LCCA was used, patch repair seems to be less economical. This can be explained by the assumption of a long service life extension or a too general judgement. Lastly, the studies assessing only through LCA highlighted UHPFRC repair for bridges as sustainable multiple times.

Furthermore, several studies indicated the value of service life prediction but also showed its complexity. The biggest advantage is the determination of a more appropriate life span that will be used in life cycle analyses. The need for more research still remains to allow further implementation. Lastly, regarding the influence of repair methods (i.e., galvanic sacrificial anodes) on the end-of-life characteristics of concrete structures, there is still not much knowledge gained. It is unclear which corrosion products are formed in the specific concrete environment and to what extent leaching will happen when concrete is reused.

This review had the objective of gathering as many relevant studies as possible to show the current state-of-the-art, so it can be used in further research work. The following conclusions (C) and recommendation (R) can be summarized:

(C1) In order to determine the most sustainable concrete repair technique, LCA and LCCA should be applied. With these methods, considering the life cycle perspective, a service life extension can be achieved with the optimal environmental and/or economical strategy.

(C2) Several studies about sustainability design frameworks are available. However, there is no research about comparing concrete repairs and rehabilitation methods through LCA and LCCA, considering all five repair principles of Standard EN1504-9.

(C3) The leaching behavior of concrete containing rest fractions of sacrificial galvanic anodes is unclear, so further research is necessary.

(C4) The assumed service life has a major influence on the results of the assessment through LCCA and LCA.

(R1) Considering current climate objectives and the need for a more circular economy, it is recommended to also take environmental performance into account, besides the technical requirements and economic performance over the structure's life cycle when selecting a certain repair.

(R2) Service life prediction should be used more in LCA and LCCA in order to take the appropriate service life (extension) into account.

Author Contributions: Writing—original draft preparation, N.R. and P.K.D.M.; writing—review and editing, M.B., B.C. and A.A.; All authors have read and agreed to the published version of the manuscript.

Funding: This research was partly funded by a post-doctoral fellowship (Postdoctoral Fellow-junior; 1207520N).

Institutional Review Board Statement: This study does not involve any studies on humans or animals.

Informed Consent Statement: Not applicable.

Acknowledgments: Research Foundation Flanders (FWO-Vlaanderen) is greatly acknowledged for supporting Dr. Matthias Buyle with a postdoctoral fellowship.

Conflicts of Interest: The authors declare no conflict of interest.

References

1. Whiteley, D.; Goethert, K.; Goodwin, F.; Golter, H.; Kennedy, J.; Wattenburg Komas, T.; Meyer, J.; Petree, M.; Smith, B.; Trepanier, S. *Sustainability for Repairing and Maintaining Concrete and Masonry Buildings*; ICRI Committee: Rosemont, IL, USA, 2014. Available online: https://cdn.ymaws.com/www.icri.org/resource/collection/1023A08D-21D0-4AE9-8F9A-5C0A111D4AC9/ICRICommittee160-Sustainability_whitepaper.pdf (accessed on 19 January 2022).
2. Confederation of Construction. *2013–2014 Annual Report: Construction and Europe [in Dutch], 2014*. Available online: https://www.constructionconfederation.be/Portals/0/Documents/Jaarverslagen/Jaarverslag-2013-2014-Confederatie-Bouw.pdf?ver=2017-09-15-161342-943 (accessed on 19 January 2022).
3. De Schepper, M.; Van den Heede, P.; Van Driessche, I.; De Belie, N. Life cycle assessment of completely recyclable concrete. *Materials* **2014**, *7*, 6010–6027. [CrossRef] [PubMed]

4. Behera, M.; Bhattacharyya, S.; Minocha, A.; Deoliya, R.; Maiti, S. Recycled aggregate from C&D waste & its use in concrete–A breakthrough towards sustainability in construction sector: A review. *Constr. Build. Mater.* **2014**, *68*, 501–516. [CrossRef]

5. Ali, M.B.; Saidur, R.; Hossain, M.S. A review on emission analysis in cement industries. *Renew. Sust. Energ. Rev.* **2011**, *15*, 2252–2261. [CrossRef]

6. Xi, F.; Davis, S.J.; Ciais, P.; Crawford-Brown, D.; Guan, D.; Pade, C.; Shi, T.; Syddall, M.; Lv, J.; Ji, L.; et al. Substantial global carbon uptake by cement carbonation. *Nat. Geosci.* **2016**, *9*, 880–883. [CrossRef]

7. European Commission. Circular economy—Overview. Available online: https://ec.europa.eu/eurostat/web/circular-economy (accessed on 23 November 2021).

8. Part 9: General principles for the use of products and systems. In *NBN EN 1504-9: Products and Systems for the Protection and Repair of Concrete Structures—Definitions, Requirements, Quality Control and Evaluation of Conformity*; European Committee for Standardization: Brussels, Belgium, 2008.

9. Jones, A.E.K.; Marsh, B.K.; Clark, L.A.; Seymour, B.P.A.M.; Long, A.M. Development of a holistic approach to ensure cement, the durability of new concrete construction. BCA Research Report, C/21. *Br. Cem. Assoc.* **1997**, *81*.

10. Wittocx, L.; Craeye, B.; Audenaert, A.; Buyle, M.; Caspeele, R.; Gruyaert, E.; Minne, P.; Dooms, B. Vademecum der Gebreken: Uitkragende Betonnen Balkons [in Dutch]. Balcon-E Odisee. 2021. Available online: https://balcon-e.odisee.be/sites/default/fil es/public/2021-09/Balcon-e%20Vademecum%20der%20gebreken_0.pdf (accessed on 19 January 2022).

11. Popov, B.N. Chapter 7—Pitting and Crevice Corrosion. In *Corrosion Engineering*; Popov, B.N., Ed.; Elsevier: Amsterdam, The Netherlands, 2015; pp. 289–325.

12. Ferreira, J.; Duarte Pinheiro, M.; de Brito, J. Economic and environmental savings of structural buildings refurbishment with demolition and reconstruction—A Portuguese benchmarking. *J. Build. Eng.* **2015**, *3*, 114–126. [CrossRef]

13. Gaspar, P.L.; Santos, A.L. Embodied energy on refurbishment vs. demolition: A southern Europe case study. *Energy Build.* **2015**, *87*, 386–394. [CrossRef]

14. Wittocx, L.; Buyle, M.; Audenaert, A.; Seuntjens, O.; Renne, N.; Craeye, B. Revamping corrosion damaged reinforced concrete balconies: Life cycle assessment and life cycle cost of life-extending repair methods. *J. Build. Eng.* **2022**, *52*, 24. [CrossRef]

15. Polder, R.B.; Peelen, W.H.A. 25 years of experience with cathodic protection of steel in concrete in the Netherlands. In *Durability of Reinforced Concrete from Composition to Protection: Selected Papers of the 6th International RILEM PhD Workshop Held in Delft, The Netherlands, July 4–5, 2013*; Andrade, C., Gulikers, J., Polder, R., Eds.; Springer International Publishing: Cham, Switzerland, 2015; pp. 69–75.

16. Qu, F.; Li, W.; Dong, W.; Tam, V.W.Y.; Yu, T. Durability deterioration of concrete under marine environment from material to structure: A critical review. *J. Build. Eng.* **2021**, *35*, 102074. [CrossRef]

17. Polder, R.B. Cathodic protection of reinforced concrete structures in the Netherlands—Experience and developments. *Heron* **1998**, *43*, 3–14.

18. Bertolini, L.; Bolzoni, F.; Pedeferri, P.; Lazzari, L.; Pastore, T. Cathodic protection and cathodic preventionin concrete: Principles and applications. *J. Appl. Electrochem.* **1998**, *28*, 1321–1331. [CrossRef]

19. Ribeiro, I.; Peças, P.; Silva, A.; Henriques, E. Life cycle engineering methodology applied to material selection, a fender case study. *J. Clean Prod.* **2008**, *16*, 1887–1899. [CrossRef]

20. Wang, J.H. 23–Lifecycle cost and performance analysis for repair of concrete tunnels. In *Eco-Efficient Repair and Rehabilitation of Concrete Infrastructures*; Pacheco-Torgal, F., Melchers, R.E., Shi, X., Belie, N.D., Tittelboom, K.V., Sáez, A., Eds.; Woodhead Publishing: Sawston, UK, 2018; pp. 637–672.

21. *ISO 14040:2006*; Environmental Management-Life Cycle Assessment-Principles and Frameworks. International Organisation for Standardization: Geneva, Switzerland, 2006.

22. Moins, B.; France, C.; Van den bergh, W.; Audenaert, A. Implementing life cycle cost analysis in road engineering: A critical review on methodological framework choices. *Renew. Sust. Energ. Rev.* **2020**, *133*, 110284. [CrossRef]

23. *Life Cycle Cost Analysis Primer*; U.S. Department of Transportation: Washington, DC, USA, 2002.

24. Matos, J.; Solgaard, A.; Linneberg, P.; Sanchez Silva, M.; Strauss, A.; Stipanovic, I.; Casas, J.; Masovic, S.; Caprani, C.; Novak, D. Life cycle cost management of concrete structures. In Proceedings of the IABSE Conference—Engineering the Past, to Meet the Needs of the Future, Copenhagen, Denmark, 25–28 June 2018; pp. 25–27.

25. Hájek, P.; Fiala, C.; Kynčlová, M. Life cycle assessments of concrete structures—A step towards environmental savings. *Struct. Concr.* **2011**, *12*, 13–22. [CrossRef]

26. Vieira, D.R.; Calmon, J.L.; Coelho, F.Z. Life cycle assessment (LCA) applied to the manufacturing of common and ecological concrete: A review. *Constr. Build. Mater.* **2016**, *124*, 656–666. [CrossRef]

27. Andrade, C.; Izquierdo, D. Benchmarking through an algorithm of repair methods of reinforcement corrosion: The repair index method. *Cem. Concr. Compos.* **2005**, *27*, 727–733. [CrossRef]

28. Carlsson Reich, M. Economic assessment of municipal waste management systems—case studies using a combination of life cycle assessment (LCA) and life cycle costing (LCC). *J. Clean Prod.* **2005**, *13*, 253–263. [CrossRef]

29. Hoogmartens, R.; Van Passel, S.; Van Acker, K.; Dubois, M. Bridging the gap between LCA, LCC and CBA as sustainability assessment tools. *Environ. Impact Assess. Rev.* **2014**, *48*, 27–33. [CrossRef]

30. Miah, J.H.; Koh, S.C.L.; Stone, D. A hybridised framework combining integrated methods for environmental Life Cycle Assessment and Life Cycle Costing. *J. Clean Prod.* **2017**, *168*, 846–866. [CrossRef]

31. Buyle, M.; Braet, J.; Audenaert, A. Life cycle assessment in the construction sector: A review. *Renew. Sust. Energ. Rev.* **2013**, *26*, 379–388. [CrossRef]
32. Janjua, S.Y.; Sarker, P.K.; Biswas, W.K. Impact of service life on the environmental performance of buildings. *Buildings* **2019**, *9*, 9. [CrossRef]
33. Morales, M.F.D.; Passuello, A.; Kirchheim, A.P.; Ries, R.J. Monte Carlo parameters in modeling service life: Influence on life cycle assessment. *J. Build. Eng.* **2021**, *44*, 103232. [CrossRef]
34. Grant, A.; Ries, R. Impact of building service life models on life cycle assessment. *Build. Res. Inf.* **2013**, *41*, 168–186. [CrossRef]
35. Val, D.; Chernin, L. Service life performance of reinforced concrete structures in corrosive environments. In *Life Cycle Civil Engineering*; CRC Press: Boca Raton, FL, USA, 2008; pp. 267–272.
36. Scope, C.; Vogel, M.; Guenther, E. Greener, cheaper, or more sustainable: Reviewing sustainability assessments of maintenance strategies of concrete structures. *Sustain. Prod. Consum.* **2021**, *26*, 838–858. [CrossRef]
37. Xie, H.-B.; Wu, W.-J.; Wang, Y.-F. Life-time reliability based optimization of bridge maintenance strategy considering LCA and LCC. *J. Clean Prod.* **2018**, *176*, 36–45. [CrossRef]
38. Navarro, I.J.; Martí, J.V.; Yepes, V. Reliability-based maintenance optimization of corrosion preventive designs under a life cycle perspective. *Environ. Impact Assess. Rev.* **2019**, *74*, 23–34. [CrossRef]
39. Choi, K.; Lee, H.W.; Mao, Z.; Lavy, S.; Ryoo, B.Y. Environmental, economic, and social implications of highway concrete rehabilitation alternatives. *J. Constr. Eng. Manag.* **2016**, *142*, 04015079. [CrossRef]
40. Wang, H. 25—Life cycle analysis of repair of concrete pavements. In *Eco-Efficient Repair and Rehabilitation of Concrete Infrastructures*; Pacheco-Torgal, F., Melchers, R.E., Shi, X., Belie, N.D., Tittelboom, K.V., Sáez, A., Eds.; Woodhead Publishing: Sawston, UK, 2018; pp. 723–738.
41. Menna Barreto, M.F.F.; Timm, J.F.G.; Passuello, A.; Dal Molin, D.C.C.; Masuero, J.R. Life cycle costs and impacts of massive slabs with varying concrete cover. *Clean Eng. Tech.* **2021**, *5*, 100256. [CrossRef]
42. Palacios-Munoz, B.; Peuportier, B.; Gracia-Villa, L.; López-Mesa, B. Sustainability assessment of refurbishment vs. new constructions by means of LCA and durability-based estimations of buildings lifespans: A new approach. *Build. Environ.* **2019**, *160*, 10. [CrossRef]
43. Vilches, A.; Garcia-Martinez, A.; Sanchez-Montañes, B. Life cycle assessment (LCA) of building refurbishment: A literature review. *Energy Build.* **2017**, *135*, 286–301. [CrossRef]
44. Chemistry LibreTexts. Pourbaix Diagrams. Available online: https://chem.libretexts.org/@go/page/183315 (accessed on 16 February 2022).
45. Lee, H.; Lee, H.-S.; Suraneni, P.; Singh, J.K.; Mandal, S. Prediction of service life and evaluation of probabilistic Life cycle Cost for surface-repaired carbonated concrete. *J. Mater. Civ. Eng.* **2020**, *32*, 04020297. [CrossRef]
46. Christodoulou, C.; Goodier, C.; Austin, S.; Webb, J.; Glass, G.K. Diagnosing the cause of incipient anodes in repaired reinforced concrete structures. *Corros. Sci.* **2013**, *69*, 123–129. [CrossRef]
47. Krishnan, N.; Kamde, D.K.; Veedu, Z.D.; Pillai, R.G.; Shah, D.; Velayudham, R. Long-term performance and life cycle cost benefits of cathodic protection of concrete structures using galvanic anodes. *J. Build. Eng.* **2021**, *42*, 102467. [CrossRef]
48. Part 2: Surface protection systems for concrete. In *NBN EN 1504-2: Products and Systems for the Protection and Repair of Concrete Structures—Definitions, Requirements, Quality Control and Evaluation of Conformity*; European Committee for Standardization: Brussels, Belgium, 2004.
49. *Conipur®Waterproofing Systems*; BASF Construction Chemicals: Ludwigshafen, Germany.
50. *European Standard EN 1504*; A Simplified, Illustrated Guide for all Involved in Concrete Repair. BASF Construction Chemicals. BASF Construction Chemicals: Ludwigshafen, Germany, 2009.
51. *The Repair and Protection of Reinforced Concrete with Sika*; Sika Services AG: Baar, Switzerland, 2018.
52. Stratfull, R. Cathodic protection of a bridge deck. *Mater Perform* **1974**, *13*, 24.
53. Rodrigues, R.; Gaboreau, S.; Gance, J.; Ignatiadis, I.; Betelu, S. Reinforced concrete structures: A review of corrosion mechanisms and advances in electrical methods for corrosion monitoring. *Constr. Build. Mater.* **2021**, *269*, 121240. [CrossRef]
54. Van den Hondel, A.; van den Hondel, H. Cathodic protection of concrete with conductive coating anodes: 25 years of experience with projects and monitoring results. In Proceedings of the MATEC Web of Conferences, International Conference on Concrete Repair, Rehabilitation and Retrofitting, Cape Town, South Africa, 19–21 November 2018; p. 6.
55. Polder, R.B.; Leegwater, G.; Worm, D.; Courage, W. Service life and life cycle cost modelling of cathodic protection systems for concrete structures. *Cem. Concr. Compos.* **2014**, *47*, 69–74. [CrossRef]
56. Wilson, K.; Jawed, M.; Ngala, V. The selection and use of cathodic protection systems for the repair of reinforced concrete structures. *Constr. Build. Mater.* **2013**, *39*, 19–25. [CrossRef]
57. Brosens, K.; Kriekemans, B.; Simpson, D. Cathodic Protection of the Historic Boerentoren (KBC Tower). Concrete Repair Bulletin, 2016. Available online: https://cdn.ymaws.com/www.icri.org/resource/resmgr/CRB/2016JanFeb/KBCTower.pdf (accessed on 10 July 2022).
58. Kamde, D.K.; Manickam, K.; Pillai, R.G.; Sergi, G. Long-term performance of galvanic anodes for the protection of steel reinforced concrete structures. *J. Build. Eng.* **2021**, *42*, 103049. [CrossRef]
59. Vector Corrosion Technologies. Vector®Galvashield®Fusion™ T2. Available online: https://www.fortius.be/content/files/3183-11961-galvashield-fusion-t2-900bd8.pdf (accessed on 25 July 2022).

60. Brueckner, R.; Cobbs, R.; Atkins, C. A review of developments in cathodic protection systems for reinforced concrete structures. In Proceedings of the MATEC Web of Conferences, Concrete Solutions 2022, Leeds, UK, 11–13 July 2022.

61. Eurostat. Glossary: Country Codes. Available online: https://ec.europa.eu/eurostat/statistics-explained/index.php?title=Glossary:Country_codes (accessed on 8 July 2022).

62. Farahani, A. Life Cycle Cost GA optimization of repaired reinforced concrete structures located in a marine environment. *SCCE* **2019**, *3*, 41–50. [CrossRef]

63. Kumar, R.; Gardoni, P. Renewal theory-based life cycle analysis of deteriorating engineering systems. *Struct. Saf.* **2014**, *50*, 94–102. [CrossRef]

64. Lepech, M.D.; Geiker, M.; Stang, H. Probabilistic design and management of environmentally sustainable repair and rehabilitation of reinforced concrete structures. *Cem. Concr. Compos.* **2014**, *47*, 19–31. [CrossRef]

65. Årskog, V.; Fossdal, S.; Gjørv, O. Life cycle assessment of repair and maintenance systems for concrete structures. In Proceedings of the Int. Workshop on Sustainable Development and Concrete Technology, Beijing, China, 20–21 May 2004; pp. 193–200.

66. Rohden, A.B.; Garcez, M.R. Increasing the sustainability potential of a reinforced concrete building through design strategies: Case study. *Case Stud. Constr. Mater.* **2018**, *9*, e00174. [CrossRef]

67. Chiu, C.-K.; Noguchi, T.; Kanematsu, M. Effects of maintenance strategies on the life cycle performance and cost of a deteriorating RC building with high-seismic hazard. *J. Adv. Concr. Technol.* **2010**, *8*, 157–170. [CrossRef]

68. Cadenazzi, T.; Lee, H.; Suraneni, P.; Nolan, S.; Nanni, A. Evaluation of probabilistic and deterministic life cycle cost analyses for concrete bridges exposed to chlorides. *Clean Eng. Tech.* **2021**, *4*, 100247. [CrossRef]

69. Shen, L.; Soliman, M.; Ahmed, S.A. A probabilistic framework for life cycle cost analysis of bridge decks constructed with different reinforcement alternatives. *Eng. Struct.* **2021**, *245*, 112879. [CrossRef]

70. Safi, M.; Sundquist, H.; Karoumi, R. Cost-efficient procurement of bridge infrastructures by incorporating life cycle cost analysis with bridge management systems. *J. Bridge Eng.* **2015**, *20*, 04014083. [CrossRef]

71. Ghodoosi, F.; Abu-Samra, S.; Zeynalian, M.; Zayed, T. Maintenance cost optimization for bridge structures using system reliability analysis and genetic algorithms. *J. Constr. Eng. Manag.* **2018**, *144*, 04017116. [CrossRef]

72. Polder, R.; Pan, Y.; Courage, W.; Peelen, W.H. Preliminary study of life cycle cost of preventive measures and repair options for corrosion in concrete infrastructure corrosion in concrete infrastructure. *Heron* **2016**, *61*, 1–13.

73. Islam, S.; Kishi, T. Effects of cover properties and repair methods on LCC estimation of reinforced concrete structure. *Internet J. Soc. Soc. Manag. Syst.* **2010**, *6*, 1–8.

74. Frangopol, D.M.; Soliman, M. Life cycle of structural systems: Recent achievements and future directions. *Struct. Infrastruct. Eng.* **2016**, *12*, 1–20. [CrossRef]

75. Van den Heede, P.; De Belie, N.; Pittau, F.; Habert, G.; Mignon, A. Life cycle assessment of Self-Healing Engineered Cementitious Composite (SH-ECC) used for the rehabilitation of bridges. In Proceedings of the 6th International Symposium on Life cycle Civil Engineering (IALCCE), Ghent, Belgium, 28–31 October 2018; pp. 2269–2275.

76. Habert, G.; Denarié, E.; Šajna, A.; Rossi, P. Lowering the global warming impact of bridge rehabilitations by using Ultra High Performance Fibre Reinforced Concretes. *Cem. Concr. Compos.* **2013**, *38*, 1–11. [CrossRef]

77. Hajiesmaeili, A.; Pittau, F.; Denarié, E.; Habert, G. Life Cycle Analysis of strengthening existing RC structures with R-PE-UHPFRC. *Sustainability* **2019**, *11*, 6923. [CrossRef]

78. Navarro, I.J.; Yepes, V.; Martí, J.V.; González-Vidosa, F. Life cycle impact assessment of corrosion preventive designs applied to prestressed concrete bridge decks. *J. Clean Prod.* **2018**, *196*, 698–713. [CrossRef]

79. Zirps, M.; Lepech, M.; Savage, S.; Michel, A.; Stang, H.; Geiker, M. Probabilistic design of sustainable reinforced concrete infrastructure repairs using sipmath. *Front. Built Environ.* **2020**, *6*, 72. [CrossRef]

80. Lepech, M.; Geiker, M.R.; Stang, H. Probabilistic design framework for sustainable repair and rehabilitation of civil infrastructure. In Proceedings of the Fib Symposium 2011, Prague, Czech Republic, 8–10 June 2011; pp. 1029–1032.

81. Ahmed, I.M.; Tsavdaridis, K.D. Life cycle assessment (LCA) and cost (LCC) studies of lightweight composite flooring systems. *J. Build. Eng.* **2018**, *20*, 624–633. [CrossRef]

82. Binder, F. Life cycle costs of selected concrete repair methods demonstrated on chloride contaminated columns. In Proceedings of the 11th International Probabilistic Workshop, Brno, Czech Republic, 6–8 November 2013.

83. Palacios-Munoz, B.; Gracia-Villa, L.; Zabalza-Bribián, I.; López-Mesa, B. Simplified structural design and LCA of reinforced concrete beams strengthening techniques. *Eng. Struct.* **2018**, *174*, 418–432. [CrossRef]

84. Turk, J.; Cotič, Z.; Mladenovič, A.; Šajna, A. Environmental evaluation of green concretes versus conventional concrete by means of LCA. *Waste Manag.* **2015**, *45*, 194–205. [CrossRef]

85. Van den Heede, P.; De Belie, N. Environmental impact and life cycle assessment (LCA) of traditional and 'green' concretes: Literature review and theoretical calculations. *Cem. Concr. Compos.* **2012**, *34*, 431–442. [CrossRef]

86. Marinković, S.; Carević, V.; Dragaš, J. The role of service life in Life Cycle Assessment of concrete structures. *J. Clean Prod.* **2021**, *290*, 125610. [CrossRef]

87. Santero, N.J.; Masanet, E.; Horvath, A. Life cycle assessment of pavements. Part I: Critical review. *Resour. Conserv. Recycl.* **2011**, *55*, 801–809. [CrossRef]

88. Walraven, J.; van der Horst, A. *FIB Model Code for Concrete Structures 2010*; Internation Federation for Structural Concrete (fib): Lausanne, Switzerland, 2013.

89. Monteiro, A.V.; Gonçalves, A. Probabilistic assessment of the depassivation limit state of reinforced concrete structures based on inspection results. *J. Build. Eng.* **2022**, *49*, 104063. [CrossRef]
90. Taffesea, W.Z.; Sistonen, E. Service Life Prediction of Repaired Structures Using Concrete Recasting Method: State-of-the-Art. *Procedia Eng.* **2013**, *57*, 1138–1144. [CrossRef]
91. De la Fuente, D.; Castano, J.; Morcillo, M. Long-term atmospheric corrosion of zinc. *Corros. Sci.* **2007**, *49*, 1420–1436. [CrossRef]
92. Thomas, S.; Birbilis, N.; Venkatraman, M.; Cole, I. Corrosion of zinc as a function of pH. *Corrosion* **2012**, *68*, 015009-1–015009-9. [CrossRef]
93. Vera, R.; Guerrero, F.; Delgado, D.; Araya, R. Atmospheric corrosion of galvanized steel and precipitation runoff from zinc in a marine environment. *J. Braz. Chem. Soc.* **2013**, *24*, 449–458. [CrossRef]
94. Appendix, J. The atmospheric corrosion chemistry of zinc. In *Atmospheric Corrosion*; Leygraf, C., Odnevall Wallinder, I., Tidblad, J., Graedel, T., Eds.; Wiley: New York, NY, USA, 2016; pp. 348–359.
95. Santana, J.; Fernández-Pérez, B.; Morales, J.; Vasconcelos, H.; Souto, R.; González, S. Characterization of the corrosion products formed on zinc in archipelagic subtropical environments. *Int. J. Electrochem. Sci.* **2012**, *7*, 12730–12741. Available online: http://www.electrochemsci.org/papers/vol12737/71212730.pdf (accessed on 10 July 2022).
96. Sagues, A.A.; Dugarte, M. Galvanic Point Anodes for Extending the Service Life of Patched Areas upon Reinforced Concrete Bridge Members. Report BD544-09, University of South Florida, USA, 2009. Available online: https://rosap.ntl.bts.gov/view/dot/16963 (accessed on 19 January 2022).
97. Diotti, A.; Plizzari, G.; Sorlini, S. Leaching behaviour of construction and demolition wastes and recycled aggregates: Statistical analysis applied to the release of contaminants. *Appl. Sci.* **2021**, *11*, 6265. [CrossRef]
98. Del Rey, I.; Ayuso, J.; Galvín, A.; Jiménez, J.R.; López, M.; García-Garrido, M. Analysis of chromium and sulphate origins in construction recycled materials based on leaching test results. *Waste Manag.* **2015**, *46*, 278–286. [CrossRef]
99. Galvín, A.P.; Ayuso, J.; García, I.; Jiménez, J.R.; Gutiérrez, F. The effect of compaction on the leaching and pollutant emission time of recycled aggregates from construction and demolition waste. *J. Clean Prod.* **2014**, *83*, 294–304. [CrossRef]
100. Nurhanim, A.; Norli, I.; Morad, N.; Khalil, H. Leaching behavior of construction and demolition waste (concrete and gypsum). *IJEE* **2016**, *7*, 203–211.
101. Chen, J.; Tinjum, J.M.; Edil, T.B. Leaching of alkaline substances and heavy metals from recycled concrete aggregate used as unbound base course. *TRR* **2013**, *2349*, 81–90. [CrossRef]
102. Trung, Z.H.; Kukurugya, F.; Takacova, Z.; Orac, D.; Laubertova, M.; Miskufova, A.; Havlik, T. Acidic leaching both of zinc and iron from basic oxygen furnace sludge. *J. Hazard. Mater.* **2011**, *192*, 1100–1107. [CrossRef]
103. Havlik, T.; Kukurugya, F.; Orac, D.; Parilak, L. Acidic leaching of EAF steelmaking dust. *World Metall. ERZMETALL* **2012**, *65*, 48–56.
104. Havlik, T.; Kukurugya, F.; Orac, D.; Vindt, T.; Miskufova, A.; Takacova, Z. Leaching of zinc and iron from blast furnace dust in sulphuric acid solutions. *Metall* **2012**, *66*, 267–270.
105. Kara De Maeijer, P.; Craeye, B.; Blom, J.; Bervoets, L. Crumb rubber in concrete—the barriers for application in the construction industry. *Infrastructures* **2021**, *6*, 116. [CrossRef]

 *infrastructures*

Article

Quantifying Anisotropic Properties of Old–New Concrete Interfaces Using X-Ray Computed Tomography and Homogenization

Guanming Zhang and Yang Lu

Department of Civil Engineering, Boise State University, Boise, ID 83706, USA; guanmingzhang@u.boisestate.edu
* Correspondence: yanglufrank@boisestate.edu; Tel.: +1-208-426-3783

Abstract: The interface between old and new concrete is a critical component in many construction practices, including concrete pavements, bridge decks, hydraulic dams, and buildings undergoing rehabilitation. Despite various treatments to enhance bonding, this interface often remains a weak layer that compromises overall structural performance. Traditional design methods typically oversimplify the interface as a homogeneous or empirically adjusted factor, resulting in significant uncertainties. This paper introduces a novel framework for quantifying the anisotropic properties of old–new concrete interfaces using X-ray computed tomography (CT) and finite element-based numerical homogenization. The elastic coefficient matrix reveals that specimens away from the interface exhibit higher values in both normal and shear directions, with normal direction values averaging 33.15% higher and shear direction values 39.96% higher than those at the interface. A total of 10 sampling units along the interface were collected and analyzed to identify the "weakest vectors" in normal and shear directions. The "weakest vectors" at the interface show consistent orientations with an average cosine similarity of 0.62, compared with an average cosine similarity of 0.23 at the non-interface, which demonstrates directional features. Conversely, the result of average cosine similarity at the interface shows randomness that originates from the anisotropy of materials. The average angle between normal and shear stresses was found to be 88.64°, indicating a predominantly orthogonal relationship, though local stress distributions introduced slight deviations. These findings highlight the importance of understanding the anisotropic properties of old–new concrete interfaces to improve design and rehabilitation practices in concrete and structural engineering.

Keywords: concrete interface; anisotropic properties; X-ray computed tomography; finite element-based numerical homogenization; mechanical weakness

Academic Editor: Patricia Kara De Maeijer

Received: 20 December 2024
Revised: 12 January 2025
Accepted: 13 January 2025
Published: 14 January 2025

Citation: Guanming Zhang and Yang Lu. Quantifying Anisotropic Properties of Old–New Concrete Interfaces Using X-Ray Computed Tomography and Homogenization. *Infrastructures* **2025**, *10*, 20. https://doi.org/10.3390/infrastructures10010020

1. Introduction

Aging civil infrastructure is becoming a critical issue worldwide, particularly in regions where structures, especially those built after World War II, are nearing or exceeding their designed lifespan. Concrete repair, such as overlays and patching, creates interfaces between old and new concrete that are often prone to mechanical weaknesses. These interfaces are heterogeneous, influenced by factors like differential shrinkage, moisture content, and surface preparation, which result in anisotropic properties. Traditional repair methods fail to adequately address these variations, as they assume homogeneity and ignore heterogeneity, which can lead to inaccurate performance predictions [1]. As infrastructure continues to age globally, enhanced methods for quantifying and improving interface performance will be essential for prolonging the lifespan of repairs and ensuring structural integrity [2].

The interface between old and new concrete is a common feature in various construction activities, such as construction joints, where fresh concrete is placed adjacent to previously poured concrete during multistage processes like building large slabs, walls, or bridge piers. It also occurs in renovation and overlay projects, where additional concrete layers are added to pavements, bridge decks, or industrial floors to restore or enhance performance. In rehabilitation efforts, this interface is critical for repairing deteriorated sections of structures like dams, tunnels, or parking garages, as well as in retrofitting buildings to strengthen or adapt them for new uses or modern code compliance. Additionally, it plays a key role in precast concrete applications, such as attaching precast panels to existing frames or foundations. Despite significant advances in bonding techniques, including the use of epoxy resins, surface roughening, and chemical bonding agents, ensuring the long-term mechanical integrity of these interfaces remains a challenge due to differential shrinkage, thermal stresses, and insufficient adhesion, particularly under cyclic loading or environmental exposure [3].

The primary cause of interface cracking between old and new concrete sections is often attributed to shrinkage cracking at the interface [4]. This type of cracking occurs due to differential volume changes when the new concrete dries and shrinks, exerting tensile stresses on the bond between the two layers. If these stresses exceed the tensile strength of the weaker interface, cracks can form [5]. Such interface cracks compromise the structural integrity, leading to potential pathways for moisture infiltration and subsequent degradation through freeze–thaw cycles, chemical reactions, or corrosion of embedded reinforcement [6]. Over time, these damages can significantly weaken the bond strength, reduce load-bearing capacity, and accelerate the overall deterioration of the composite structure [7]. Therefore, it is essential to study the mechanical properties of the old–new concrete interfaces.

Many researchers have made contributions to the old–new concrete interfaces, including qualification of interface properties, influencing factors of interface strength, and establishment of constitutive models with emphasis on the concrete interface or transition zone. Aaleti and Sritharan utilize slant shear test to investigate the behavior at the interface of ultra-high-performance concrete (UHPC), and findings show that adequate shear transfer is achieved with a surface roughness of at least 10^{-3} m, irrespective of concrete strength or curing conditions [8]. Ahmed and Aziz studied the shear performance of dry and epoxied joints in precast concrete segmental bridges, finding that epoxied joints offer 25–28% higher shear capacity than dry joints but fail in a brittle manner [4]. Austin, Robins, and Pan reviewed shear bond strength testing methods for concrete repairs, emphasizing the importance of surface preparation and material compatibility. Their work underscores the need for multidimensional approaches to accurately understand adhesion in cementitious repairs [9]. Bentz et al. and Beushausen et al. explore the effects of substrate moisture state on the bond strength and interfacial microstructure of repair materials. The difference is that the former uses neutron and X-ray radiography, while the latter uses the conventional shear test in the laboratory [10,11]. The bond properties between new and old concrete are tested using conventional macroscopic testing methods, and the influence of different factors on the bond strength is investigated [12–14]. Nanoindentation tests and numerical simulations are used in the research of Xiao et al. They focus on the mechanical properties and stress–strain behavior and find that the new mortar matrix and ITZs significantly impact MRAC's mechanical performance, particularly under uniaxial compression and tension [15]. Overall, previous studies have often treated the interface as a homogeneous and isotropic entity for simplicity in calculation and analysis, overlooking its heterogeneous characteristics at the microscopic scale.

Traditional design and test methodologies either assume a homogeneous interface or apply a reduction factor based on empirical data. However, these assumptions introduce uncertainties, as they fail to account for the complex mechanical interactions at the interface [12]. The presence of voids, cracks, and discontinuities at the old–new concrete interface leads to anisotropic behavior that cannot be captured by traditional design approaches. This anisotropy affects the strength, stiffness, and durability of the structure. Moreover, the lack of reliable methods to quantify this weakness from a solid physics and mechanical basis further complicates the problem. This issue is becoming more prominent against the background of aging infrastructure [1].

Micromechanics is an analytical and computational methodology used to study the mechanical behavior of materials at the microstructural level. It examines the interactions between multiple phases—such as grains, inclusions, and interfaces—within a material and investigates how these interactions influence the material's macroscopic mechanical properties. By focusing on microscale features like inclusions, voids, and grain boundaries, micromechanics provides insights into how these elements collectively affect the material's overall performance. This approach enables the prediction of bulk material properties by incorporating intricate microstructural details, effectively bridging the gap between microscale phenomena and macroscopic behavior. Micromechanics has been successfully applied to predict effective properties such as electrical conductivity, thermal conductivity, and elastic moduli [2].

However, traditional micromechanics methods are primarily designed for materials with standard geometries, making their application to random geometries challenging. Randomness in both geometry and the spatial distribution of material phases poses a significant long-term challenge for the application of micromechanics theory to heterogeneous microstructures. Among the various approaches within micromechanics, numerical homogenization has emerged as a state-of-the-art technique for calculating the effective properties of stochastic heterogeneous microstructures. In this work, FEM-based numerical homogenization is conducted due to its advanced computational capability to account for microscale variations. This method provides accurate macroscopic property predictions that capture the complexity and randomness of heterogeneities within the material [16].

Recent advancements in imaging technologies, such as X-ray computed tomography (CT), and computational modeling techniques, like digital image correlation (DIC) and finite element modeling (FEM), present promising solutions for characterizing the interfaces between different phases in concrete materials [17]. X-ray CT provides a non-destructive method to obtain high-resolution 3D images of the internal structure of concrete, allowing for a detailed examination of the distribution of aggregates, voids, and the cement paste matrix [18]. These images offer valuable microstructural data that can be directly integrated into computational models for further analysis. When combined with homogenization techniques, such as those used in micromechanical modeling, these advanced imaging and computational methods enable a more accurate prediction of the mechanical behavior of concrete interfaces under various loading conditions [19]. By employing homogenization, the microstructural data can be used to create effective macroscopic material properties that reflect the heterogeneous nature of concrete, improving the reliability of models predicting concrete performance [20]. Accurate characterization of these interfaces is critical for designing durable, cost-effective repairs, particularly in aging infrastructure subjected to repeated loading and environmental exposure [21]. The integration of X-ray CT data with computational models, therefore, represents a significant advancement in the field of concrete material characterization and repair design, enabling more efficient and precise engineering solutions.

With the research gap in quantifying the anisotropic properties of old–new concrete interfaces, this study aims to propose a method for quantifying the mechanical properties of the old–new concrete interface within a heterogeneous model. By combining X-ray CT scanning technology, homogenization techniques, and data analysis, a new framework is developed to offer deeper insights into the heterogeneous characteristics at the old–new concrete interface. This approach contributes to the understanding of concrete interfaces and holds the potential for advancing future repair and rehabilitation strategies in concrete structures.

2. Methodology

The methodology can be expressed as Figure 1.

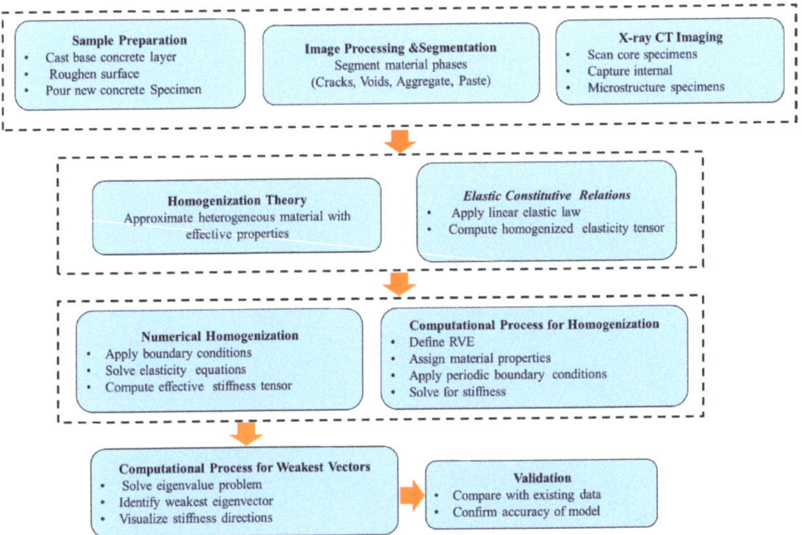

Figure 1. Schematic representation of the methodological framework.

As shown in Figure 1, a comprehensive framework is implemented to analyze the mechanical properties of concrete specimens in this study, focusing on the old–new concrete interface. Samples are meticulously prepared to replicate real-world conditions, including casting, surface texturing, and controlled curing. Image processing identifies key material phases, while X-ray CT imaging non-destructively captures the internal microstructure, revealing voids, cracks, and aggregate distribution critical to mechanical behavior. Homogenization theory bridges microscale features with macroscopic behavior, and numerical homogenization solves elasticity equations to determine the effective stiffness tensor. A representative volume element (RVE) with periodic boundary conditions ensures realistic simulations, while eigenvalue analysis identifies the weakest vectors to highlight potential failure directions.

This paper integrates X-ray CT imaging and numerical homogenization to assess the anisotropic properties of the old–new concrete interface, which consists of the following steps.

2.1. Sample Preparation

Concrete specimens are prepared to replicate typical conditions found at old–new concrete interfaces in construction [22]. The specimens have a diameter of 2.54×10^{-2} m and a length of 2.032×10^{-1} m.

The concrete used in this study was a standard mixture designed for structural applications, with a strength grade of C30, in accordance with Ref. [23] for ready-mixed concrete and Ref. [24] for Portland cement. It consisted of Portland cement (Class 42.5), fine aggregate (natural sand with a maximum particle size of 4 mm), coarse aggregate (crushed limestone with a maximum particle size of 20 mm), and potable water. The water-to-cement ratio was set to 0.45 to ensure a target compressive strength of 30 MPa at 28 days. A plasticizer admixture was added at 1.5% of the cement weight to improve workability. The consistency of the fresh concrete was determined using Abram slump test according to Ref. [25], which resulted in a slump value of approximately 75 mm, indicating a rather dry mix with lower workability suitable for the casting process.

The mixture design is summarized in Table 1:

Table 1. Mixture design for concrete.

Ingredient	Specification	Quantity (per m^3)	Density (kg/m^3)
Cement	Portland cement (Class 42.5)	371 kg	3150
Fine Aggregate	Natural sand, max particle size 4 mm	742 kg	2650
Coarse Aggregate	Crushed limestone, max size 20 mm	1166 kg	2700
Water	Potable water	166.95 kg	1000
Admixture	Plasticizer	1.5% by cement weight	~1000 (liquid-based)

The preparation process begins by casting a base layer of concrete to represent the existing structure. After casting, the base layer undergoes an initial curing period of 28 days. This curing allows the base concrete to harden and develop its strength, representing the aged structure in practice. Once the base concrete has gained enough strength, its surface is carefully treated to simulate true construction joint bonding conditions. The surface is roughened using mechanical tools to expose the aggregates and increase mechanical interlocking. These surface treatments mimic field practices used to improve the bond strength between old and new concrete.

After surface preparation, a fresh layer of concrete is mixed and poured over the treated surface to form a composite specimen. The new concrete is placed carefully, ensuring thorough compaction, especially at the interface, to minimize air pockets and potential weak zones. To adequately consolidate the fresh concrete, a mechanical vibrator (Model XYZ, frequency 50 Hz) was used for a duration of 30 s per batch. For smaller or hard-to-reach areas, hand tamping with a steel rod (diameter 10^{-3} m) was performed to ensure proper compaction. Specimens were then maintained under controlled water bath curing conditions for a specified period of 28 days, allowing hydration and bonding comparable to field conditions. Once the curing was complete, cylindrical cores were drilled perpendicular to the interface to capture a cross-section of the old–new boundary.

2.2. X-Ray CT Imaging

X-ray CT scans are performed on the cores to capture the internal microstructure of the old–new concrete interface [26], highlighting the distribution and morphology of voids, microcracks, aggregates, and the cement matrix.

As shown in Figure 2, voids at the interface are irregularly shaped and unevenly distributed, often forming micro-porous zones that reduce bond strength. Microcracks, frequently oriented along stress concentration paths, vary in width and alignment, contributing to directional mechanical weaknesses. Aggregate particles near the interface show partial embedding, with occasional gaps or weak bonds, while the cement matrix exhibits differences in density and fracture patterns due to variations in hydration and compaction between old and new concrete layers. The interfacial transition zones (ITZs), thin layers surrounding aggregates, display higher porosity and weaker properties compared to the bulk matrix. These microstructural heterogeneities lead to an anisotropic mechanical response, with stress transfer and stiffness varying by direction. The 55-micrometer resolution of X-ray CT scans provides detailed data for quantifying these features, forming a foundation for homogenization analysis. The observed microstructural variations will be further validated in Section 3.

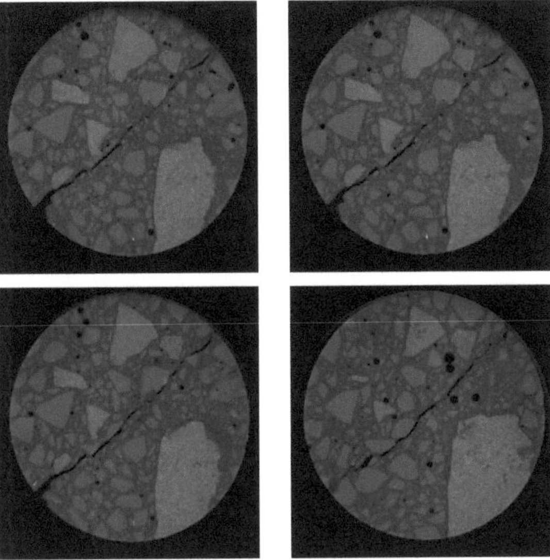

Figure 2. X-ray CT micrographs with morphological information of concrete.

2.3. Image Processing and Segmentation

The CT images are processed using image segmentation techniques to differentiate between material phases, including cracks, voids, aggregate, and paste. This step is crucial for generating accurate 3D models of the interface for further analysis [27]. The CT images undergo segmentation to differentiate the primary material phases, including aggregates, paste, voids, and cracks, enabling detailed morphological analysis of the old–new concrete interface. Initial preprocessing involves contrast enhancement and Gaussian filtering to reduce noise and sharpen phase boundaries. A threshold-based binary mask is created to isolate regions of interest, followed by morphological operations to remove noise and refine phase boundaries. Shrink-wrap iterations are applied to enforce circular constraints around the interface, mimicking the geometry observed in actual cross-sections.

Subsequently, grayscale intensity ranges corresponding to each phase are defined for segmentation. Aggregates are segmented within the high-intensity range, paste occupies an intermediate range, and voids and cracks are identified in the lower range. Connected component analysis is performed to classify voids and cracks based on aspect ratios, ensuring precise differentiation of these critical features. The segmented images are stacked to construct comprehensive 3D models for quantifying anisotropic properties, linking microstructural variations to macroscopic mechanical behavior.

As shown in Figure 3, the image on the left represents the segmented version of the original image (Figure 2). The segmented image consists of four grayscale values corresponding to the four material phases: cracks (15), voids (69), paste (129), and aggregate (255). Multiple consecutive segmented images are stacked to create a 3D digital representation of the four-phase concrete. Samples are then extracted at both the interface and non-interface regions to generate multiple subcubes, each measuring $40 \times 40 \times 40$ pixels. The image on the right of Figure 3 illustrates a sampling practice extracting a subcube along the interface region. This method effectively preserves the key features of the multiphase material, such as the distinct characteristics of cracks, voids, paste, and aggregate, while ensuring that the resulting data are sufficiently detailed to provide accurate and reliable predictions of the material's macroscopic properties. It is physically based and represents sensory information from the real world, maintaining the critical microstructural information and capturing the complexity of interface regions. This approach enhances the precision of the analysis without oversimplifying the internal structure of the concrete.

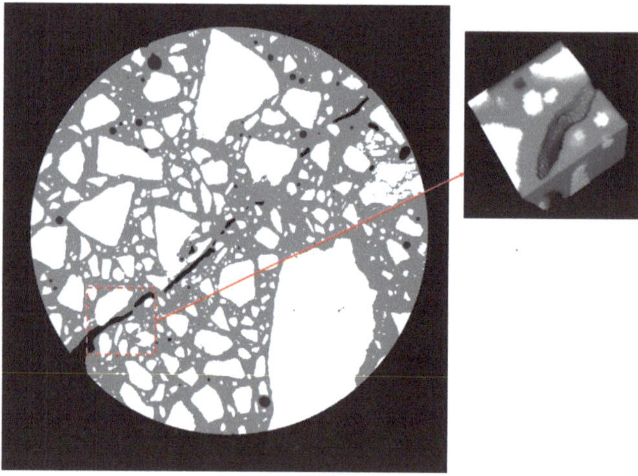

Figure 3. Segmented X-ray CT image depicting four-phase material composition.

In the model development process for this study, a simplified tetrahedral mesh was generated to represent the heterogeneous material microstructure, following the approach described by Ref. [28], which involves topological and geometric data processing to create meshes with consistent interfaces between aggregates and cement paste.

2.4. Finite Element-Based Numerical Homogenization

It is noteworthy that homogenization theory approximates a heterogeneous material as an equivalent homogeneous material with effective properties [21]. The fundamental principle is to establish a relationship between the microscopic behavior of individual phases and the macroscopic response of the composite material [27]. The effective properties are typically represented by the elastic modulus tensor, which relates stress and strain on the macroscopic scale.

The method of asymptotic numerical homogenization with periodic boundary conditions is one of the most effective approaches to obtaining the effective homogenized stiffness matrix of a heterogeneous elastic material. In this finite element-based approach, six boundary conditions are applied to a representative volume element (RVE), which is modeled with detailed micro-/mesoscale multi-phasic constituents, to calculate the homogenized elastic properties of the RVE [29].

The elastic behavior of a homogeneous or pseudo-homogeneous material is described by the linear elastic constitutive law:

$$\sigma = C{:}\varepsilon \tag{1}$$

where

σ is the stress tensor.

C is the elastic modulus tensor (homogenized elasticity tensor).

ε epsilon is the strain tensor.

In multiphase materials, the effective elastic modulus tensor C can be computed using periodic boundary conditions and a systematic approach involving multiple load cases [26].

The first step in the computational process is to define a representative volume element (RVE) that captures the essential features of the microstructure. The RVE should be large enough to represent the statistical behavior of the material while being small enough for computational feasibility.

Once the RVE is defined, material properties must be assigned to each constituent phase. These properties include Young's modulus, Poisson's ratio, and other relevant mechanical properties, which can vary significantly between different phases.

Boundary conditions play a crucial role in the homogenization process. Periodic boundary conditions are often applied to ensure consistency across the RVE. This involves establishing relationships between the displacements and tractions on corresponding boundaries. The periodic boundary condition can be expressed as follows:

$$u_d(x) = u_s(x + L) \tag{2}$$

where

u_d and u_s are the displacement vectors at the destination and source boundaries, respectively.

L is the position vector connecting the source and destination boundaries.

Additionally, the continuity of tractions across adjacent RVE boundaries is enforced:

$$T_1 = T_2 \tag{3}$$

where T_1 and T_2 are the traction vectors at the adjacent boundaries—these conditions simulate the behavior of homogeneous materials and are crucial for calculating the equivalent elastic modulus matrix.

To compute the homogenized elasticity tensor C, six different load cases are applied, covering standard deformation modes in three axial and three shear directions. Each load case prescribes a non-zero component of the average strain tensor while keeping the others at zero. The average traction vector for each load case is used to construct C.

The effective stiffness tensor can be obtained by solving the following relation in each load case:

$$T = C{:}\,\varepsilon \tag{4}$$

where T is the average traction vector corresponding to the applied strain.

The elastic constitutive law for a homogeneous or pseudo-homogeneous material is written as follows:

$$\begin{bmatrix} \sigma_{avg,11} \\ \sigma_{avg,12} \\ \sigma_{avg,13} \\ \sigma_{avg,14} \\ \sigma_{avg,15} \\ \sigma_{avg,16} \end{bmatrix} = D_{avg} \cdot \begin{bmatrix} \epsilon_{avg,11} \\ \epsilon_{avg,12} \\ \epsilon_{avg,13} \\ \epsilon_{avg,14} \\ \epsilon_{avg,15} \\ \epsilon_{avg,16} \end{bmatrix} \tag{5}$$

where D_{avg} is the homogenized elasticity tensor. To find the components of the homogenized elasticity tensor using the periodic boundary conditions, six load cases are required. In each load case, one strain component is prescribed while the others are zero.

After obtaining the heterogeneous elastic matrix, this work proposes a novel method to assess the anisotropic properties of old–new concrete, addressing the current gap in accurate quantification. This approach provides valuable insight for better understanding the interface between old and new concrete. The core of the analysis lies in solving the eigenvalues and eigenvectors of the material, identifying its stiffness properties in different directions, and particularly focusing on the "weakest" eigenvector associated with the smallest eigenvalue. This analysis plays a key role in understanding the material's response under various stress and strain conditions, especially in the evaluation of elastic properties of multiphase or complex materials.

Given a 6×6 constitutive matrix C, it can be represented as follows:

$$
\begin{bmatrix}
C_{11} & C_{12} & C_{13} & C_{14} & C_{15} & C_{16} \\
C_{21} & C_{22} & C_{23} & C_{24} & C_{25} & C_{26} \\
C_{31} & C_{32} & C_{33} & C_{34} & C_{35} & C_{36} \\
C_{41} & C_{42} & C_{43} & C_{44} & C_{45} & C_{46} \\
C_{51} & C_{52} & C_{53} & C_{54} & C_{55} & C_{56} \\
C_{61} & C_{62} & C_{63} & C_{64} & C_{65} & C_{66}
\end{bmatrix}
\tag{6}
$$

This matrix C characterizes the material's anisotropic stiffness properties, which is crucial for understanding how the material responds to stress and strain in different directions.

2.5. Weakest Vector Index

The homogenization method effectively addresses the issue of macroscopic representation of the elastic modulus in heterogeneous models. To gain a deeper understanding and evaluation of the directional properties of the elastic modulus at the old–new concrete interface under heterogeneous conditions, this study introduces a new index in this subsection called the "weakest vector."

The introduction of this index begins with the definition of the eigenvector, and the eigenvalue problem is solved as follows:

$$
Cv = \lambda v \tag{7}
$$

where λ represents the eigenvalue, and v is the corresponding eigenvector. The eigenvalue λ represents the stiffness of the material in the direction defined by v. Larger eigenvalues indicate higher stiffness, while smaller eigenvalues represent lower stiffness or higher compliance.

After obtaining all the eigenvalues and eigenvectors, the smallest eigenvalue λ_{min} and its corresponding eigenvector v_{weak} are extracted. This direction, v_{weak}, is referred to as the "weakest direction", representing the direction in which the material is most susceptible to deformation. This is mathematically expressed as follows:

$$
\lambda_{min} = min(\lambda_1, \lambda_2, \ldots, \lambda_6), v_{weak} = vi, where \ \lambda_i = \lambda_{min} \tag{8}
$$

To visualize stiffness in different directions, the eigenvectors are scaled by their eigenvalues and plotted in 3D space. The length of each eigenvector corresponds to its eigenvalue, illustrating the material's relative stiffness in that direction. The weakest eigenvector is highlighted with a distinctive color, with annotations indicating its eigenvalue and index.

This analytical method is crucial for assessing the material's response under complex stress conditions and identifying potential weaknesses. Studying the weakest stiffness direction provides deeper insights, enabling engineers and researchers to better assess the structural performance and safety of the material.

2.6. Validation

The Self-Consistent Scheme (SCS) is a well-established micromechanical model used to predict the effective elastic properties of heterogeneous materials [30]. It assumes that each phase in the composite behaves as though embedded in an effective medium, which is itself an unknown property to be determined. Current research has demonstrated the validity of comparing the Self-Consistent Scheme (SCS) with FEM analysis for characterizing composite material properties, showing that FEM-based calculations can be reliably verified using SCS [31].

The iterative process ensures that the strain field within each inclusion satisfies the self-consistency condition. The effective stiffness tensor C_{eff} is computed by solving the following relation:

$$C_{eff} = C_m + \sum_{i=1}^{n} v_i (C_i - C_m) A_i \tag{9}$$

where C_m is the stiffness tensor of the matrix, C_i is the stiffness tensor of inclusion i, v_i is the volume fraction of phase i, and A_i is the strain concentration factor derived from the Eshelby tensor.

As shown in Figure 4, to validate our adopted finite element-based homogenization method, this work applied the SCS to a model containing two phases: a matrix with Young's modulus $E_m = 7 \times 10^{10}$ Pa and Poisson's ratio $v_m = 0.25$ and spherical inclusions with $E_i = 2 \times 10^{11}$ Pa and $v_i = 0.3$. The inclusion, with a fraction of 11.31%, is randomly distributed within a cubic domain with an edge length of 10^{-3} m. The material properties and phase geometry are consistent across both the SCS and finite element models.

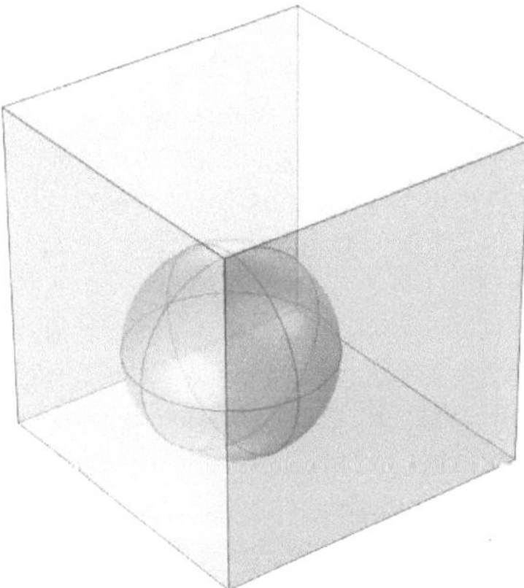

Figure 4. Geometry input for SCS and FEM methodologies.

The effective stiffness tensors derived from the SCS and FEM approaches are compared below. For brevity, the two 6×6 matrices are as follows:

As shown in (10) and (11), the results reveal overall consistency between the two methods, particularly along the diagonal terms (C_{11}, C_{22}, C_{33}), where the relative differences are less than 2% (e.g., $C_{11} = 241.29$ from SCS vs. 237.14 from FEM, yielding a relative error of 1.7%). Off-diagonal terms (C_{12}, C_{13}, etc.) and shear components (C_{44}, C_{55}, C_{66}) show minor discrepancies, with maximum absolute differences of approximately 4×10^9 Pa in shear moduli. The observed negative off-diagonal values in the elastic matrix can be attributed to the intrinsic heterogeneity of the composite material and the complex interfacial interactions between the two phases, reflecting the inherent anisotropic mechanical coupling and physical behavior of the system.

$$\begin{bmatrix} 241.29 & 101.82 & 101.82 & 0.00 & 0.00 & 0.00 \\ 101.82 & 241.29 & 101.82 & 0.00 & 0.00 & 0.00 \\ 101.82 & 101.82 & 241.29 & 0.00 & 0.00 & 0.00 \\ 0.00 & 0.00 & 0.00 & 70.63 & 0.00 & 0.00 \\ 0.00 & 0.00 & 0.00 & 0.00 & 70.63 & 0.00 \\ 0.00 & 0.00 & 0.00 & 0.00 & 0.00 & 70.63 \end{bmatrix} \tag{10}$$

$$\begin{bmatrix} 237.14 & 97.65 & 97.65 & -0.29 & -0.46 & -0.63 \\ 97.65 & 237.14 & 97.65 & -0.37 & -0.21 & -0.49 \\ 97.65 & 97.65 & 237.14 & -0.41 & -0.43 & -0.36 \\ -0.29 & -0.37 & -0.41 & 69.06 & 0.82 & 0.61 \\ -0.46 & -0.21 & -0.43 & 0.82 & 69.06 & 0.78 \\ -0.63 & -0.49 & -0.36 & 0.61 & 0.78 & 69.06 \end{bmatrix} \tag{11}$$

Compared with statistical homogeneity, the validation confirms the robustness of the finite element-based homogenization approach combined with our selected component properties. The FEM results demonstrate its capability to accurately predict the effective elastic moduli of multiphase, heterogeneous materials with random microstructures, validating its use in homogenization studies for random composite materials.

3. Results and Discussion

The results of numerical evaluation reveal a strong correlation between the quantified anisotropy of the elastic modulus and the microstructural features of the concrete, particularly in the interface regions. The stiffness tensors exhibit notable directional dependence, with reduced stiffness observed perpendicular to the interface. This reduction is closely tied to microstructural characteristics identified through X-ray CT scans, such as the distribution and morphology of voids, microcracks, aggregates, and the cement matrix.

Voids at the interface are irregularly shaped and unevenly distributed, forming microporous zones that weaken bond strength. Microcracks, aligned along stress concentration paths, vary in width and orientation, introducing directional weaknesses in mechanical response. Aggregates near the interface show partial embedding with occasional gaps or weak bonds, while the cement matrix displays variations in density and fracture patterns due to differences in hydration and compaction between old and new concrete layers. The interfacial transition zones (ITZs), thin, porous layers surrounding aggregates, further contribute to directional stiffness variability. These microstructural heterogeneities, quantified at a resolution of 5.5×10^{-5} m using X-ray CT, underlie the observed anisotropic behavior.

The findings emphasize the critical role of interface heterogeneity in governing mechanical behavior, challenging traditional design assumptions that neglect anisotropic effects. The current reliance on empirical reduction factors fails to capture the true mechanical response of the interface, potentially leading to inaccuracies in structural performance estimation. These microstructural observations provide a robust basis for homogenization analysis, bridging the gap between detailed features and macroscopic anisotropy.

3.1. Results of Elastic Constitutive Matrix

Constitutive relations in linear elasticity are given by the generalized Hooke law, which linearly relates the stress and strain tensors to the elasticity and/or compliance tensors. The coefficients of these linear relations are the elastic and/or compliance moduli, as calculated in this subsection [32].

The elements of matrix C reveal key insights into the material's anisotropic properties at the old–new concrete interface. The diagonal terms C_{11}, C_{22}, and C_{33} represent the material's stiffness along the x, y, and z axes, indicating its resistance to axial deformation. Higher values highlight the primary load-bearing capabilities. The off-diagonal terms C_{12}, C_{13}, and C_{23} capture anisotropic coupling, where stress in one direction induces strain in another, reflecting the transitional properties of the old–new concrete interface. The shear terms C_{44}, C_{55}, and C_{66} describe resistance to shear deformation in the yz, xz, and xy planes, providing insights into interfacial bonding and structural cohesion under shear loads [28].

Using the homogenized model derived from numerical simulations, the effective elastic constitutive matrix C is calculated based on segmented X-ray CT images. In this paper, a total of 18 samples are calculated with a size of $40 \times 40 \times 40$ for each sample. According to the sampling position, it can be divided into interface position samples and non-interface position samples.

The matrix C is determined by applying six prescribed load cases, each inducing a unique strain component. This approach enables us to obtain a complete 6×6 matrix of stiffness coefficients, capturing the interactions between stress and strain in multiple directions. Selected components of the calculated C matrix are presented in the following tables.

As shown in (12), the elasticity matrix of the non-interface position reveals that the diagonal elements represent axial stiffness along the principal directions (x, y, z), with higher values indicating greater stiffness. In the non-interface position matrix, the high values in axial directions, especially $C_{11} = 31{,}937.09$, $C_{22} = 31{,}193.96$, and $C_{33} = 35{,}270.82$, suggest that this region is generally stiffer due to better compaction and fewer microstructural voids in the old–new concrete interface. This uniform axial stiffness across principal directions is typical in regions where the concrete is well-bonded.

$$
\begin{bmatrix}
31{,}937.09 & 9165.88 & 9207.04 & 5.61 & 142.71 & 89.06 \\
9165.88 & 31{,}193.96 & 9332.21 & 256.75 & 22.84 & 182.96 \\
9207.04 & 9332.21 & 35{,}270.82 & 404.62 & 198.80 & 100.80 \\
5.61 & 256.75 & 404.62 & 11{,}518.02 & 126.12 & 28.12 \\
142.71 & 22.84 & 198.80 & 126.12 & 11{,}437.41 & 30.72 \\
89.06 & 182.96 & 100.80 & 28.12 & 30.72 & 11{,}141.91
\end{bmatrix}
\tag{12}
$$

As shown in (13), in contrast, the interface position matrix shows lower values across the diagonal, particularly $C_{11} = 18{,}970.84$ and $C_{22} = 14{,}562.60$, suggesting that the material is more compliant or less stiff in these directions. This is likely due to weaker bonding or microstructural defects, such as voids or cracks.

$$
\begin{bmatrix}
18{,}970.84 & 6042.31 & 5695.41 & -26.32 & 527.61 & 4527.15 \\
6042.31 & 14{,}562.60 & 4885.64 & -388.06 & 128.25 & 4420.14 \\
5695.41 & 4885.64 & 33314.93 & -391.05 & 734.93 & 1998.46 \\
-26.32 & -388.06 & -391.05 & 7625.52 & 2883.81 & 131.13 \\
527.61 & 128.25 & 734.93 & 2883.81 & 8721.25 & -25.82 \\
4527.15 & 4420.14 & 1998.46 & 131.13 & -25.82 & 7737.81
\end{bmatrix}
\tag{13}
$$

Examining the off-diagonal elements reveals insights into shear and coupling effects between different axes. Higher off-diagonal values typically suggest stronger bonding and resistance to deformation in off-axis directions, while lower or negative values may imply structural weaknesses or anisotropy. In the non-interface position matrix, many off-diagonal elements are relatively small, with values close to zero or low magnitudes (e.g.,

$C_{14} = 5.61$, $C_{13} = 9207.04$), indicating minimal shear coupling or anisotropy effects between the principal axes. This consistency aligns with a well-bonded, isotropic-like region.

In the interface position matrix, however, the off-diagonal terms are more variable, with some significantly negative values (e.g., $C_{14} = -26.32$, $C_{24} = -388.06$) and others with larger magnitudes (e.g., $C_{15} = 527.61$, $C_{35} = 734.93$). The presence of negative values suggests internal stresses or localized anisotropy, likely due to structural weaknesses at the interface. These irregularities indicate greater anisotropic behavior and less cohesive bonding between phases in the interface position.

Overall, the non-interface position has consistently higher values in both axial and off-diagonal terms, indicating a generally stiffer and more isotropic behavior. This reflects a stronger bond at the old–new concrete interface, with minimal voids or cracks. In contrast, the interface position matrix shows significant variability, with lower and negative values in the off-diagonal terms, indicating anisotropic behavior where certain directions are more compliant. These variations are likely due to defects such as cracks, voids, or weak bonding regions, making this position more susceptible to deformation in specific directions.

In practical terms, the non-interface position is likely to resist load application more effectively in all directions, maintaining structural integrity under stress. The interface position, however, suggests potential failure zones, particularly when loads are applied along or perpendicular to planes where defects reduce stiffness.

The calculated constitutive matrix reveals significant anisotropy due to the heterogeneous nature of the old–new concrete interface. The variation in stiffness across different directions, with distinct values for C_{11}, C_{22}, and C_{33}, highlights the directional stiffness characteristics of the interface. This anisotropic behavior arises from the presence of voids, microcracks, and bonding inconsistencies within the interface region.

Additionally, the off-diagonal terms indicate substantial coupling between axial and shear deformations, a typical feature in composite materials with uneven bonding. These shear coupling effects underscore the complex response of the interface when subjected to multi-directional loading, reflecting its non-uniform and transitional nature. Together, these observations characterize the old–new concrete interface as a zone of variable stiffness and intricate mechanical behavior [33].

3.2. Calculation and Visualization of Weakest Vectors

As described in Section 2, the process of calculating the "weakest vector" in a constitutive matrix analysis begins with the definition of eigenvalues and eigenvectors.

To obtain the eigenvalues λ_i and corresponding eigenvectors of C, the matrix equation is solved:

$$C \times v_i = \lambda_i \times v_i \qquad (14)$$

where each λ_i is an eigenvalue, and each v_i is the corresponding eigenvector. This computation produces a diagonal matrix of eigenvalues and a matrix of eigenvectors, where each column in the eigenvector matrix corresponds to an eigenvector associated with an eigenvalue in the eigenvalue matrix. Extracting the eigenvalues as a vector allows for further analysis of the material's directional properties.

The "weakest vector" is identified as the eigenvector associated with the smallest eigenvalue of C. This smallest eigenvalue represents the minimum response or stiffness in the direction of its corresponding eigenvector, making it the "weakest" direction in the material. Specifically, this work identifies the minimum eigenvalue $\lambda\text{min} = \min\{\lambda_1, \lambda_2, \ldots, \lambda_6\}$ and finds the index idxweakest $= \arg\min\{\lambda_1, \lambda_2, \ldots, \lambda_6\}$ to locate the corresponding eigenvector vweakest $=$ vidxweakest.

To visualize the eigenvalues and eigenvectors, each eigenvector can be scaled proportionally to its eigenvalue, with length and thickness adjusted to show the relative magnitude of each eigenvalue. The scaling factor for each eigenvector is calculated as follows:

$$\lambda_i^{\text{scaled}} = 1 + 4 \cdot \frac{\lambda_i - \min(\{\lambda\})}{(\{\lambda\}) - \min(\{\lambda\})} \tag{15}$$

This proportional scaling highlights variations among eigenvalues, enabling a clear comparison of the material's characteristic directions. The "weakest vector", marked distinctly, stands out as the vector of minimal response, signifying the direction in which the material exhibits the least resistance. This analysis can provide valuable insights into material behavior under different forces and is widely applicable in fields like materials science and mechanical engineering, where understanding directional responses to stress and strain is essential.

As illustrated in Figures 5 and 6, to investigate the anisotropic properties of the old–new concrete interface, the weakest vectors are visualized. The first three components of each weakest vector correspond to the direction of minimum resistance under normal stress and are represented by a red arrow with a solid line. Conversely, the last three components of the weakest vector indicate the direction of minimum resistance under shear stress, denoted by a blue arrow with a dashed–dotted line.

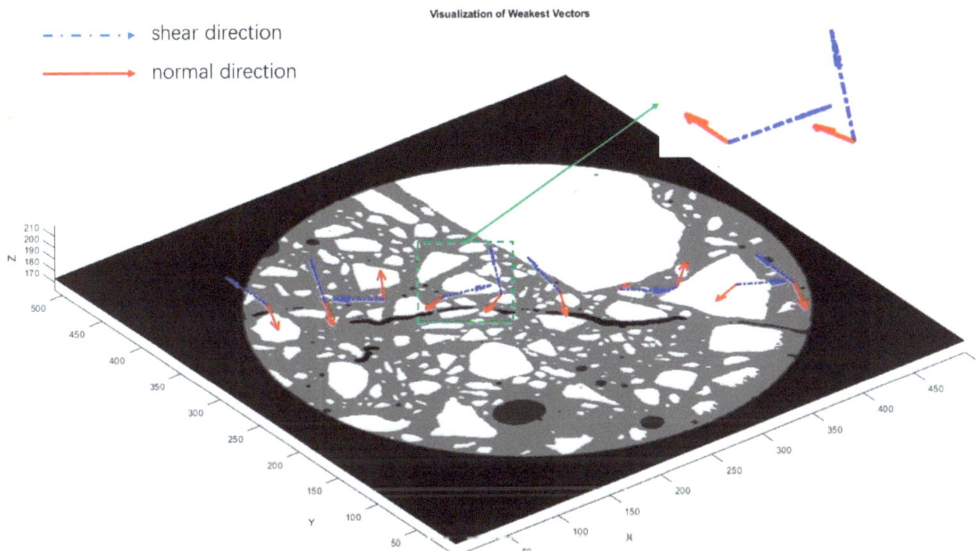

Figure 5. Three-dimensional visualization of weakest vectors in old–new concrete (solid lines indicate normal direction; dashed–dotted lines indicate shear direction).

The angles between the weakest normal stress direction and the corresponding weakest shear stress direction are calculated for ten specific vector pairs. The computed angles are as follows: 101.38°, 85.68°, 73.11°, 96.41°, 80.26°, 91.82°, 72.37°, 85.27°, 103.42°, and 96.68°. These values indicate a variation in the relative orientation between the normal and shear stress components across the sampled pairs. The mean angle is determined to be 88.64°, reflecting a predominantly orthogonal relationship with some deviation due to local stress distributions.

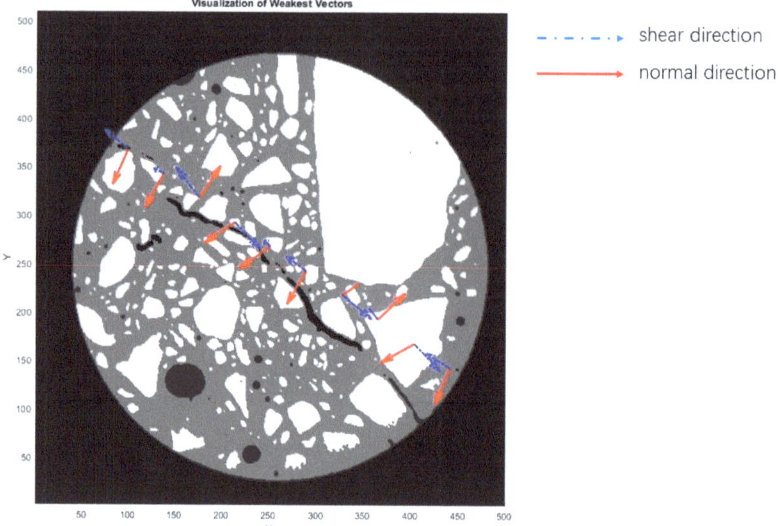

Figure 6. Two-dimensional visualization of weakest vectors in old–new concrete (solid lines represent normal direction; dashed–dotted lines represent shear direction).

3.3. Similarity Analysis of Weakest Vectors

Based on the visualization of the weakest vectors, it is found that there are similarities between these vectors, which might show the properties of the interface of old–new concrete. To analyze the anisotropic properties of the interface deeply, the similarity is quantified in this subsection.

As shown in (16), based on the calculated elastic matrix, we can obtain the "weakest vectors" of interface positions. Each column represents a vector. This work resamples 10 cubic specimens at the interface, each measuring $40 \times 40 \times 40$, at various interface positions. For each sample, the "weakest vectors" are computed, representing the direction of minimal material resistance. These "weakest vectors" are then assembled into a matrix, termed the "weakest vector matrix". This matrix has dimensions of 6×10, where each column corresponds to the vector of an individual sample.

$$
\begin{bmatrix}
-0.1079 & -0.1036 & 0.0142 & -0.0296 & -0.0752 & -0.1148 & 0.0126 & 0.0596 & -0.0309 & -0.0916 \\
-0.2310 & -0.1894 & 0.0211 & -0.0193 & -0.0508 & -0.1945 & 0.0085 & 0.0556 & -0.0172 & -0.2131 \\
-0.0004 & 0.0151 & 0.0012 & 0.0092 & 0.0134 & 0.0076 & -0.0049 & -0.0024 & 0.0057 & -0.0080 \\
-0.5979 & -0.2465 & -0.6190 & 0.6719 & -0.1405 & -0.5026 & 0.7869 & -0.5626 & 0.6900 & -0.6347 \\
0.4888 & 0.1254 & 0.7744 & -0.6686 & 0.1604 & 0.3620 & -0.6160 & 0.7053 & -0.7074 & 0.3722 \\
0.5819 & 0.9363 & -0.1289 & 0.3165 & 0.9727 & 0.7518 & -0.0323 & -0.4236 & 0.1487 & 0.6361
\end{bmatrix}
\tag{16}
$$

The calculation process implements a method to compute a cosine similarity matrix for a set of vectors, each representing a "weakest vector". Cosine similarity is a measure used to evaluate the directional alignment between two non-zero vectors by calculating the cosine of the angle between them. This measure ranges from -1 to 1, where a value of 1 indicates perfect alignment in the same direction, -1 indicates alignment in opposite directions, and 0 indicates orthogonality or complete lack of alignment.

Initially, the vectors are defined and organized into a matrix, where each column corresponds to an individual vector. A square matrix is then created to store the cosine similarity values, with dimensions equal to the number of vectors. The calculation of cosine similarity between each unique pair of vectors, including each vector with itself, follows by retrieving vectors a and b and computing the similarity using the formula:

$$\cos\_\mathrm{sim}(a,b) = \frac{a \cdot b}{||a|| \times ||b||} \tag{17}$$

where $a \cdot b$ denotes the dot product of the two vectors, and $||a||$ and $||b||$ are their magnitudes. The computed similarity is stored symmetrically in the matrix since cosine similarity is a symmetric measure.

As shown in (18), further analysis is then performed on the cosine similarity matrix. The average cosine similarity is calculated by taking the absolute values of the matrix entries, excluding the diagonal elements, and averaging them. The average similarity at non-interface positions is 0.23, compared to 0.62 at interface positions. This indicates that the "weakest vectors" at the interface demonstrate greater directional consistency. This metric indicates the general alignment across the set of vectors. Additionally, the variance of these similarity values is computed to reveal the diversity in orientation. In this case, the variance is 0.11, which shows that these "weakest vectors" have minimal fluctuation. The low variance suggests homogeneity in alignment. This phenomenon is consistent with the results observed in X-ray CT images. The interface between the new and old concrete has obvious directionality as a whole, which leads to the convergence of the weakest normal stress and shear stress in direction. However, due to the local heterogeneity, the similarity of the weakest vector only reaches 0.62. It can be inferred that if the size of the sampling unit is larger or the heterogeneity of the interface position is lower, the similarity will increase.

$$\begin{bmatrix}
1.0000 & 0.8084 & 0.6672 & -0.5367 & 0.7482 & 0.9723 & -0.7937 & 0.4153 & -0.6645 & 0.9907 \\
0.8084 & 1.0000 & 0.1235 & 0.0537 & 0.9831 & 0.9221 & -0.3045 & -0.1862 & -0.1130 & 0.8485 \\
0.6672 & 0.1235 & 1.0000 & -0.9752 & 0.0837 & 0.4888 & -0.9596 & 0.9510 & -0.9949 & 0.5933 \\
-0.5367 & 0.0537 & -0.9752 & 1.0000 & 0.1095 & -0.3346 & 0.9298 & -0.9865 & 0.9850 & -0.4673 \\
0.7482 & 0.9831 & 0.0837 & 0.1095 & 1.0000 & 0.8786 & -0.2422 & -0.2272 & -0.0625 & 0.7853 \\
0.9723 & 0.9221 & 0.4888 & -0.3346 & 0.8786 & 1.0000 & -0.6459 & 0.2019 & -0.4842 & 0.9840 \\
-0.7937 & -0.3045 & -0.9596 & 0.9298 & -0.2422 & -0.6459 & 1.0000 & -0.8622 & -0.9529 & -0.7523 \\
0.4153 & -0.1862 & 0.9510 & -0.9865 & -0.2272 & 0.2019 & -0.8622 & 1.0000 & -0.9529 & 0.3329 \\
-0.6645 & -0.1130 & -0.9949 & 0.9850 & -0.0625 & -0.4842 & 0.9734 & -0.9529 & 1.0000 & -0.6003 \\
0.9907 & 0.8485 & 0.5933 & -0.4673 & 0.7853 & 0.9840 & -0.7523 & 0.3329 & -0.6003 & 1.0000
\end{bmatrix} \tag{18}$$

Finally, the process identifies, for each vector, the maximum cosine similarity with any other vector, excluding itself, thereby indicating the closest directional alignment among the set. According to computation, the average value of maximum cosine similarity is 0.99, showing that every "weakest vector" has at least one other vector that has the same direction as itself. This phenomenon further confirms that the "weakest vectors" at the interface between new and old concrete have a high similarity in direction.

3.4. Discussion

The heterogeneous nature of the old–new concrete interface is a critical factor contributing to its anisotropic behavior in many rehabilitation projects. Due to differential shrinkage, thermal expansion, and the inherent incompatibilities between the old and new concrete layers, the interface exhibits directional variations in properties such as stiffness, strength, and crack resistance [3,4]. These variations are not uniformly distributed across the interface, leading to localized regions where the mechanical properties differ significantly from those of surrounding areas. This heterogeneity, when not adequately quantified, can result in unpredictable behavior under load, potentially compromising the integrity of the structure over time. Therefore, elastic moduli matrix, "weakest vectors", and similarity analysis are conducted to qualify the heterogeneity of the old–new concrete interface.

This section first shows a 6×6 calculated elastic moduli matrix of interface position and non-interface position based on finite element-based numerical homogenization [19,21,29]. As shown in (19), We compare the calculated elastic modulus coefficients with the results from Ref. [19], which reveals similar characteristics in the computed data. Specifically, the matrix exhibits a high degree of alignment, with significantly higher values along the main diagonal. This comparison indirectly supports the validity of the simulation results obtained in this study.

$$\begin{bmatrix} 36.605 & 8.425 & 8.379 & 0.069 & -0.183 & -0.048 \\ 8.425 & 33.015 & 8.216 & 0.022 & -0.024 & -0.018 \\ & & 32.927 & 0.008 & -0.047 & -0.005 \\ & & & 12.571 & -0.046 & -0.027 \\ & \text{Symmetric} & & & 12.517 & 0.005 \\ & & & & & 12.366 \end{bmatrix} \tag{19}$$

As for the results obtained in this study, the elasticity matrix comparison of interface and non-interface highlights significant differences between the non-interface and interface positions in terms of stiffness and anisotropy. In the non-interface region, the diagonal elements exhibit high and uniform values (e.g., C_{11} = 31,937.09, C_{22} = 31,193.96, C_{33} = 35,270.82), indicating greater axial stiffness and isotropic-like behavior due to better compaction and fewer voids. In contrast, the interface region shows lower stiffness in most directions (e.g., C_{11} = 18,970.84, C_{22} = 14,562.60), with notable anisotropy reflected in the relatively high C_{33} = 33,314.93, suggesting directional dependency caused by microstructural defects. Additionally, the off-diagonal elements further reveal differences in shear coupling and anisotropy; the non-interface region exhibits small values close to zero, reflecting minimal coupling and isotropic behavior, whereas the interface region displays greater variability with negative values (e.g., C_{14} = -26.32) and higher magnitudes (e.g., C_{35} = 734.93), indicative of internal stresses, localized anisotropy, and weaker bonding. These findings underscore the critical role of microstructural integrity, with non-interface regions being stiffer and more isotropic, while interface regions contribute to compliance and directional dependency due to their inherent weaknesses.

To effectively quantify the anisotropic properties of the concrete interface, this work innovatively proposed and identified the "weakest vector" index based on the research summary of the progress and gap of existing research [2,6,8,9]. These vectors represent directions in which the interface is most susceptible to failure, either due to weaker bonding, higher porosity, or pre-existing cracks. The identification of these weakest vectors is crucial for understanding how the interface will respond to external forces, as regions aligned with these vectors are more likely to experience failure under stress. The "weakest vector" analysis identifies the direction of minimum stiffness in a material by analyzing eigenvalues and eigenvectors of its stiffness matrix. This vector represents the material's weakest response under stress. Visualizing the weakest normal and shear stress directions as distinct arrows highlights the material's directional properties, particularly in anisotropic regions like the old–new concrete interface. The analysis reveals variations in the relative orientation of normal and shear stress components, reflecting the interface's complex stress behavior. This approach provides valuable insights into material anisotropy and stress distribution, enhancing the understanding of structural properties at critical interfaces. By analyzing these vectors, engineers can pinpoint critical regions where localized damage may initiate, thus providing valuable information for optimizing repair strategies and ensuring the long-term durability of the structure.

Once the "weakest vectors" are identified, it becomes necessary to quantify the degree of similarity between different regions of the interface. The similarity analysis of "weakest vectors" reveals the directional properties of the old–new concrete interface [2,34]. By calculating the cosine similarity of these vectors, the analysis quantifies their alignment and homogeneity. The average cosine similarity of 0.62 suggests a relatively high degree of alignment, indicating that the weakest vectors share similar directional tendencies. The weakest normal stress direction and the weakest shear stress direction of each sample are nearly orthogonal, which is physically reasonable. Furthermore, the low variance of 0.11 highlights minimal fluctuation in these alignments, signifying uniformity across samples. Each weakest vector has at least one other vector closely aligned with it, as demonstrated by the high average maximum cosine similarity of 0.99. These findings confirm that the weakest vectors exhibit consistent directional properties, reflecting the anisotropic nature of the interface.

The quantitative analysis of anisotropic properties in the old–new concrete interface requires a nuanced understanding of the directional dependence of the material's mechanical behavior.

Existing studies have demonstrated the existence of anisotropy at the concrete interface [8], but there is a lack of methods to quantify it. Traditional methods often oversimplify this complexity, but by incorporating more advanced techniques [7,16,17], a more accurate representation of the anisotropic nature of the interface can be achieved. These approaches enable a deeper insight into how varying material properties and structural irregularities at the interface influence the overall behavior of the composite material under different loading conditions. Furthermore, the introduction of similarity analysis not only explores the directional characteristics of the weakest vector but also eliminates the randomness of the weakest vector indicator in the calculation results. In other words, similar results are obtained on multiple samples, which can prove the effectiveness and rationality of the "weakest vector" index.

4. Conclusions

This study presents a novel framework for quantifying the anisotropic properties and micromechanical behavior of old–new concrete interfaces through X-ray CT and finite element-based numerical homogenization. The results highlight significant anisotropy in the interface, arising from microcracks, voids, and other heterogeneities, which challenge the assumptions of traditional design approaches that rely on homogeneity or empirical reduction factors. By modeling the homogenization of 4-phase concrete materials and generating the elastic moduli matrix, this study identifies the weakest directional vectors, offering a detailed analysis of the interface's heterogeneous properties. These findings provide critical insights into the internal microstructure of concrete, with important implications for improving the design, renovation, and reconstruction of concrete structures.

1. The elasticity matrix highlights significant differences between the interface and non-interface regions of old–new concrete. Non-interface regions exhibit higher axial stiffness (e.g., C_{11} = 31,937.09) and near-isotropic behavior due to better compaction and fewer defects, while interface regions show reduced stiffness (e.g., C_{11} = 18,970.84) and pronounced anisotropy caused by voids and microcracks;

2. Off-diagonal terms in the interface matrix, such as negative values (C_{14} = −26.32) and larger magnitudes (C_{35} = 734.93), reflect weaker bonding and irregular stress transfer. These findings underscore the interface's microstructural heterogeneity and directional compliance, which reduce load-bearing capacity and highlight the critical role of anisotropy in interface behavior;

3. The "weakest vectors" at the old–new concrete interface are investigated, representing directions of minimal resistance to normal and shear stress. The average angle of $88.64°$ between the weakest normal stress and weakest shear stress reflects a predominantly orthogonal relationship with some deviation due to local stress distributions;

4. Cosine similarity analysis reveals higher directional consistency at the interface, with an average similarity of 0.62 compared to 0.23 at non-interface positions, which is consistent with the results observed in X-ray CT images.

Author Contributions: Conceptualization, Y.L.; methodology, G.Z. and Y.L.; software, G.Z.; validation, G.Z.; formal analysis, G.Z.; investigation, G.Z.; resources, G.Z.; data curation, G.Z.; writing—original draft preparation, G.Z.; writing—review and editing, Y.L.; visualization, G.Z.; supervision, Y.L.; project administration, Y.L. All authors have read and agreed to the published version of the manuscript.

Funding: This research received no external funding.

Data Availability Statement: The datasets presented in this article are not readily available because the data are part of an ongoing PhD study. Requests to access the datasets should be directed to guanmingzhang@u.boisestate.edu.

Conflicts of Interest: The authors have no conflicts of interest.

References

1. Brown, R.E.; Willis, H.L. The economics of aging infrastructure. *IEEE Power Energy Mag.* **2006**, *4*, 36–43. [CrossRef]
2. Li, H. Microstructural Quantification, Property Prediction, and Stochastic Reconstruction of Heterogeneous Materials Using Limited X-Ray Tomography Data. Doctoral Dissertation, Arizona State University, Tempe, AZ, USA, 2017.
3. Liu, H.; Zou, H.; Zhang, J.; Zhang, J.; Tang, Y.; Zhang, J.; Xiao, J. Interface bonding properties of new and old concrete: A review. *Front. Mater.* **2024**, *11*, 1389785. [CrossRef]
4. Zhang, J.; Li, J.; Zhao, Y.; Wang, S.; Guan, Z. Concrete Cover Cracking and Reinforcement Corrosion Behavior in Concrete with New-to-Old Concrete Interfaces. *Materials* **2023**, *16*, 5969. [CrossRef]
5. Golewski, G.L. The phenomenon of cracking in cement concretes and reinforced concrete structures: The mechanism of cracks formation, causes of their initiation, types and places of occurrence, and methods of detection—A review. *Buildings* **2023**, *13*, 765. [CrossRef]
6. Kayondo, M.; Combrinck, R.; Boshoff, W.P. State-of-the-art review on plastic cracking of concrete. *Constr. Build. Mater.* **2019**, *225*, 886–899. [CrossRef]
7. Zhang, T.; Rahman, M.A.; Peterson, A.; Lu, Y. Novel Damage Index-Based Rapid Evaluation of Civil Infrastructure Subsurface Defects Using Thermography Analytics. *Infrastructures* **2022**, *7*, 55. [CrossRef]
8. Aaleti, S.; Sritharan, S. Quantifying bonding characteristics between UHPC and normal-strength concrete for bridge deck application. *J. Bridge Eng.* **2019**, *24*, 04019041. [CrossRef]
9. Austin, S.; Robins, P.; Pan, Y. Shear bond testing of concrete repairs. *Constr. Build. Mater.* **1999**, *13*, 249–258. [CrossRef]
10. Bentz, D.P.; De la Varga, I.; Muñoz, J.F.; Spragg, R.P.; Graybeal, B.A.; Hussey, D.S.; Jacobson, D.L.; Jones, S.Z.; LaManna, J.M. Influence of substrate moisture state and roughness on interface microstructure and bond strength: Slant shear vs. pull-off testing. *Constr. Build. Mater.* **2020**, *235*, 117491. [CrossRef]
11. Beushausen, H.; Höhlig, B.; Talotti, M. The influence of substrate moisture preparation on bond strength of concrete overlays and the microstructure of the OTZ. *Cem. Concr. Res.* **2016**, *91*, 2–11. [CrossRef]
12. Carbonell Muñoz, M.A.; Harris, D.K.; Ahlborn, T.M.; Froster, D.C. Bond performance between ultrahigh-performance concrete and normal-strength concrete. *J. Mater. Civ. Eng.* **2014**, *26*, 04014023. [CrossRef]
13. Courard, L.; Piotrowski, T.; Garbacz, A. Near-to-surface properties affecting bond strength in concrete repair. *Constr. Build. Mater.* **2014**, *65*, 496–505. [CrossRef]
14. Diab, A.M.; Abd Elmoaty, A.M.; Tag Eldin, M.R. Slant shear bond strength between self-compacting concrete and old concrete. *Constr. Build. Mater.* **2020**, *234*, 117387. [CrossRef]
15. Xiao, J.; Li, W.; Corr, D.J.; Shah, S.P. Effects of interfacial transition zones on the stress-strain behavior of modeled recycled aggregate concrete. *Cem. Concr. Res.* **2013**, *52*, 82–99. [CrossRef]
16. Boukhatem, B.; Kenai, S.; Tagnit-Hamou, A.; Ghrici, M. Application of new information technology on concrete: An overview. *J. Civ. Eng. Manag.* **2011**, *17*, 248–258. [CrossRef]

17. Vicente, M.A.; González, D.C.; Mínguez, J. Recent advances in the use of computed tomography in concrete technology and other engineering fields. *Micron* **2019**, *118*, 22–34. [CrossRef] [PubMed]
18. Fukuda, D.; Nara, Y.; Kobayashi, Y.; Maruyama, M.; Koketsu, M.; Hayashi, D.; Kaneko, K. Investigation of self-sealing in high-strength and ultra-low-permeability concrete in water using micro-focus X-ray CT. *Cem. Concr. Res.* **2012**, *42*, 1494–1500. [CrossRef]
19. Luo, Q.; Liu, D.; Qiao, P.; Zhou, Z.; Zhao, Y.; Sun, L. Micro-CT-based micromechanics and numerical homogenization for effective elastic property of ultra-high performance concrete. *Int. J. Damage Mech.* **2020**, *29*, 45–66. [CrossRef]
20. Wang, R.; Wu, H.; Zhao, M.; Liu, Y.; Chen, C. The classification and mechanism of microcrack homogenization research in cement concrete based on X-ray CT. *Buildings* **2022**, *12*, 1011. [CrossRef]
21. Sanahuja, J.; Toulemonde, C. Numerical homogenization of concrete microstructures without explicit meshes. *Cem. Concr. Res.* **2011**, *41*, 1320–1329. [CrossRef]
22. Moharekpour, M.; Liu, P.; Schmidt, J.; Oeser, M.; Jing, R. Evaluation of design procedure and performance of continuously reinforced concrete pavement according to AASHTO design methods. *Materials* **2022**, *15*, 2252. [CrossRef]
23. *ASTM C94/C94M-24c*; Standard Specification for Ready-Mixed Concrete. ASTM International: West Conshohocken, PA, USA, 2024.
24. *ASTM C150/C150M-20*; Standard Specification for Portland Cement. ASTM International: West Conshohocken, PA, USA, 2020.
25. *ASTM C143/C143M-20*; Standard Test Method for Slump of Hydraulic-Cement Concrete. ASTM International: West Conshohocken, PA, USA, 2020.
26. du Plessis, A.; Boshoff, W.P. A review of X-ray computed tomography of concrete and asphalt construction materials. *Constr. Build. Mater.* **2019**, *199*, 637–651. [CrossRef]
27. Denisiewicz, A.; Kuczma, M.; Kula, K.; Socha, T. Influence of boundary conditions on numerical homogenization of high performance concrete. *Materials* **2021**, *14*, 1009. [CrossRef]
28. Lu, Y.; Garboczi, E.J. Bridging the gap between random microstructure and 3D meshing. *J. Comput. Civ. Eng.* **2014**, *28*, 04014007. [CrossRef]
29. Wu, T.; Temizer, I.; Wriggers, P. Computational thermal homogenization of concrete. *Cem. Concr. Compos.* **2013**, *35*, 59–70. [CrossRef]
30. Mercier, S.; Molinari, A. Homogenization of elastic–viscoplastic heterogeneous materials: Self-consistent and Mori-Tanaka schemes. *Int. J. Plast.* **2009**, *25*, 1024–1048. [CrossRef]
31. Feng, Y.; Yong, H.; Zhou, Y. A concurrent multiscale framework based on self-consistent clustering analysis for cylinder structure under uniaxial loading condition. *Compos. Struct.* **2021**, *266*, 113827. [CrossRef]
32. Jiang, D. Stress, Strain, and Elasticity. In *Continuum Micromechanics: Theory and Application to Multiscale Tectonics*; Springer: Cham, Switzerland, 2023; pp. 57–96.
33. Deng, F.; Xu, L.; Cao, C.; Chi, Y. Effect of cellulose nanofiber addition on the microstructure characterization and nano-mechanical behavior of interfacial transition zones in recycled concrete. *J. Mater. Res. Technol.* **2024**, *33*, 7572–7585. [CrossRef]
34. Nemat-Nasser, S.; Lori, M.; Datta, S.K. Micromechanics: Overall Properties of Heterogeneous Materials. *ASM J. Appl. Mech.* **1996**, *63*, 561. [CrossRef]

 *infrastructures*

Article

Mechanical Properties of Pervious Recycled Aggregate Concrete Reinforced with Sackcloth Fibers (SF)

Arissaman Sangthongtong, Noppawan Semvimol, Thitima Rungratanaubon, Kittichai Duangmal and Panuwat Joyklad

1   Department of Environmental Science, Faculty of Environment, Kasetsart University, Bangkok 10900, Thailand
2   Research and Innovation Development Unit for Infrastructure and Rail Transportation Structural System (RIDIR), Faculty of Engineering, Srinakharinwirot University, Nakhonnayok 26120, Thailand
*   Correspondence: noppawan.sem@ku.th

Abstract: The excessive production of construction waste is a significant concern as it requires proper disposal and may become economically unfeasible. Reusing construction waste in producing new concrete can substantially reduce the disposal requirements of construction waste. In addition, this results in a sustainable solution for the rapidly depleting natural resources of concrete. Pervious concrete may contain up to 80% coarse aggregates and could be an exceptional host for reusing construction waste. This study aimed to investigate the mechanical properties of pervious concrete constructed with natural and recycled aggregates. The substandard properties of recycled aggregates were improved by adding natural fibers from sackcloth. This study presents an experimental program on 45 samples of pervious concrete with air void ratios and the size of coarse aggregates as the parameters of interest. The compressive strength of the pervious concrete decreased by increasing the air void ratio regardless of the size of the aggregates. The type of aggregates did not influence the permeability of pervious concrete, and the maximum temperature in pervious concrete increased as the quantity of air void ratios increased. The decrease in compressive strength was 40–60% as the void ratio was increased from 10–30% for all types of concrete mixes, such as natural and recycled aggregates. The permeability of small-size aggregates with 10% designed air void ratios for natural and recycled aggregates with sackcloth was 0.705 cm/s.

Keywords: pervious concrete; recycled aggregates; permeability; compressive strength; temperature

Citation: Arissaman Sangthongtong, Noppawan Semvimol, Thitima Rungratanaubon, Kittichai Duangmal and Panuwat Joyklad Mechanical Properties of Pervious Recycled Aggregate Concrete Reinforced with Sackcloth Fibers (SF). *Infrastructures* **2023**, *8*, 38. https://doi.org/10.3390/ infrastructures8020038

Academic Editor: Patricia Kara De Maeijer

Received: 23 January 2023
Revised: 14 February 2023
Accepted: 15 February 2023
Published: 18 February 2023

1. Introduction

In recent years, the reuse of construction waste in new construction works has been a topic of interest [1–8]. The use of recycled aggregates as an alternative to natural aggregates in the production of concrete addresses two key issues: (1) It eases the handling and proper disposal requirements of construction waste, and (2) the demand for the rapidly depleting natural resources may be reduced [9–14]. It has been estimated that the annual use of concrete will increase to 18 billion tons by the year 2050 [15]. Thus, effective ways of minimizing the excessive use of depleting natural resources for concrete production must be devised.

Porous Portland cement concrete (PPCC), also known as pervious concrete, is differentiated from Portland cement concrete (PCC) due to the presence of a large volume of air voids [16]. PPCC mainly finds its applications in permeable pavements and infiltration beds [17]. One of the best ways to utilize recycled aggregates could be in the production of PPCC. This is because PPCC comprises up to 80% of coarse aggregates with little or no fine aggregates [18]. Recycled aggregates typically possess lower density, higher porosity, and higher water absorption characteristics than natural aggregates. This may be attributed to the mortar adhered to the surface of recycled aggregates [19–21]. Consequently, recycled aggregates are mainly recommended for non-structural applications [22,23]. Thus, the use of recycled aggregates in porous pavements is subjected to the mechanical properties of the resulting mix as the pavements are designed to bear traffic loads. Natural sources, such as solar radiation, air temperature, and wind, are the catalyst to vary pavement temperatures periodically. The properties of comprising constituents in PPCC also influence the pavement temperature. As a result, pavements may exhibit different temperatures under the same environmental conditions. Hassan et al. [24] concluded that the temperature at the top and bottom surface of pavements increased with the increase in air void ratios. Albedo is a property that characterizes the amount of heat absorbed by the pavement. Li et al. [25] found that an increased albedo resulted in a decreased heat absorption characteristic of the pavement. Studies have shown that PPCC has lower albedo than PCC and demonstrates higher heat absorption characteristics [26,27]. It has been suggested that pavement temperature affects the surrounding environment, and an increased temperature variation may disturb the comfort of human bodies [28]. A strong correlation between air temperature and surface temperature has been observed [29].

Although in the past, many studies have been conducted on the mechanical properties of pervious concrete with natural stone aggregates [30–38]; however, no studies were found on mechanical properties of pervious recycled aggregate concrete reinforced with sackcloth fibers (SF). The mechanical properties of concrete constructed with recycled aggregates differ substantially from those constructed with natural aggregates. The present study aims to investigate the presence of recycled aggregates in pervious concrete, focusing on mechanical properties, such as compressive strength, permeability, and temperature variation inside the resulting PPCC. By acknowledging the substandard characteristics of the concrete constructed with recycled aggregates, natural fibers originating from sackcloth (sackcloth fibers SF) with a weight of 3% of the cement content were added to the concrete mix in the presence of recycled aggregates. Therefore, the mechanical properties of PPCC with natural aggregates, recycled aggregates, and recycled aggregates strengthened with sackcloth natural fibers were investigated in the present study.

2. Materials and Methods

2.1. Test Matrix

In this study, three groups of specimens were tested. Each group was comprised of fifteen specimens. Specimens in Group 1 were constructed with pervious Portland cement concrete with 100% natural aggregates, whereas 100% of natural coarse aggregates in Group 2 specimens were replaced with recycled aggregates. The recycled aggregates were obtained from locally available construction waste. Specimens in Group 3 were also comprised of 100% recycled coarse aggregates. In Thailand, usually sackcloth bags are used for the curing of concrete at the construction site. In this study, used sackcloth bags were collected and cut into small lengths to develop natural sackcloth fibers (SF). The sackcloth fibers (SF) were added to the concrete mix to enhance the mechanical properties of the PPCC mix. The weight of the natural sackcloth fibers was kept as 3% of the weight of the cement used. The air void ratios in each group varied from 10% to 30%, with an increment of 5%. Thus, five samples in each group were constructed with air void ratios of 5%, 10%, 15%, 20%, and 25% each. Further, three samples were tested corresponding to each air void ratio. Three samples corresponding to each void ratio were differentiated depending on the size of coarse aggregates. Three sizes of coarse aggregates were used, namely large (L), medium (M), and small (S). The large coarse aggregates referred to the size retained on sieve #3/8 inches. The medium size referred to the aggregates with 50% passing through the sieve #3/8 inches and 50% retained on no. 4 sieve, whereas small aggregates were the ones that passed 100% from the no. 4 sieve. The details of the test specimens are presented in Table 1. Specimen name in each group was differentiated depending on the type of coarse aggregates and the presence of sackcloth. Natural, recycled, and recycled aggregates strengthened with sackcloth were recognized as "NA", "RA", and "RA-SF", respectively. Further, specimens within each group were differentiated depending on the air void contents. For instance, Vv10 represented a specimen with a 10% air void ratio.

Table 1. Details of test specimens.

ID	Coarse Aggregate Type			
	Natural (%)	Recycled (%)	Recycled with Sackcloth (%)	Air Void Ratio (%)
NA-Vv10	100	-	-	10
NA-Vv15	100	-	-	15
NA-Vv20	100	-	-	20
NA-Vv25	100	-	-	25
NA-Vv30	100	-	-	30
RA-Vv10	-	100	-	10
RA-Vv15	-	100	-	15
RA-Vv20	-	100	-	20
RA-Vv25	-	100	-	25
RA-Vv30	-	100	-	30
RA-Vv10-SF	-	-	100	10
RA-Vv15- SF	-	-	100	15
RA-Vv20- SF	-	-	100	20
RA-Vv25- SF	-	-	100	25
RA-Vv30- SF	-	-	100	30

All specimens were constructed using a water-to-cement ratio of 0.30. The volume of cement paste was constant at 20% in all specimens. The mix details are provided in Table 2. Further, type-I Portland cement was used in the construction of all specimens. Locally available natural aggregates from Saraburi province in Thailand were used to construct Group 1 specimens. The recycled aggregates were acquired from recycled structural concrete with compressive strength ranging from 23.5 MPa to 40.0 MPa. Various tests were performed to obtain the mechanical properties of coarse aggregates. Figure 1 shows the process of obtaining recycled aggregates from the structural concrete waste. A typical sackcloth is shown in Figure 2.

Table 2. Details of mix components (kg).

ID	Cement (kg)	Water	Fiber	Coarse Aggregates		Total Weight
				NA	RA	
NA-Vv10	279	112	-	1771	-	2162
NA-Vv15	279	112	-	1645	-	2036
NA-Vv20	279	112	-	1518	-	1909
NA-Vv25	279	112	-	1392	-	1783
NA-Vv30	279	112	-	1265	-	1656
RA-Vv10	279	112	-	-	1771	2162
RA-Vv15	279	112	-	-	1645	2036
RA-Vv20	279	112	-	-	1518	1909
RA-Vv25	279	112	-	-	1392	1783
RA-Vv30	279	112	-	-	1265	1656
RA-Vv10-SF	276	110	3	-	1771	2160
RA-Vv15-SF	276	110	3	-	1645	2034
RA-Vv20-SF	276	110	3	-	1518	1907
RA-Vv25-SF	276	110	3	-	1392	1781
RA-Vv30-SF	276	110	3	-	1265	1654

Figure 1. Acquisition of recycled aggregates in process.

Figure 2. Typical sackcloth.

2.2. Unit Weight and Water Absorption

The unit weight and water absorption of the aggregates were determined by following the recommendations of ASTM C29 [39]. Table 3 presents the details of the process to determine the unit weight and water absorption characteristics of natural and recycled aggregates. Three trials were performed for both natural and recycled aggregates. The average values of unit weight and water absorption are reported in Table 3. The average unit weight of natural coarse aggregates was 1493.9 kg/m^3, whereas the unit weight of recycled aggregates was estimated at 1358.2 kg/m^3, which is 9.81% lower than the unit of natural aggregates. Similarly, the water absorption of recycled aggregates was 4.52%, which was 511% higher than the water absorption of natural aggregates.

Table 3. Unit weight and water absorption characteristics of natural and recycled aggregates.

Type of Coarse Aggregate	Natural Aggregate			Recycle Aggregate		
Weight of Measuring Cylinder (g)	2435.62					
Weight of Cylinder and Water (g)	5142.29					
Weight of Water (g)	2706.67					
Volume of Measuring Cylinder (m^3)	0.00302					
Weight of Cylinder + Sample (kg)	6954.2	6897.53	6984.78	6535.2	6474.5	6598.5
Weight of Sample Alone	4518.6	4461.91	4549.16	4099.6	4038.9	4162.9
Unit Weight of Sample (kg/m^3)	1496.7	1477.91	1506.81	1357.9	1337.8	1378.9
Avg. Unit Weight of Sample (kg/m^3)	1493.9			1358.2		
Weight of Sample in SSD (g)	2010.3	2020.65	2014.39	2035.7	2042.6	2056.4
Weight of Sample in Water (g)	1184.6	1201.25	1194.58	1358.5	1379.7	1345.6
Weight of Sample in Air (g)	1996.3	2005.36	1999.26	1945.7	1958.5	1965.1
Oven Dry Bulk Specific Gravity	2.42	2.447	2.439	2.87	2.95	2.76
SSD Bulk Specific Gravity	2.43	2.466	2.457	3.01	3.08	2.89
Apparent Specific Gravity	2.46	2.494	2.485	3.31	3.38	3.17
Percentage of Absorption (%)	0.70	0.76	0.76	4.63	4.29	4.64
Avg. Percentage of Absorption (%)	0.74			4.52		

2.3. Sieve Analysis of Coarse Aggregates

The determination of fineness modulus through sieve analysis results in the average size of the aggregates. Sieve analysis was performed in accordance with ASTM C136/C136M-19 [40]. Figure 3a,b presents the sieve analysis results performed on recycled and natural aggregates, respectively. The fineness modulus of recycled and natural aggregates was estimated at 6.003 and 6.350, respectively. This suggests that the average size of both recycled and natural aggregates was between 4.75 mm and 10 mm.

2.4. Resistance to Abrasion and Impact

The relative quality of natural and recycled aggregates was determined by performing the Los Angeles test in accordance with ASTM C131/C131M-20 [41]. The results of the Los Angeles test performed on recycled and natural aggregates are shown in Table 4. The abrasion values of natural and recycled aggregates were 23.45% and 30.86%, respectively. This suggests that recycled aggregates exhibited a 31.60% higher abrasion tendency than natural aggregates.

Table 4. Unit weight and water absorption characteristics of natural and recycled aggregates.

Type of Coarse Aggregate	Natural Aggregate			Recycle Aggregate		
	1	2	3	1	2	3
Number of Spheres	12	12	12	12	12	12
Weight of Sample (g)	5000.21	5000.58	5000.12	5001.04	5000.48	5000.13
Weight of Sample in Pan (g)	3874.62	3786.94	3822.19	3465.47	3418.69	3487.77
Abrasion (%)	22.511	24.270	23.558	30.705	31.633	30.246
Avg. Abrasion (%)	23.446			30.861		

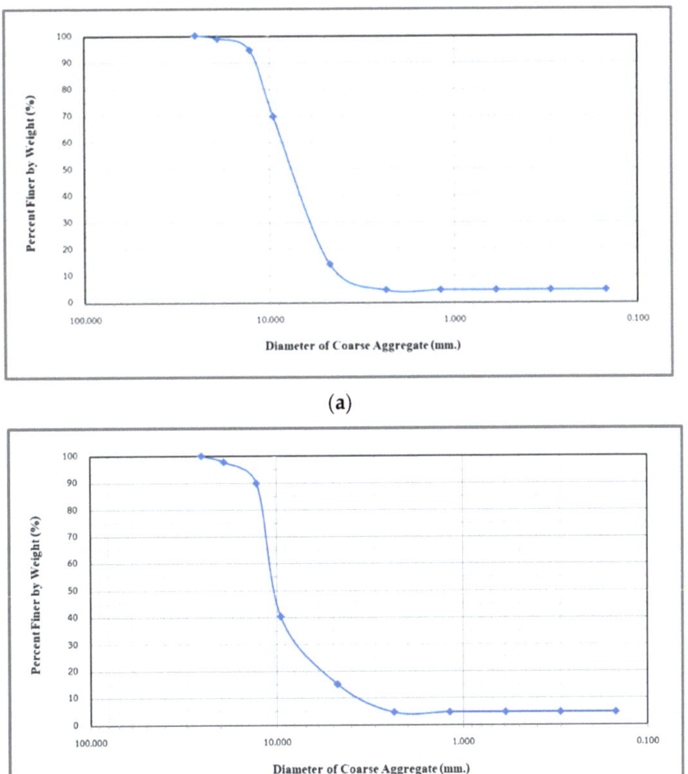

(**a**)

(**b**)

Figure 3. Results of sieve analysis on (**a**) recycled coarse aggregates and (**b**) natural coarse aggregates.

2.5. Temperature Measurement

The present study investigated the temperature of porous concrete by using thermo-couples. To replicate the actual conditions of pavement, specimens were embedded inside the ground, as shown in Figure 4. The temperature at the top surface, mid-height, and bottom of specimens was monitored. The temperature corresponding to the mid-height represented the temperature inside the specimen. Thermocouple wires were attached to the outer surfaces at the top and bottom of specimens. Thermocouples were connected to a logger to record continuous temperature measurements. The measured temperatures reported in this study are the average of the temperatures recorded at three positions.

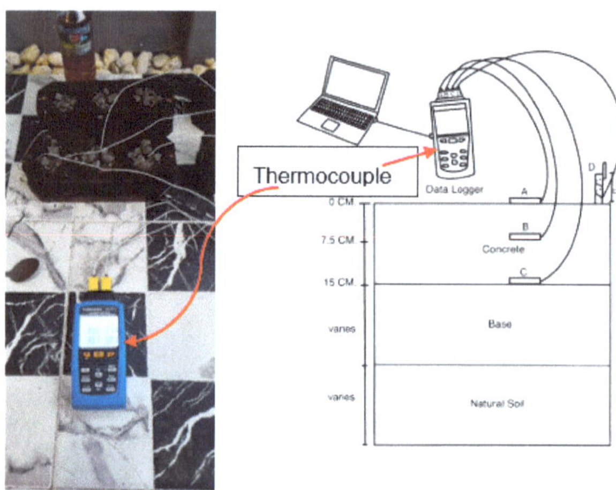

Figure 4. Temperature measurement scheme.

3. Results

3.1. Compressive Strength Results

Compressive strength for each specimen was determined on the 7th and 28th day of casting, as per the recommendations of ASTM C39/C39M [42]. A summary of the results of compressive strength is presented in Table 5. The typical failure modes of specimens are shown in Figure 5. All specimens experienced sudden failure due to the crushing of concrete. The aggregates started peeling off the specimen with a gentle sound near the ultimate load. The crushing resulted in the separation of coarse aggregates that is considerably different from the splitting and cone formations in PCC. A similar failure was observed in pervious concrete elsewhere [43].

Table 5. Unit weight and water absorption characteristics of natural and recycled aggregates.

	Compressive Strength (KSC)					
■ Sample	S (Days)		M (Days)		L (Days)	
	7	28	7	28	7	28
■ NA-Vv10	54.177	59.863	84.547	94.610	33.383	37.147
■ NA-Vv15	51.517	57.570	79.697	87.783	33.570	37.433
■ NA-Vv20	41.223	45.840	77.090	84.953	26.707	29.710
■ NA-Vv25	29.420	32.920	67.203	74.633	18.223	20.103
■ NA-Vv30	23.733	26.037	56.820	62.330	16.870	18.550
■ RA-Vv10	57.357	64.700	90.283	101.340	39.237	43.763
■ RA-Vv15	56.913	63.627	89.360	98.297	40.153	44.813

Table 5. *Cont.*

	Compressive Strength (KSC)					
■ Sample	S (Days)		M (Days)		L (Days)	
	7	28	7	28	7	28
■ RA-Vv20	44.420	49.750	86.607	95.277	35.623	39.453
■ RA-Vv25	28.513	31.680	70.437	78.667	23.743	26.403
■ RA-Vv30	28.913	32.130	61.533	68.703	18.123	19.997
■ RA-Vv10-SF	62.647	70.417	93.630	105.080	43.533	49.100
■ RA-Vv15-SF	59.290	66.083	93.290	103.617	41.793	46.717
■ RA-Vv20-SF	48.823	53.987	89.333	99.230	37.757	41.853
■ RA-Vv25-SF	36.457	41.017	74.337	83.720	24.590	27.633
■ RA-Vv30-SF	30.603	33.777	62.507	68.923	18.303	20.310

Figure 5. Ultimate failure modes of cube specimens.

The variation of the compressive strength of test specimens with air void ratios for different aggregate sizes is shown in Figure 6. It is evident that the compressive strength decreased as the air void ratio increased, regardless of the size of the aggregates. Aggregates corresponding to small and large size referred to the size retained on the #3/8-inch and no. 4 sieves, respectively. Hence, a uniform grading was achieved in small- and large-size aggregates. On the contrary, medium-size aggregates referred to the combination of sizes with 50% retained on the #3/8-inch sieve and remaining retained on the no. 4 sieve. As a result, a better grading was achieved in medium-size aggregates. This is reflected in the compressive strength results of the medium-size aggregates as compared to the large- and small-size aggregates, as shown in Figure 6. It is observed that the compressive strength of medium-size aggregates was highest among small- and large-size aggregates at all air void ratios and irrespective of the type of aggregates. For uniformly graded aggregates, the strength of PPCC decreased as the size of coarse aggregates increased. A similar trend was also observed in existing studies on pervious concrete [44–46]. Zhong and Wille [47] suggested that an increased aggregate size increased the pore size. As a larger pore size reduces the number of pores per unit volume for a constant porosity, the bonding area of the coarse aggregates reduces, which results in a lower compressive strength.

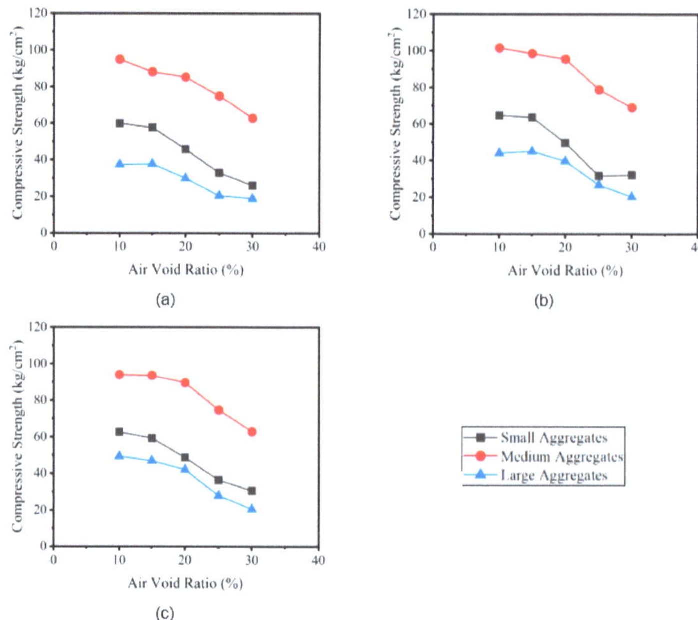

Figure 6. Variation of compressive strength (at 28th day) of PPCC with an air void ratio: (**a**) natural aggregate PPCC, (**b**) recycled aggregate PPCC, and (**c**) recycled aggregate PPCC strengthened with sackcloth fibers.

3.2. Permeability Results

An adequate permeability is desired from pervious concrete when used in pavements. Permeability tests were performed by following the recommendations of ASTM D5084-16a [48]. The calculated air void ratios and permeabilities are presented in Table 6 for different aggregate sizes. Figure 7 shows a relationship between the permeability and air void ratios for different aggregate sizes and types. It is interesting to observe that the type of aggregates did not influence the permeability of PPCC. For instance, the permeability of small-size aggregates with 10% designed air void ratios for natural, recycled, and recycled aggregates with sackcloth was 0.705 cm/s. It was found that the permeability increased with the air void ratios in all aggregate types and sizes. Further, the permeability of large-size aggregates was highest for a specific specimen type. The permeability of small- and medium-size aggregates did not differ considerably. This can be attributed to the better grading of medium-size aggregates as compared to that of small- and large-size aggregates.

3.3. Temperature Variation of PPCC

The temperature variation of PPCC was recorded for 24 h starting from 9:00 A.M. Table 7 presents the hourly temperature measurements for specimens with 10% air void ratios. Figure 8 plots the temperature variation at different time intervals. It is evident that the temperature variation trend was identical in all aggregate types and sizes. The upper bound on temperature was created by large-size aggregates, whereas the temperature in small- and medium-sized aggregate PPCC could not be separated. A similar observation on temperature measurement in PPCC was also reported elsewhere [28]. The maximum temperature was recorded at 15:00 h for all specimen types. The minimum temperature was reported around 6:00 h for all specimen types. It is interesting to observe that the increase in temperature was sudden beyond 6:00 h for specimens with large-size aggregate. Figure 9 provides the temperature variation of specimens with void ratios ranging from

15% to 30%. The temperature variation trend was identical for all void ratios with large-size aggregate specimens creating the upper and lower bounds of the recorded temperature.

Table 6. Unit weight and water absorption characteristics of natural and recycled aggregates.

■ Sample	Air Void Ratio (%)			Permeability (cm/s)		
	S	M	L	S	M	L
■ NA-Vv10	7.844	7.920	9.415	0.705	0.702	1.019
■ NA-Vv15	12.949	12.845	14.176	0.995	0.997	1.294
■ NA-Vv20	16.386	16.555	18.101	1.276	1.268	1.464
■ NA-Vv25	21.388	21.350	24.305	1.455	1.444	1.609
■ NA-Vv30	26.352	26.465	28.686	1.615	1.611	1.779
■ RA-Vv10	7.456	7.465	9.250	0.705	0.700	1.018
■ RA-Vv15	12.652	12.540	14.403	0.995	0.996	1.294
■ RA-Vv20	16.559	16.487	18.563	1.276	1.268	1.463
■ RA-Vv25	21.597	21.538	23.914	1.457	1.447	1.611
■ RA-Vv30	27.426	27.229	28.241	1.621	1.613	1.777
■ RA-Vv15-SF	12.310	12.437	14.275	0.996	0.995	1.294
■ RA-Vv20-SF	16.328	16.372	19.153	1.274	1.268	1.464
■ RA-Vv25-SF	21.260	21.394	24.441	1.456	1.446	1.609
■ RA-Vv30-SF	26.238	26.363	28.456	1.634	1.611	1.777

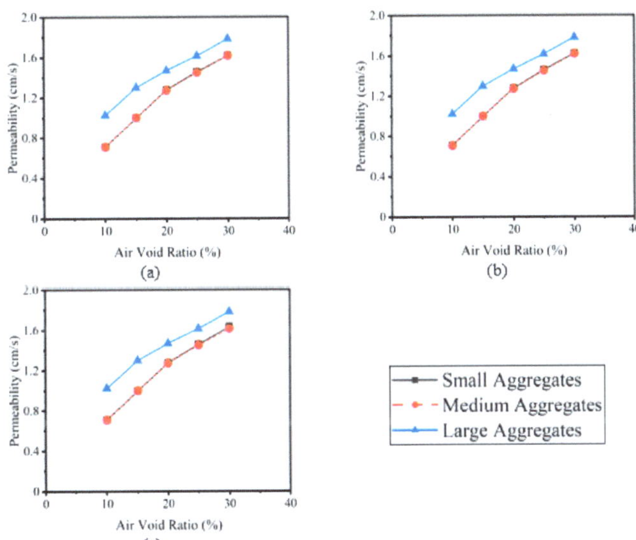

Figure 7. Variation of permeability of PPCC with an air void ratio: (**a**) natural aggregate PPCC, (**b**) recycled aggregate PPCC, and (**c**) recycled aggregate PPCC strengthened with sackcloth fibers.

Table 7. Unit weight and water absorption characteristics of natural and recycled aggregates.

Time	Vv10 (°C)								
	NA-S	RA-S	RA-SF-S	NA-M	RA-M	RA-SA-M	NA-L	RA-L	RA-SA-L
09:00	31.638	31.784	32.087	31.748	31.740	32.120	32.110	32.107	32.569
10:00	32.267	32.201	32.100	32.258	32.215	32.136	32.593	32.544	32.620
11:00	33.544	33.521	33.501	33.561	33.510	33.535	33.856	33.853	34.141
12:00	36.568	36.473	37.217	36.549	36.454	37.271	37.842	37.848	37.969
13:00	40.323	40.460	40.571	40.388	40.552	40.508	41.605	41.621	41.811
14:00	42.304	42.391	42.018	42.503	42.406	42.052	42.669	42.667	42.862
15:00	43.084	43.091	43.235	43.010	43.187	43.274	43.651	43.635	43.789
16:00	42.556	42.497	42.789	42.510	42.532	42.718	43.492	43.414	43.485
17:00	40.895	40.949	40.974	40.909	40.984	40.948	42.876	42.802	42.836
18:00	38.932	38.998	38.901	38.975	38.988	38.947	39.899	39.990	40.211
19:00	36.936	36.947	37.189	37.019	36.958	37.225	37.453	37.560	37.664
20:00	34.945	34.888	35.120	35.111	34.908	35.213	35.624	35.648	35.756
21:00	33.277	33.383	33.867	33.223	33.376	33.809	33.129	33.144	33.276
22:00	31.914	31.860	32.016	31.937	31.993	32.134	32.675	32.601	32.739
23:00	30.710	30.684	30.883	30.743	30.668	30.886	31.270	31.256	31.347
0:00	29.920	29.917	29.980	29.959	29.996	29.997	29.973	29.925	30.062
1:00	29.178	29.267	29.339	29.106	29.261	29.336	29.241	29.263	29.370
2:00	28.634	28.620	28.894	28.696	28.660	28.820	28.736	28.779	28.836
3:00	28.024	28.020	27.973	28.114	28.186	27.906	27.992	27.894	27.943
4:00	27.551	27.531	27.793	27.506	27.563	27.779	27.167	27.226	27.365
5:00	26.974	26.921	26.889	26.994	26.939	26.874	26.869	26.843	26.951
6:00	26.550	26.598	26.609	26.605	26.530	26.674	26.345	26.337	26.447
7:00	27.282	27.243	27.568	27.246	27.285	27.580	30.275	30.269	30.364
8:00	28.151	28.197	28.541	28.121	28.247	28.513	31.147	31.156	31.264
9:00	29.550	29.641	29.670	29.546	29.678	29.547	32.198	32.211	32.354

The variation of peak temperature with air void ratios is shown in Figure 10. It is interesting to observe that the maximum temperature of PPCC increased with the air void ratio. This trend was the same in all specimen types, irrespective of the type and size of the aggregates.

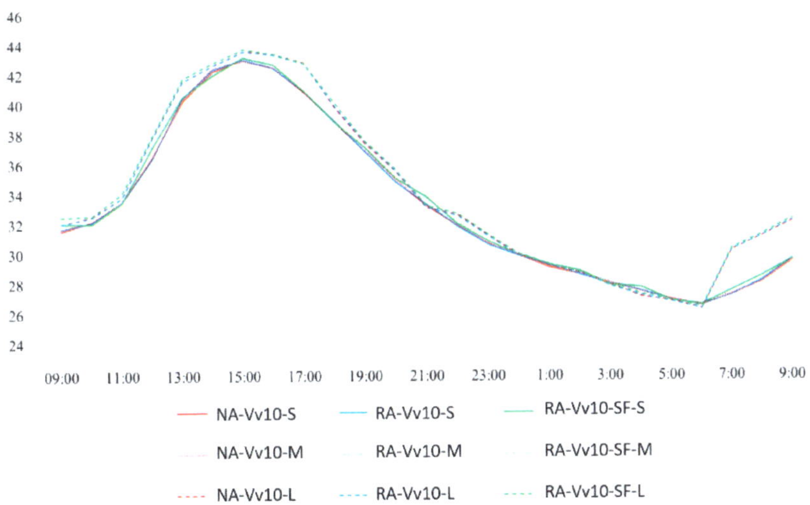

Figure 8. Variation of temperature in specimens with a 10% air void ratio.

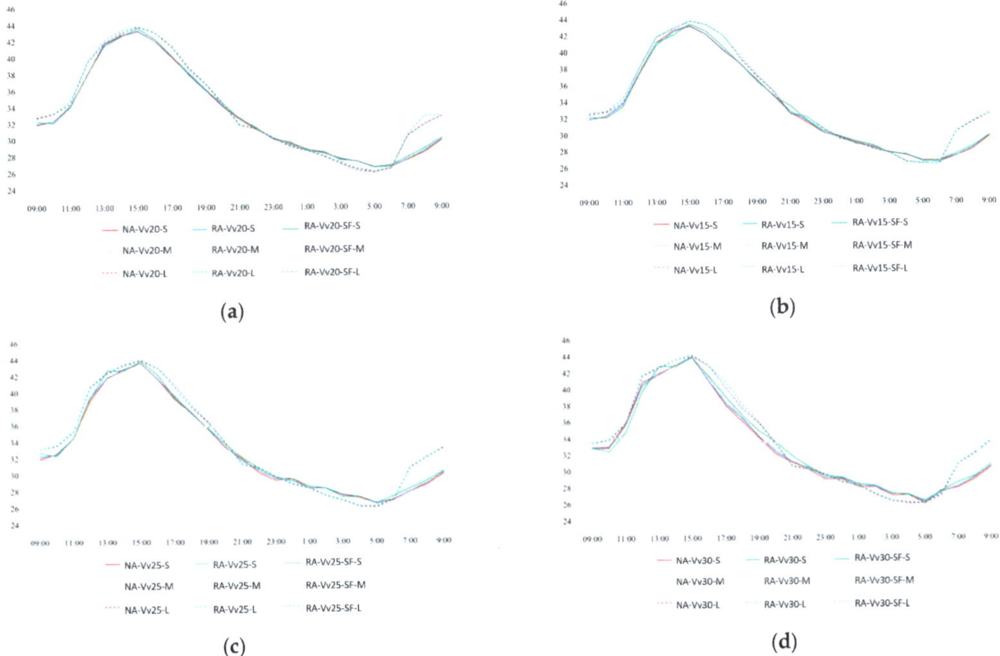

Figure 9. Temperature variation in PPCC specimens with (**a**) a 15% air void ratio, (**b**) a 20% air void ratio, (**c**) a 25% air void ratio, and (**d**) a 30% air void ratio.

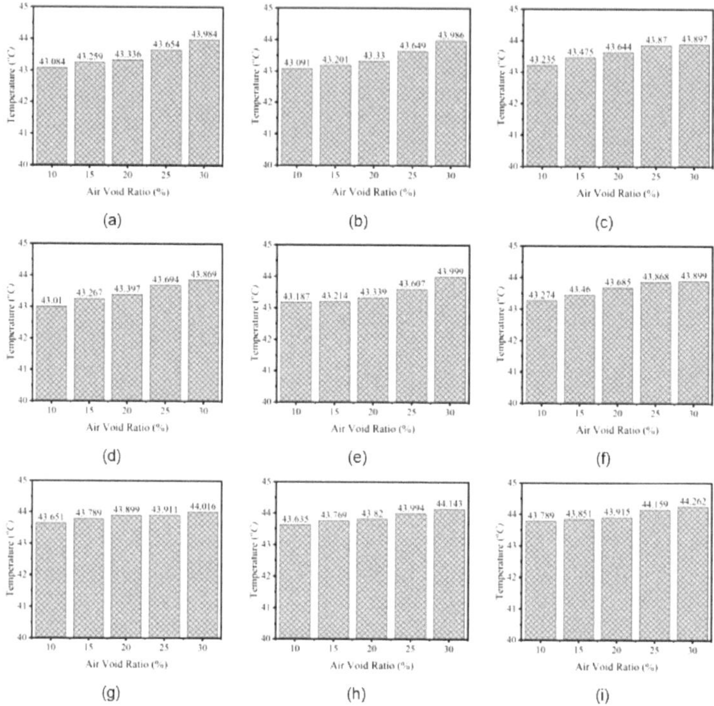

Figure 10. Effect of air void ratios on the maximum temperature of PPCC: (**a**) NA-S, (**b**) RA-S, (**c**) RA-SF-S, (**d**) NA-M, (**e**) RA-M, (**f**) RA-SF-M, (**g**) NA-L, (**h**) RA-L, and (**i**) RA-SF-L.

4. Conclusions

This study aimed to investigate the mechanical properties of pervious concrete constructed with natural and recycled aggregates. The substandard properties of recycled aggregates were improved by adding natural fibers from sackcloth. This study presents an experimental program on 45 samples of pervious concrete with air void ratios and the size of coarse aggregates as the parameters of interest. Experimental results in terms of compressive strength, permeability, and temperature variation were reported. The following important conclusions can be deduced.

1. The compressive strength decreased as the air void ratio increased, irrespective of the size of aggregates. Three sizes of aggregates were tested, namely small, medium, and large. The compressive strength of PPCC specimens with large-size aggregates was consistently lower than that of specimens with small-size aggregates. A better grading of aggregates was used in medium-size aggregates as compared to the uniform grading in small and large-size aggregates. As a result, the compressive strength of PPCC specimens with medium-size aggregates was the highest.

2. The type of aggregates did not influence the permeability of PPCC. For instance, the permeability of small-size aggregates with 10% designed air void ratios for natural, recycled, and recycled aggregates with sackcloth was 0.705 cm/s. It was found that the permeability increased with air void ratios in all aggregate types and sizes.

3. The temperature variation trend was identical in all aggregate types and sizes. The upper bound on temperature was created by large-size aggregates, whereas the temperature in small- and medium-sized aggregate PPCC could not be separated.

4. The maximum temperature of PPCC increased as the quantity of air void ratios increased. This trend was same in all specimen types, irrespective of the type and size of aggregates.

Author Contributions: Conceptualization, A.S., N.S., T.R., K.D. and P.J.; Methodology, A.S., N.S., T.R., K.D. and P.J.; Resources, T.R.; Writing–original draft, A.S., N.S., T.R. and K.D; Writing–review & editing, A.S., N.S. and K.D. All authors have read and agreed to the published version of the manuscript.

Funding: This research received no external funding.

Data Availability Statement: Not applicable.

Acknowledgments: This research was successfully accomplished with the assistance of government officials' officers of the Directorate of civil engineering, Royal Thai Air Force that has facilitated access to the sackcloth testing tool, including thanks to the Department of Civil Engineering, Nawaminthakasatriyathirat Air Force Academy for the funding and the location that includes the compression testing tool for cube concrete blocks. The researcher would like to thank everyone for this opportunity.

Conflicts of Interest: The authors declare no conflict of interest.

References

1. Poon, C.S.; Kou, S.C.; Lam, L. Use of Recycled Aggregates in Molded Concrete Bricks and Blocks. *Constr. Build. Mater.* **2002**, *16*, 281–289. [CrossRef]
2. Tabsh, S.W.; Abdelfatah, A.S. Influence of Recycled Concrete Aggregates on Strength Properties of Concrete. *Constr. Build. Mater.* **2009**, *23*, 1163–1167. [CrossRef]
3. Thomas, C.; Setién, J.; Polanco, J.A.; Alaejos, P.; Sánchez De Juan, M. Durability of Recycled Aggregate Concrete. *Constr. Build. Mater.* **2013**, *40*, 1054–1065. [CrossRef]
4. Gao, C.; Huang, L.; Yan, L.; Kasal, B.; Li, W. Behavior of Glass and Carbon FRP Tube Encased Recycled Aggregate Concrete with Recycled Clay Brick Aggregate. *Compos. Struct.* **2016**, *155*, 245–254. [CrossRef]
5. Nováková, I.; Mikulica, K. Properties of Concrete with Partial Replacement of Natural Aggregate by Recycled Concrete Aggregates from Precast Production. *Procedia Eng.* **2016**, *151*, 360–367. [CrossRef]
6. Hamad, B.S.; Dawi, A.H. Sustainable Normal and High Strength Recycled Aggregate Concretes Using Crushed Tested Cylinders as Coarse Aggregates. *Case Stud. Constr. Mater.* **2017**, *7*, 228–239. [CrossRef]
7. Reza, F.; Wilde, W.J.; Izevbekhai, B. Sustainability of Using Recycled Concrete Aggregates as Coarse Aggregate in Concrete Pavements. *Transp. Res. Rec. J. Transp. Res. Board* **2018**, *2672*, 99–108. [CrossRef]
8. Ohemeng, E.A.; Ekolu, S.O. Comparative Analysis on Costs and Benefits of Producing Natural and Recycled Concrete Aggregates: A South African Case Study. *Case Stud. Constr. Mater.* **2020**, *13*, e00450. [CrossRef]
9. Noguchi, T.; Kitagaki, R.; Tsujion, M. Minimizing Environmental Impact and Maximizing Performance in Concrete Recycling. *Struct. Concr.* **2011**, *12*, 36–46. [CrossRef]
10. Zeybek, Ö.; Özkılıç, Y.O.; Karalar, M.; Çelik, A.İ.; Qaidi, S.; Ahmad, J.; Burduhos-Nergis, D.D.; Burduhos-Nergis, D.P. Influence of replacing cement with waste glass on mechanical properties of concrete. *Materials* **2022**, *15*, 7513. [CrossRef]
11. Karalar, M.; Bilir, T.; Çavuşlu, M.; Özkılıç, Y.O.; Sabri, M.M.S. Use of recycled coal bottom ash in reinforced concrete beams as replacement for aggregate. *Front. Mater.* **2022**, *9*, 1064604. [CrossRef]
12. Qaidi, S.; Najm, H.M.; Abed, S.M.; Özkılıç, Y.O.; Al Dughaishi, H.; Alosta, M.; Sabri, M.M.S.; Alkhatib, F.; Milad, A. Concrete containing waste glass as an environmentally friendly aggregate: A review on fresh and mechanical characteristics. *Materials* **2022**, *15*, 6222. [CrossRef]
13. Çelik, A.İ.; Özkılıç, Y.O.; Zeybek, Ö.; Karalar, M.; Qaidi, S.; Ahmad, J.; Burduhos-Nergis, D.D.; Bejinariu, C. Mechanical Behavior of Crushed Waste Glass as Replacement of Aggregates. *Materials* **2022**, *15*, 8093. [CrossRef]
14. Karalar, M.; Özkılıç, Y.O.; Aksoylu, C.; Sabri MM, S.; Alexey, N.B.; Sergey, A.S.; Evgenii, M.S. Flexural behavior of reinforced concrete beams using waste marble powder towards application of sustainable concrete. *Front. Mater.* **2022**, *701*. [CrossRef]
15. Meng, T.; Zhang, J.; Wei, H.; Shen, J. Effect of Nano-Strengthening on the Properties and Microstructure of Recycled Concrete. *Nanotechnol. Rev.* **2020**, *9*, 79–92. [CrossRef]
16. Lian, C.; Zhuge, Y.; Beecham, S. The Relationship between Porosity and Strength for Porous Concrete. *Constr. Build. Mater.* **2011**, *25*, 4294–4298. [CrossRef]
17. Beecham, S. Water Sensitive Urban Design: A Technological Assessment. *J. Stormwater Ind. Assoc.* **2003**, *17*, 5–13.
18. Aamer Rafique Bhutta, M.; Hasanah, N.; Farhayu, N.; Hussin, M.W.; Tahir, M.B.M.; Mirza, J. Properties of Porous Concrete from Waste Crushed Concrete (Recycled Aggregate). *Constr. Build. Mater.* **2013**, *47*, 1243–1248. [CrossRef]
19. Li, X. Recycling and Reuse of Waste Concrete in China: Part I. Material Behaviour of Recycled Aggregate Concrete. *Resour. Conserv. Recycl.* **2008**, *53*, 36–44. [CrossRef]
20. Etxeberria, M.; Vázquez, E.; Marí, A.; Barra, M. Influence of Amount of Recycled Coarse Aggregates and Production Process on Properties of Recycled Aggregate Concrete. *Cem. Concr. Res.* **2007**, *37*, 735–742. [CrossRef]

21. Henry, M.; Pardo, G.; Nishimura, T.; Kato, Y. Balancing Durability and Environmental Impact in Concrete Combining Low-Grade Recycled Aggregates and Mineral Admixtures. *Resour. Conserv. Recycl.* **2011**, *55*, 1060–1069. [CrossRef]
22. Courard, L.; Michel, F.; Delhez, P. Use of Concrete Road Recycled Aggregates for Roller Compacted Concrete. *Constr. Build. Mater.* **2010**, *24*, 390–395. [CrossRef]
23. Leite, F.D.C.; Motta, R.D.S.; Vasconcelos, K.L.; Bernucci, L. Laboratory Evaluation of Recycled Construction and Demolition Waste for Pavements. *Constr. Build. Mater.* **2011**, *25*, 2972–2979. [CrossRef]
24. Hassn, A.; Aboufoul, M.; Wu, Y.; Dawson, A.; Garcia, A. Effect of Air Voids Content on Thermal Properties of Asphalt Mixtures. *Constr. Build. Mater.* **2016**, *115*, 327–335. [CrossRef]
25. Li, H.; Harvey, J.T.; Holland, T.J.; Kayhanian, M. The Use of Reflective and Permeable Pavements as a Potential Practice for Heat Island Mitigation and Stormwater Management. *Environ. Res. Lett.* **2013**, *8*, 015023. [CrossRef]
26. Haselbach, L.; Boyer, M.; Kevern, J.T.; Schaefer, V.R. Cyclic Heat Island Impacts on Traditional versus Pervious Concrete Pavement Systems. *Transp. Res. Rec. J. Transp. Res. Board* **2011**, *2240*, 107–115. [CrossRef]
27. Zhang, R.; Jiang, G.; Liang, J. The Albedo of Pervious Cement Concrete Linearly Decreases with Porosity. *Adv. Mater. Sci. Eng.* **2015**, *2015*, 746592. [CrossRef]
28. Hu, L.; Li, Y.; Zou, X.; Du, S.; Liu, Z.; Huang, H. Temperature Characteristics of Porous Portland Cement Concrete during the Hot Summer Session. *Adv. Mater. Sci. Eng.* **2017**, *2017*, 2058034. [CrossRef]
29. Benrazavi, R.S.; Binti Dola, K.; Ujang, N.; Sadat Benrazavi, N. Effect of Pavement Materials on Surface Temperatures in Tropical Environment. *Sustain. Cities Soc.* **2016**, *22*, 94–103. [CrossRef]
30. Karalar, M.; Özkılıç, Y.O.; Deifalla, A.F.; Aksoylu, C.; Arslan, M.H.; Ahmad, M.; Sabri, M.M.S. Improvement in bending performance of reinforced concrete beams produced with waste lathe scraps. *Sustainability* **2022**, *14*, 12660. [CrossRef]
31. Çelik A, İ.; Özkılıç, Y.O.; Zeybek, Ö.; Özdöner, N.; Tayeh, B.A. Performance assessment of fiber-reinforced concrete produced with waste lathe fibers. *Sustainability* **2022**, *14*, 11817. [CrossRef]
32. Amin, M.N.; Khan, K.; Saleem, M.U.; Khurram, N.; Niazi MU, K. Influence of mechanically activated electric arc furnace slag on compressive strength of mortars incorporating curing moisture and temperature effects. *Sustainability* **2017**, *9*, 1178. [CrossRef]
33. Zeybek, Ö.; Özkılıç, Y.O.; Çelik, A.İ.; Deifalla, A.F.; Ahmad, M.; Sabri Sabri, M.M. Performance evaluation of fiber-reinforced concrete produced with steel fibers extracted from waste tire. *Front. Mater.* **2022**, *692*. [CrossRef]
34. Beskopylny, A.N.; Shcherban', E.M.; Stel'makh, S.A.; Meskhi, B.; Shilov, A.A.; Varavka, V.; Evtushenko, A.; Özkılıç, Y.O.; Aksoylu, C.; Karalar, M. Composition Component Influence on Concrete Properties with the Additive of Rubber Tree Seed Shells. *Appl. Sci.* **2022**, *12*, 11744. [CrossRef]
35. Shcherban', E.M.; Stel'Makh, S.A.; Beskopylny, A.N.; Mailyan, L.R.; Meskhi, B.; Shilov, A.A.; Chernil'Nik, A.; Özkılıç, Y.O.; Aksoylu, C. Normal-Weight Concrete with Improved Stress–Strain Characteristics Reinforced with Dispersed Coconut Fibers. *Appl. Sci.* **2022**, *12*, 11734. [CrossRef]
36. Khan, K.; Ahmad, W.; Amin, M.N.; Aslam, F.; Ahmad, A.; Al-Faiad, M.A. Comparison of Prediction Models Based on Machine Learning for the Compressive Strength Estimation of Recycled Aggregate Concrete. *Materials* **2022**, *15*, 3430. [CrossRef]
37. Amin, M.N.; Iqtidar, A.; Khan, K.; Javed, M.F.; Shalabi, F.I.; Qadir, M.G. Comparison of machine learning approaches with traditional methods for predicting the compressive strength of rice husk ash concrete. *Crystals* **2021**, *11*, 779. [CrossRef]
38. Sakib, N.; Bhasin, A.; Islam, M.K.; Khan, K.; Khan, M.I. A review of the evolution of technologies to use sulphur as a pavement construction material. *Int. J. Pavement Eng.* **2021**, *22*, 392–403. [CrossRef]
39. *ASTM C29/C29M-17a*; Standard Test Method for Bulk Density ("Unit Weight") and Voids in Aggregate. ASTM International: West Conshohocken, PA, USA, 2017.
40. *ASTM C136/C136M-19*; Standard Test Method for Sieve Analysis of Fine and Coarse Aggregates. ASTM International: West Conshohocken, PA, USA, 2020.
41. *ASTM C131/C131M-20*; Standard Test Method for Resistance to Degradation of Small-Size Coarse Aggregate by Abrasion and Impact in the Los Angeles Machine. ASTM International: West Conshohocken, PA, USA, 2020.
42. *ASTM ASTM C39/C39M-21*; Standard Test Method for Compressive Strength of Cylindrical Concrete Specimens 1. ASTM International: West Conshohocken, PA, USA, 2018.
43. Yu, F.; Sun, D.; Wang, J.; Hu, M. Influence of Aggregate Size on Compressive Strength of Pervious Concrete. *Constr. Build. Mater.* **2019**, *209*, 463–475. [CrossRef]
44. Yang, J.; Jiang, G. Experimental Study on Properties of Pervious Concrete Pavement Materials. *Cem. Concr. Res.* **2003**, *33*, 381–386. [CrossRef]
45. Bhutta, M.A.R.; Tsuruta, K.; Mirza, J. Evaluation of High-Performance Porous Concrete Properties. *Constr. Build. Mater.* **2012**, *31*, 67–73. [CrossRef]
46. Huang, B.; Wu, H.; Shu, X.; Burdette, E.G. Laboratory Evaluation of Permeability and Strength of Polymer-Modified Pervious Concrete. *Constr. Build. Mater.* **2010**, *24*, 818–823. [CrossRef]

47. Zhong, R.; Wille, K. Compression Response of Normal and High Strength Pervious Concrete. *Constr. Build. Mater.* **2016**, *109*, 177–187. [CrossRef]
48. *ASTM D5084-16a*; Standard Test Methods for Measurement of Hydraulic Conductivity of Saturated Porous Materials Using a Flexible Wall Permeameter. ASTM International: West Conshohocken, PA, USA, 2016.

Review

An Overview of Methods to Enhance the Environmental Performance of Cement-Based Materials

Daniel Suarez-Riera, Luciana Restuccia, Devid Falliano, Giuseppe Andrea Ferro, Jean-Marc Tuliani and Matteo Pavese et al.

1   Department of Structural, Geotechnical and Building Engineering, Politecnico di Torino, Corso Duca degli Abruzzi 24, 10129 Torino, Italy; daniel.suarez@polito.it (D.S.-R.); luciana.restuccia@polito.it (L.R.); devid.falliano@polito.it (D.F.); giuseppe.ferro@polito.it (G.A.F.)
2   Department of Applied Science and Technology, Politecnico di Torino, Corso Duca degli Abruzzi 24, 10129 Torino, Italy; jeanmarc.tulliani@polito.it (J.-M.T.); matteo.pavese@polito.it (M.P.)
*   Correspondence: luca.lavagna@polito.it

Abstract: Urbanization and demographic growth have led to increased global energy consumption in recent years. Furthermore, construction products and materials industries have contributed significantly to this increase in fossil fuel use, due to their significant energy requirements, and consequent environmental impact, during the extraction and processing of raw materials. To address this environmental problem, architectural design and civil engineering are trying to implement strategies that enable the use of high-performance materials while minimizing the usage of energy-intensive or toxic and dangerous building materials. These efforts also aim to make buildings less energy-consuming during their useful life. Using waste materials, such as Construction and Demolition Waste (CdW), is one of the most promising approaches to address this issue. In recent years, the European Union (EU) has supported recovery strategies focused on using CdW, as they account for more than 30% of the total waste production in the EU. In this regard, reuse techniques—such as incorporating concrete fragments and bricks as road floor fillers—have been the subject of targeted scientific research. This review will outline various strategies for producing green cement and concrete, particularly emphasizing the reuse of Construction and Demolition Waste (CdW).

Keywords: construction and demolition waste; sustainability; recycling strategies; waste management; cement

Citation: Daniel Suarez-Riera, Luciana Restuccia, Devid Falliano, Giuseppe Andrea Ferro, Jean-Marc Tuliani and Matteo Pavese et al. . An Overview of Methods to Enhance the Environmental Performance of Cement-Based Materials. *Infrastructures* **2024**, *9*, 94. https://doi.org/10.3390/infrastructures9060094

Academic Editor: Patricia Kara De Maeijer

Received: 30 April 2024
Revised: 7 June 2024
Accepted: 8 June 2024
Published: 11 June 2024

1. Introduction

Cement is a fundamental component of construction material used globally; specifically, Ordinary Portland Cement (OPC) is the primary building material for worldwide housing and infrastructure, making it incredibly significant. It is a binding agent that holds together the aggregates, such as sand and gravel, to form mortar or concrete, the world's most widely used building material [1]. According to the International Energy Agency (Figure 1), the global demand for cement was estimated to be 4.3 Gt in 2021, and is expected to continuously increase by 2030 (30% more respect 2020), driven primarily by the construction of infrastructure and housing in developing countries [2]. Concrete is a ubiquitous material that has played a crucial role in modern construction and infrastructure development. Its versatility, durability, and low cost have made it a go-to building material for many applications, from high-rise buildings and bridges to sidewalks and retaining walls. However, its impact on the built environment and the environment is multifaceted, with both positive and negative consequences.

On the one hand, concrete structures are designed to withstand the forces of nature, such as storms and floods, which makes them safer and more reliable than other materials. This means that people who live and work in buildings made from concrete are better protected from the effects of natural disasters [3,4]. In addition, its use has significantly impacted urbanization, since more extensive and complex structures have created larger cities and urban areas. Conversely, the growth of urban areas has also had negative consequences, such as increased air pollution, the destruction of wetlands, traffic congestion, and the loss of green spaces and other habitats. Furthermore, cement production is one of the leading sources of greenhouse gas (GHG) emissions, contributing about 8% of global carbon dioxide (CO_2) emissions [5,6]. Its production requires substantial heat, using up to 4 GJ of energy per ton of clinker. To produce 1000 kg of cement, roughly 120 kg of coal, with an energy content of 27.5 MJ/kg [7], is utilized, while electricity consumption ranges between 90 and 120 kWh/ton [8]. Therefore, there is a need to develop more sustainable and eco-friendly alternatives to traditional cement and cement-based materials.

Various measures are being taken to reduce the energy and carbon footprint and enhance sustainability to mitigate the adverse impact of cement-based materials on the environment. An encouraging advancement is the employment of carbon capture technology in cement production, which can substantially decrease the emission of carbon dioxide into the atmosphere [9]. Another option is the replacement of significant amounts of Portland cement with less impacting materials, reducing the effects of mining and protecting biodiversity while using natural resources efficiently. It is also crucial to replace raw materials such as limestone and clay with waste from other industrial activities that can be used in cement-making, and would otherwise be sent to landfills. This would result in "green cement" with a lower amount of clinker, an innovative and sustainable solution offering a more environmentally friendly alternative to OPC [10]. It is produced using innovative manufacturing processes and raw materials that significantly reduce energy consumption and GHG emissions compared to conventional OPC. Alternative raw materials can be used instead of traditional ones, including contaminated soil, waste from road cleaning, and other materials containing iron, aluminum, and silica. Some specific examples of such waste materials are coal fly ash and blast furnace slag [11,12]. The demand for green cement is rapidly increasing due to the urgent need to reduce the construction industry's energy and carbon footprint. This industry is one of the most energy-demanding, and thus contributes heavily to global GHG emissions, accounting for approximately 40% of total energy-related CO_2 emissions [13]. Finally, the environmental impact of cement-based materials can be mitigated through sustainable production practices and the use of alternative materials. Recycling concrete waste, bricks [14] and clay into fine and coarse aggregates reduces landfill waste and the extraction of natural raw materials, promoting eco-friendly development. High-quality recycled coarse aggregates can replace up to 100% of natural aggregates in concrete, improving environmental performance. Methods like acid treatment and CO_2 curing enhance the quality of recycled aggregates. Recycled sand from coarse aggregates offers a cheaper and sustainable alternative to natural sand, supporting high-value recycling. This innovation improves the mechanical strength and durability of recycled mortar, boosting sustainable construction practices [15].

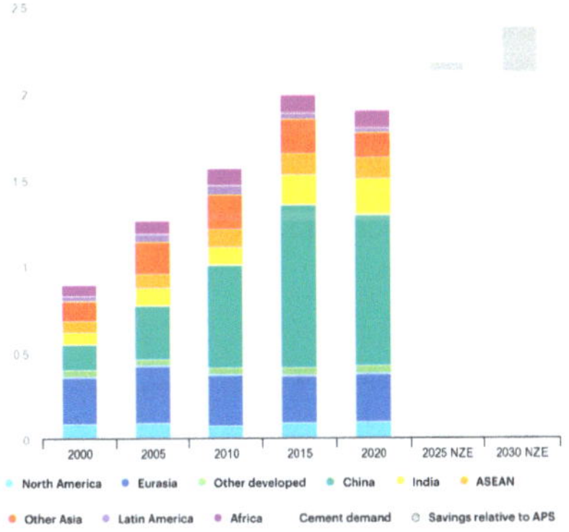

Figure 1. The global demand for cement [16].

2. Carbon Sequestration in Cement-Based Materials

Carbon sequestration in the cement industry is a complex and perplexing topic, with many factors at play, since cement production is one of the largest contributors to carbon dioxide emissions [17]. This makes it a prime target for carbon sequestration efforts, which aim to capture and store carbon dioxide emissions safely and permanently.

At its core, carbon sequestration involves capturing carbon dioxide emissions before they are released into the atmosphere and storing them to prevent their release. This can take many forms in the cement industry, including carbon capture and storage (CCS) technology. It involves capturing carbon dioxide emissions from cement production and storing them underground or elsewhere [18,19]. However, while carbon sequestration may seem like a straightforward solution to the carbon emissions problem in the cement industry, many challenges and uncertainties make it a highly complex and challenging endeavor. For example, the effectiveness of CCS technology in the cement industry has yet to be thoroughly tested or proven, and there are concerns about the safety and sustainability of storing carbon dioxide underground [20]. Moreover, the costs associated with carbon sequestration in the cement industry are significant, and it is unclear who will bear these costs. CCS encounters various technical and financial impediments that must be surmounted to enable its widespread implementation. A crucial financial challenge is that CCS is not profitable, necessitating substantial capital investment. In addition to financial barriers, there are also significant technical challenges associated with CCS, such as the uncertainty surrounding long-term CO_2 leakage rates.

Furthermore, some countries may not have an adequate geological storage capacity for CCS, which may increase transportation and injection costs, particularly for offshore storage. This limitation applies to several countries, including the UK, Norway, Singapore, Brazil, and India [21,22]. Cement producers may be reluctant to invest in CCS technology due to the high costs and uncertainties. At the same time, governments and other stakeholders may be hesitant to provide the necessary funding and support [22]. Furthermore, various technical and logistical challenges are associated with carbon sequestration in the cement industry. For example, capturing and storing carbon dioxide requires significant energy and resources, which can offset the emissions reductions achieved through sequestration. In addition, the process of storing carbon dioxide underground can be highly complex and requires careful monitoring and management to prevent leaks or other environmental risks. Despite these challenges, many experts believe that carbon sequestration is essential for reducing carbon emissions in cement and other carbon-intensive sectors. With suitable investments and policies in place, it may be possible to overcome the technical, financial, and logistical hurdles associated with carbon sequestration and achieve meaningful emissions reductions. However, it is crucial to recognize that the cement industry has no one-size-fits-all solution to carbon sequestration. Depending on the specific circumstances and challenges individual cement producers and regions face, different approaches may be required. For example, some producers may find switching to alternative cement production methods that produce fewer emissions more cost-effective. In contrast, others may need to rely on CCS technology to achieve emissions reductions. Ultimately, the success of carbon sequestration in the cement industry will depend on various factors, including technological advancements, policy support, and public awareness and engagement. While there are no easy answers or quick fixes, the proper calculation of economic and environmental costs for carbon sequestration will be crucial in the choice of transitioning to a low-carbon economy [23].

3. Green Cement

Green cement is a term referring to cement obtained through innovative manufacturing processes and/or using raw materials that significantly reduce energy consumption and GHG emissions compared to traditional Portland cement. There are currently several methodologies for making "green cement", such as the use of pozzolanic materials (fly ash, slag, etc.) as a replacement for Portland cement [24], and in this work we focus our attention on: geopolymer binders and calcium sulfoaluminate (CSA) cement and cementitious materials with improved self-healing ability.

3.1. Geopolymer Binders

Geopolymer binders are a wide class of binders, generally referred to as alkali-activated materials. They are characterized by the presence of a highly alkaline solution, typically sodium hydroxide or sodium silicate, where aluminosilicates dissolve. These materials have been gaining popularity in recent years since they need less energy than Portland cement for their production. In fact, while Portland cement production requires around 1450 °C, alkali-activated materials require lower temperatures for clay calcination, and commonly use wastes from other industrial process as raw material; for instance, ground granulated blast furnace slag (GGBFS), fly ashes [25], stone muds [26] and even municipal solid waste incineration residues [27]. This can reduce the amount of waste going to landfills and reduces the need for new materials to be mined or extracted. The production of GGBFS is estimated at around 300 million tons per year [28], which is very large even if still much lower than what would be needed to completely substitute Portland cement. Thus, a widespread substitution of OPC with geopolymer-based binders would require using very large amounts of extracted raw materials.

Nevertheless, geopolymer binders seem very interesting due to their interesting set of properties (Figure 2). First, they present good durability, avoiding some of the issues of OPC, for instance alkali-aggregate reaction, which can cause it to break down over time. They also have a higher resistance to chemicals and can be used in harsh environments, such as in the construction of chemical plants or wastewater treatment facilities. Geopolymers also have the potential to reduce economic and environmental costs, since their production is less energy-intensive than traditional cements and requires a lower amount of fossil fuels. The possibility of using waste materials that would otherwise be disposed of further reduces the need for new raw materials and leads to cost savings for the construction industry, and potentially even for consumers [29]. However, some studies [30] suggest that a standard concrete is less expensive than one based on geopolymers, even if this last presents a lower environmental impact.

One of the challenges facing the widespread adoption of geopolymer binders is the lack of standardization and regulation. Traditional Portland cement is regulated and standardized by organizations such as ASTM International and the European Committee for Standardization, ensuring consistency and quality. On the other hand, geopolymer binders do not have the same standardization level, making it difficult for contractors and engineers to use them in construction projects. However, efforts are underway to standardize and regulate them. In 2019, the International Union of Laboratories and Experts in Construction Materials, Systems, and Structures (RILEM) published guidelines for testing and characterizing geopolymer binders. These guidelines aim to provide a standardized approach to the testing and characterization of these materials, which can help to ensure consistency and quality in their use [31].

At the beginning of the development of geopolymers, another important issue was the high temperature needed for the curing. This issue was partially solved using calcium-containing waste materials, which can react at low temperature thanks to the presence of alkali as an activator. This is similar to what happens in Type III cements, where slag is used as a supplementary cement material, and may require activation by portlandite formed by the hydration of Portland cement. Anyway, some issues related to workability and shrinkage are still yet to be solved in order to allow a significant commercial penetration of geopolymer binders.

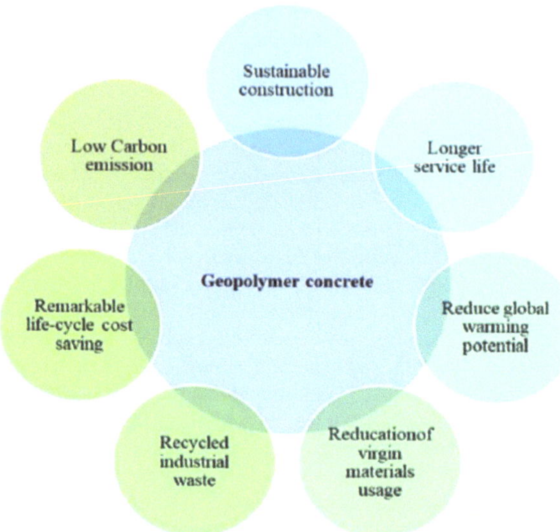

Figure 2. Usefulness of geopolymer concrete in construction [32].

3.2. Calcium Sulfoaluminate Cement

Another approach that was explored in recent years to reduce the energy footprint of concrete regards the use of alternative cements, based on less energy-intensive raw materials. One possibility is to use Calcium Sulfoaluminate Cement (CSA cement), a class of specialty cements that are composed of calcium sulfoaluminate ($4CaO \cdot 3Al_2O_3 \cdot CaSO_4$), dicalcium silicate ($2CaO \cdot SiO_2$) and gypsum ($CaSO_4 \cdot 2H_2O$) [33,34]. CSA cement is a type of hydraulic cement first developed in the late 1950s [35]. It is made starting from a mixture of calcium sulfate, alumina, and limestone. Unlike OPC, which requires long curing times to achieve its strength, CSA cement can set and harden rapidly, often within hours. This property makes it an attractive option for construction projects that require quick turnaround times. In addition to its rapid setting time, CSA cement offers several other benefits over OPC.

For one, its production seems to emit less CO_2 than Portland cement [36,37]. This is due to the specific composition of CSA cement, which requires less limestone as a raw material, and to a lower clinkering temperature, at 1250–1350 °C instead of 1450 °C for Portland cement. Ren and coworkers [38] suggest that the higher costs of CSA cement with respect to Portland are related to the higher costs of alumina-containing raw materials, and that the use of waste material can lower both the cost and the environmental impact of the CSA cement. It must be considered that Chinese LCA normalization was used in these papers, and that scale factors also contribute to its higher cost with respect to Portland cement. In any case, CSA cement seems to be able to significantly reduce the environmental impact of concrete [39].

Another advantage of CSA cement is its high early strength. It can achieve up to 50 MPa strength within 24 h of casting. This property makes it ideal for projects requiring rapid construction or bearing heavy loads soon after casting. For instance, CSA cement has been used in constructing runways and other infrastructure projects where quick turnaround times are essential; additionally, it has good resistance to chemical attacks, making it suitable for projects in harsh environments [40]. Despite these advantages, CSA cement is not yet widely used in construction. One reason is that it is still more expensive than Portland cement, as discussed above. Moreover, it is not yet as well-understood as OPC, and its properties can vary depending on the specific mixtures used. This variability makes it difficult to predict its behavior in different applications. Another big challenge facing its extensive adoption is the disputable availability of raw materials. In contrast to OPC, which uses readily available materials such as limestone, clay, and gypsum, CSA cement requires bauxite or another alumina-bearing mineral, or waste materials of suitable composition, which may not be widely available in every part of the world. It must be remembered that the produced amount of Portland cement is enormous and that every alternative must be considered in relation to the availability of the resources used for its production.

3.3. Cementitious Materials with Improved Self-Healing Ability

The high annual cost of maintenance and the growing concern about the safety and sustainability of infrastructure in Europe have increased interest in the development of self-healing cementitious materials and preventative repair methods. The appearance of small cracks (less than 300 microns in size) in concrete is inevitable, but does not necessarily cause the collapse of structures. However, small cracks weaken the functionality, accelerate degradation, and reduce the service life and sustainability of such structures. The enormous development that concrete technology has undergone in recent decades has led to the development of materials with extremely low porosity, but has not prevented the intrinsic risk of cracking. On the contrary, high-performance concretes are even more fragile and sensitive to cracking in a short time, compared to concretes that have a lower compressive strength. This behavior has given rise to the study of methodologies that can heal these fissures, which can be divided into passive methods, applied manually after an inspection and allowing only superficial cracks to be sealed, and active methods, incorporated into fresh concrete and therefore allowing both internal cracks and surface cracks to be healed. The latter techniques are also called self-healing techniques (Figure 3).

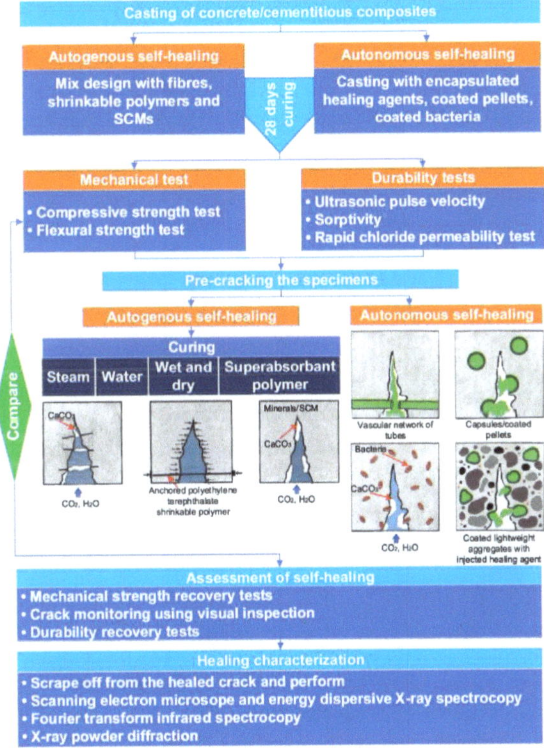

Figure 3. Self-healing framework [41].

The autogenous ability of concrete to repair itself is based on traditional constituents of the cement matrix, but can also be induced by targeted additions to the mix. This phenomenon was observed for the first time in 1836 by the French Academy of Sciences when the autogenous repair of concrete was highlighted in pipes, water retention structures, etc. [42]. The mechanisms that contribute to the autogenous repair of a crack, when it has formed and is exposed to water, are mainly twofold: the continuous hydration of anhydrous cement grains and the precipitation of calcium carbonate crystals ($CaCO_3$) on the edges of the crack, following the chemical reaction between the calcium ions (Ca^{2+}) of the cement matrix and the carbonate ions (CO_3^{2-}) available in water, or carbon dioxide from air in contact with the damaged area. The autogenous repair induced by the continuous hydration of anhydrous cement grains is very useful, as the new hydration products have similar mechanical properties as the primary C-S-H gel that are in any case higher than those of calcium carbonate precipitates. However, the conditions for the formation of secondary C-S-H are different because the nucleation and growth take place on the edges of the cracks, and not in the bulk of the cement paste, as well as the fact that the water/cement (w/c) ratio could be much higher in the case of water arriving from the external environment. Autogenous repair is effective for small cracks between 10 and 100 microns, sometimes up to 200 microns, but only in the presence of water. The type of cement seems not to be important, but the clinker content determines the release of Ca^{2+} ions and the subsequent ability of the matrix to form calcium carbonate-based precipitates. On the contrary, the silicate additions have an effect depending on their nature and quantity in the mixture, related to the characteristic pozzolanic reactions with the Portlandite, which affects the duration of the self-repair mechanisms [42–44]. Furthermore, concretes with high mechanical resistance, prepared with a low w/c and a high binder content, contain many anhydrous cement grains that can potentially produce significant quantities of C-S-H. Finally, the age of the concrete also has an influence, as for short times, more anhydrous grains are available, while for longer times the formation of $CaCO_3$ as a self-healing agent will prevail. To conclude, autogenous repair can be favored by specific additions, such as the addition of blast furnace slag and fly ash, or porosity-reducing additives acting through crystallization (the so-called crystalline admixtures), which react with water to form insoluble precipitates (based on modified C-S-H and a calcium-based hydrated compound) in the pores and superabsorbent polymers with the ability to swell in the presence of liquids (swelling up to 1000 g/g).

The autonomous repair of concrete is based on the incorporation of various microcapsules (<1 mm), whose rupture releases a repairing agent contained in it so as to seal the crack. Various polymeric capsules have been studied (based on: urea, melamine or phenol and formaldehyde [45–47], polyurethane [48] resin, polystyrene [49], polyvinyl alcohol [50], acrylates [51], or even pig gelatine/gum acacia [52] or silica [51]) containing various repairing agents (epoxy, polyurethane or acrylic resin, calcium silicate, colloidal silica or calcium sulpho-aluminate). However, polymeric shells and epoxy-based cargos are the most widely investigated systems. Microcapsules significantly affect the viscosity of the fresh mix, while after 28 d of curing, the compressive strength and the elastic modulus show a consistent decrease with the increasing concentration of microcapsules [52]. The literature survey shows that the mechanical recovery rate of cracked samples is roughly proportional to the content of the microcapsules [43].

Macrocapsules based on fibers (with external diameter 1 mm and length 100 mm) or glass tubes (with external diameter 3 mm and length 100 mm) filled with cyanoacrylate resin or sodium silicate have been also studied [53,54]. However, glass tubes may be subjected to the silica alkali reaction. To avoid this inconvenience, ceramic capsules [55], extruded EVA [56], PLA, PMMA, PEG [57], or cement tubes [58] with external diameters up to 8.4 mm and lengths up to 5 cm, have also been tested. When incorporated in a self-compacting concrete, cementitious macrocapsules (with a of 1.6 vol%) showed a compressive strength that was not significantly influenced by the presence of the tubes, whatever their orientation with respect to the load direction [59]. The average sealing efficiency ranged from 54 to

74% for the samples containing cementitious capsules. However, although promising, these technologies based on the micro- and macroencapsulation of repairing agents present the limit of the actual durability of the encapsulated repairing agents over time. Finally, microcapsules are also limited to one-time use, contrarily to macrocapsules.

Thus, thirty years ago [60], vascular networks (VN) based on hollow fibers were proposed: they operate according to the same healing mechanism as capsule-based systems, but with the advantage of a continuous external supply of a healing agent to the damaged zones within concrete. Moreover, the delivery of a healing agent through a vascular network can be done under pressure, further increasing the efficiency of repair, and with the potential to allow multiple healing cycles. Recently [61], thanks to the ability of additive manufacturing to produce complex geometries, ductile-porous 3D-printed VN were investigated. Load regains up to 56% and stiffness recovery up to 91% were achieved with a polyurethane resin as the healing agent. Combining traditional fabrication techniques like extrusion or injection molding for manufacturing linear pipes and additive manufacturing to produce branched parts seems a good compromise to save time and money in view of the possible large-scale diffusion of VN in the future.

A last very interesting solution for the autonomous repair of concrete involves the use of ureolytic bacteria capable of decomposing urea into ammonium/ammonia (Equation (1)) and carbonate ions. If calcium ions are available in sufficient quantities, then calcium carbonate precipitates (Equation (2)) [62]:

$$CO(NH_2)_2 + 2H_2O \xrightarrow{\text{ureolytic bacteria}} 2NH_4^+ + CO_3^{2-} \tag{1}$$

$$Ca^{2+} + CO_3^{2-} \rightarrow CaCO_3 \tag{2}$$

Theoretically, one mole of calcium carbonate can be formed if one mole of urea is present. However, the process is strongly controlled by the enzyme that produces urease (the kinetics of the reaction are 10^{14} times faster when it is present).

The precipitation of calcium carbonate is also induced by the reduction in nitrates in oxygen-poor environments via denitrifying microorganisms (Equation (3)):

$$5HCOO^- + 2NO_3^- \rightarrow N_2 + 3HCO_3^- + H_2O \tag{3}$$

The self-healing of concrete due to the precipitation of $CaCO_3$ by bacteria is based on the following parameters: adequate pH value and specific nutrients for bacterial cells. For nitrate-reducing bacteria, calcium formate and calcium nitrate are used as nutrients. After 56 days of immersion in water, the sealing of cracks with a width of up to 480 ± 16 μm could be observed [63]. Nitrate reduction can also lead to the production of nitrite ions, known corrosion inhibitors of steel reinforcement (Equation (4)):

$$2HCOO^- + 2NO_3^- + 2H^+ \rightarrow 2CO_2 + H_2O + 2NO_2^- \tag{4}$$

The bacteria *Pseudomonas aeruginosa* and *Diaphorobacter nitroreducens* can survive in mortars if inserted into expanded clay granules, diatomaceous earth or activated carbon, for example. The company Basilisk, a spin-off of Delft University in the Netherlands, markets some bacteria of the *Bacilli* genus capable of producing spores that can remain dormant for up to 200 years [64]. When fed with calcium lactate, they precipitate calcium carbonate (Equation (5)):

$$CaC_6H_{10}O_6 + 6O_2 \rightarrow CaCO_3 + 5CO_2 + 5H_2O \tag{5}$$

The carbon dioxide from the reaction can in turn react with portlandite to produce additional calcium carbonate. Aerobic metabolic degradation requires the presence of oxygen; therefore, a limited availability of this gas decreases the quantity of precipitated calcium carbonate. However, the absence of oxygen slows down the rate and risks of corrosion of steel reinforcement, and the presence of aerobic bacteria will still prolong the service life of reinforced concrete structures in environments favorable to corrosion.

3.4. Potential Benefits, Applications and Challenges

As discussed in the previous paragraphs, two possible "green" cements are being envisaged in order to reduce the environmental impact related to Portland cement production. They could be used in a wide range of applications, including concrete, mortar, and grout, making them a versatile alternative to traditional cement. It is important to consider the specific characteristics of these cements in order to understand their potential advantages and drawbacks in construction and building applications.

Durability and chemical resistance seem to be improved with respect to OPC, allowing their use also in harsh environments, while the high strength of CSA and high-calcium geopolymers makes them interesting for structural applications. This reinforces the interest in these materials, which was initially based mostly on their lower embodied energy and carbon footprint, and the possibility of using waste materials, either for the cement paste preparation (in the case of geopolymers) or as raw materials during the clinkering process (in the case of CSA). Workability issues are to be considered, but the two cements cover both the short and long setting ranges, such that for fast curing applications, CSA would be preferred, while geopolymers are preferred in slow curing.

On the whole, the use of green cement could simultaneously benefit society, the environment and the construction industry, thanks to reduced energy consumption and carbon footprint, improved waste materials use and an improvement in circularity. Another significant issue raised in particular for geopolymers is the health and safety of cement industry workers, but high-calcium geopolymers allow one to avoid the use of dangerous highly alkaline solutions, obtaining acceptable materials from the safety point of view.

The use of these green cements in reinforced concrete is still hampered by the scarce knowledge about the corrosion behavior of steel inside both CSA and geopolymer-based concrete [65,66]. New types of fibers, either natural or artificial, are being proposed, but the possibility of substituting steel into reinforced concrete is not yet scientifically substantiated, in particular due to the difficulty of preparing reinforcements that are continuous, strong, tough, and durable [67,68].

Another important issue related to green cement is that, currently, its production is typically more expensive than traditional cement due to the scale factor, and the requirement of alternative raw materials and innovative manufacturing processes. Specific resources may not always be available, secondary raw materials from waste have often varying and non-homogeneous compositions, and legislation must be amended to allow for the recycling of waste in construction materials. Processing plants must be built and upscaled, which is difficult due to the competitiveness of traditional cement, which is a cheap and well-known material with a very standardized production process. All these issues make green cement less attractive to cost-conscious builders and developers who prioritize cost over sustainability. The request to reduce the construction industry's energy consumption and carbon footprint has increased the demand for green cement, but currently, it remains a niche product, even if the global market for environmentally sustainable cement is expected to expand significantly in the coming years, driven by rising interest in sustainable and environmental friendly materials. In fact, a significant push toward the use of this cement comes from governments and regulatory bodies worldwide, which are promoting its use through policies and incentives. For example, the European Union aims to reduce GHG emissions from the construction sector by 60% by 2050, creating more demand for green cement [69].

4. Construction and Demolition Waste—CDW

Sustainable construction practices have emerged as a response to the significant natural resource consumption associated with traditional building and construction technologies. These practices aim to repurpose industrial waste and by-products to minimize construction's environmental impact and protect valuable resources. The waste materials resulting from construction and demolition operations are collectively referred to as CDW. Due to their enormous volume, generating adverse environmental and economic effects, they are regarded as one of the more significant challenges in the construction industry. Indeed, the escalating amount of waste generated and its disposal process negatively impacts the environment and society; this category of waste represents a significant portion of global waste, accounting for between 30% and 40% of the total solid waste, with a global net use rate between 20% and 30% [70]. In the world context, the United States' recovery rate is approximately 70%, while in China, it remains low, at less than 5% [71]. In the United States, total CDW was estimated to be 600 million tons in 2018 [72], while China generates approximately 2.4 billion tons of CDW every year, which accounts for roughly 40% of the total urban waste produced in the country [73]. The rapid urbanization in China has led to increased CDW generation, resulting in significant pressure placed on waste management systems and a severe "garbage siege" phenomenon prevalent in many urban areas [74]. The low utilization of CDW can be attributed to several factors, including the lack of reuse and recycling design for buildings, insufficient recovery facilities in some areas, and low demand for some materials due to regulatory restrictions. The competitiveness of CDW recycling can be improved through intrusive measures, such as increasing raw material prices or imposing taxes, but also by simply establishing end-of-waste criteria for specific CDW fractions. In the European Union, for example, approximately 3 billion tons of waste are generated each year, with one-third of this amount originating from construction and demolition activities [75,76], with an average recovery rate of almost 50%. Nevertheless, it varies significantly among member states, ranging from 10% to 80%. For instance, Italy has a recovery rate of almost 80%, France at 48%, Spain at around 40%, and Germany at 34% [77].

4.1. CDW Composition

Depending on the source and separation methods, CDW primarily comprises inert mineral materials with varying amounts of other components. However, definitions and compositions of CDW can vary from state to state. It can be broadly classified into five categories: metal, concrete and mineral, wood, miscellaneous, and unsorted mixed fractions [78]. More specifically, these waste materials may include concrete, bricks, tiles, ceramics, wood, glass, plastic, bituminous mixtures and tars, ferrous and non-ferrous metals, soils, stones, insulation materials, gypsum-based materials (such as plasterboards), chemicals, waste electronic and electrical equipment (WEEE), packaging materials, and hazardous substances. Hazardous substances commonly found in building materials include asbestos, lead-based paints, phenols, polychlorinated biphenyls (PCBs), and polycyclic aromatic hydrocarbons (PAHs). Many substances, such as insulation, roofs, tiles, and fire-resistant sealing, are often used in conjunction with concrete to complete the structure and finishes. The composition of CDW varies significantly depending on factors such as local typology, construction techniques, climate conditions, economic activities, and technological advancements in the area [79]. Additionally, the composition of CDW changes over time due to aging buildings and low-quality structures built between the 1960s and 70s, which are now reaching the end of their lifespan and require demolishing [80]. Therefore, defining a standard composition representative of a large region is one of the biggest challenges to face. Figure 4 depicts the average composition of CDW.

in eco-mortars is around 50%, beyond which the decrease in strength becomes significant. Finally, the influence of the saturation state and replacement percentage or fraction of natural sand with recycled sand on the properties of mortars led to greater water absorption and smaller slumps, but not better mechanical properties, which were superior when adding dried RS, as this absorbs water only during the preparation and curing stages [96]. Regarding the substitution fraction, it was observed that the compressive strength of mortars with RS decreased linearly as the replacement percentage of RS increased.

As a general remark on the substitution of recycled sand in mortar and concrete, it seems that the typical effect is a decrease in mechanical properties. This is reasonable, since RS is less pure and more porous than standard sand. However, the fact that the pores of the sand can absorb part of the water can alter the results, giving the impression that the mechanical properties increase due to the use of RS, when in fact the effect is a reduction in water available for cement hydration, i.e., the reduction in water-to-cement ratio.

6. Improving the Microstructural Properties of Recycled Aggregates

As discussed in the previous paragraph, mortars made with recycled aggregates have poor mechanical performance. In fact, mixed recycled aggregates from CDW contain various constituents, such as natural aggregates, cement, bricks, tiles, glass, small amounts of metal, and other minor organic and inorganic impurities. This mixed composition contributes to lowering the performance of cement-based materials containing recycled aggregates. Due to the presence of CDW, compressive strength, as well as tensile and shear strength, are reduced due to higher porosity, crushing index, micro-cracks in the interfacial transition zones, contamination, and variances in quality.

The presence of micro-cracks in the interfacial transition zones due to non-homogenous recycled aggregates can result in the penetration of harmful reactive substances such as sulfate ions, which can react with the hydration products of the cement [97,98]. This reaction produces gypsum and ettringite, further weakening the recycled concrete aggregate due to the higher volume of these reaction products applying internal stresses. Therefore, improving the microstructural and mechanical properties of recycled concrete aggregate has become crucial to enhancing its applicability and usefulness when producing recycled concrete [99–101].

The existing literature indicates that there are six major methods available to enhance the properties of recycled aggregates. These methods can be categorized into two groups: the "improve by removing" category, whereby weaker parts of the recycled aggregate, for instance cracked cement zones, are removed by chemical and thermal processes; and the "improve by adding" category, where the aggregate is reinforced by the addition of mineral admixtures, or by self-healing, carbonation, sequential mixing, or fortification by coating and infiltration processes [102–104].

The chemical approach involves the use of strong acids, such as hydrochloric (HCl) and sulfuric acid (H_2SO_4), to dissolve certain hydration products in the cement. This method effectively removes loose and cracked mortar from the recycled aggregates (RA), reducing water absorption and improving concrete performance [105,106]. However, Tam et al. [106] showed that using acid-treated recycled concrete aggregate (RCA) in concrete allows for a maximum replacement of only 30% of the natural coarse aggregate. Furthermore, strong acids pose safety risks and introduce harmful chemicals into the concrete. In the thermal approach, instead, the RA is heated to a high temperature of over 400 °C to remove the hydration products and weaken the residual mortar, which is then mechanically removed from the natural aggregate. Not surprisingly, this method requires a significant amount of thermal energy and may produce fine powders that attach to the surface of the RA, negatively impacting its quality. The mechanical rubbing process can also cause new micro-cracks to form, further weakening the RA [107].

In the "improve by adding" category, instead, one way to improve the RA quality is to strengthen the adhered mortar. Shi et al. [103] investigated using both pozzolan slurry (including silica fume, nano-SiO_2, and fly ash slurries) and CO_2 treatment as enhancement methods for RCA. Their findings show that concrete made with treated RCA had a compressive strength that was increased from 17% to 55%. Polymer emulsions also effectively reduced RCA water absorption by 5% to 30% [108]. Zhan et al. [109] used a carbonation process that enhanced the properties of the recycled aggregates. With this method, the water absorption of RCA was reduced by 20%, down to 24%. Furthermore, the durability of concrete made with treated RCA was significantly improved compared to untreated RCA. In addition, mineral precipitation, which leverages the activity of bacteria to precipitate calcium carbonate on the surface of the RCA [110], can also significantly reduce water absorption by 13%, down to 17%, and enhance the microstructure of the RCA. On the other hand, this method is costly and not practical for widespread use.

From the coating and infiltration point of view, as stated by Tam et al. [111], a limited number of research studies are available that focus on identifying methods to improve the microstructural properties of recycled aggregates. In their review, they explain that research studies in this area have been superficial and incomplete, leaving an extensive knowledge gap that requires further investigation and filling. Given the abundance of available chemical varieties, there is potential for developing numerous chemicals and solutions to treat and improve the microstructural properties of recycled concrete aggregate. Figure 5 provides an illustration summarizing the improvement methods.

An interesting approach to improving the RA involves crystallization technology. This method, known as crystalline waterproofing, is widely spreading in concrete applications, and involves active substances that react with the hydration products or dehydrated cement particles in the concrete to produce additional reactants in the form of crystals [112–115]. These crystals then effectively block off the pores in the concrete, decreasing its overall permeability [116,117]. Recent studies have shown that the use of this additive does not contribute to the improvement of concrete compression strength, but significantly increases durability [118]. The increase in durability also appears to be due to the closure of capillary pores resulting from the formation of ettringite on crack surfaces [119]. No effects on workability have been detected [120]. This innovative approach offers a promising alternative for reducing porosity and improving the waterproofing properties of concrete structures, and could be applied to RA to partially fill the present pores.

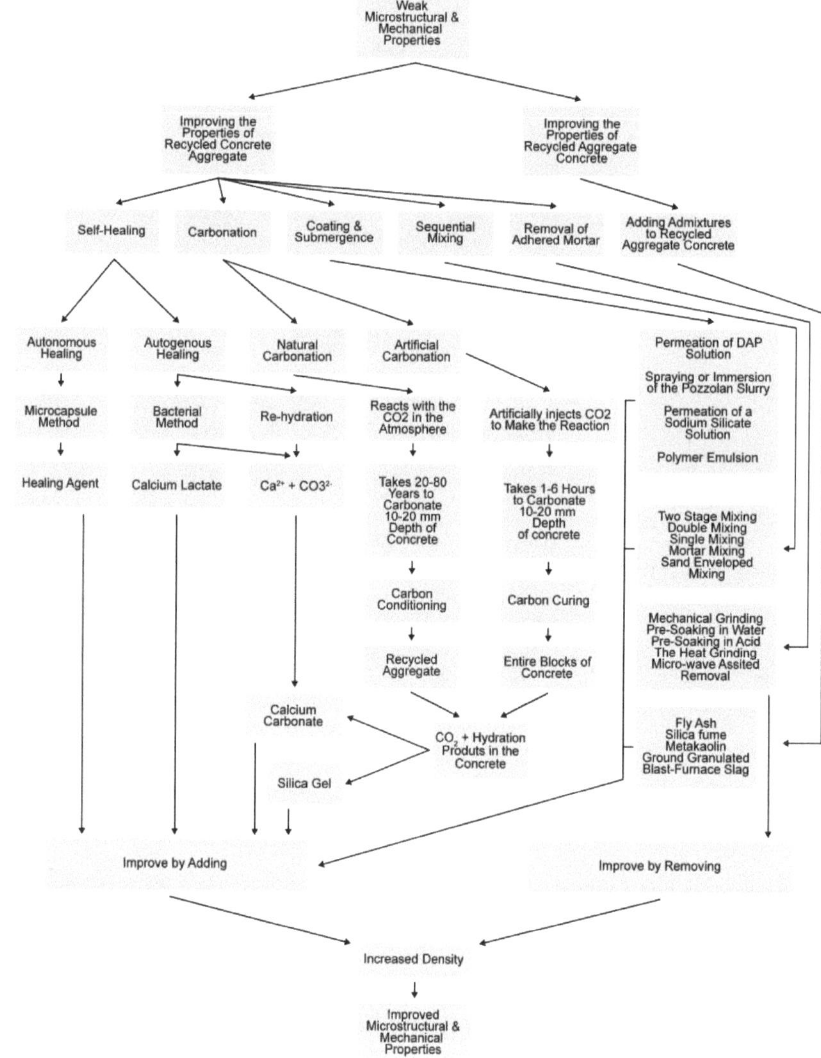

Figure 5. An overview of the enhancement techniques and their interconnections [111].

7. Conclusions

This overview emphasizes the need to address the environmental impacts of construction materials used in sustainable urban development, given the industry's high energy consumption and carbon emissions.

- Sustainable improvements in cement-based materials can be pursued via two strategies: using eco-friendly cements and fully exploiting Construction and Demolition Wastes (CDW) as aggregates, whilst considering techniques to improve recycled aggregates' properties.
- Architectural and civil engineering efforts to promote high-performance, eco-friendly materials are crucial. The EU's support for CDW recovery strategies highlights the importance of managing construction waste.

- The potential use of green cement and crystallizing agents to enhance cement sustainability. Green cement reduces energy use and carbon footprint, unlike traditional Portland cement. Crystallizing agents improve concrete durability and self-healing, reducing maintenance needs. Their use enhances both the environmental performance and longevity of cement-based materials.
- Using CDW, green cement, and crystallizing agents offers a path toward a greener construction industry. Continued research and collaboration are essential to expand these strategies and create a sustainable built environment for future generations.

Author Contributions: Conceptualization, D.S.-R., L.R. and D.F.; methodology, L.L., G.A.F., J.-M.T. and M.P.; validation, M.P., J.-M.T. and D.F.; formal analysis, D.S.-R., L.R., L.L. and D.F.; investigation, D.S.-R. and M.P.; resources, L.L. and J.-M.T.; data curation, L.L., G.A.F., L.R., M.P.; writing—original draft preparation, D.S.-R., D.F. and L.L.; writing—review and editing, D.S.-R., J.-M.T., D.F., M.P. and L.L.; visualization, D.S.-R.; supervision, G.A.F., J.-M.T., L.R. and M.P.; project administration, L.R. and G.A.F.; funding acquisition, L.L. and J.-M.T. All authors have read and agreed to the published version of the manuscript.

Funding: This research received no external funding.

Data Availability Statement: No new data were created or analyzed in this study. Data sharing is not applicable to this article.

Conflicts of Interest: The authors declare no conflicts of interest.

References

1. Mehta, P.; Monteiro, P. *Concrete: Microstructure, Properties, and Materials*; McGraw-Hill Education: New York, NY, USA, 2014; ISBN 0-07-179787-4.
2. Key Facts & Figures. Available online: https://cembureau.eu/about-our-industry/key-facts-figures/ (accessed on 3 April 2024).
3. Hajek, P. Concrete Structures for Sustainability in a Changing World. *Procedia Eng.* **2017**, *171*, 207–214. [CrossRef]
4. Hassler, U.; Kohler, N. Resilience in the Built Environment. *Build. Res. Inf.* **2014**, *42*, 119–129. [CrossRef]
5. Izumi, Y.; Iizuka, A.; Ho, H.-J. Calculation of Greenhouse Gas Emissions for a Carbon Recycling System Using Mineral Carbon Capture and Utilization Technology in the Cement Industry. *J. Clean. Prod.* **2021**, *312*, 127618. [CrossRef]
6. Galvez-Martos, J.-L.; Schoenberger, H. An Analysis of the Use of Life Cycle Assessment for Waste Co-Incineration in Cement Kilns. *Resour. Conserv. Recycl.* **2014**, *86*, 118–131. [CrossRef]
7. Sarawan, S.; Wongwuttanasatian, T. A Feasibility Study of Using Carbon Black as a Substitute to Coal in Cement Industry. *Energy Sustain. Dev.* **2013**, *17*, 257–260. [CrossRef]
8. Zhang, L.; Mabee, W.E. Comparative Study on the Life-Cycle Greenhouse Gas Emissions of the Utilization of Potential Low Carbon Fuels for the Cement Industry. *J. Clean. Prod.* **2016**, *122*, 102–112. [CrossRef]
9. Meng, D.; Unluer, C.; Yang, E.-H.; Qian, S. Carbon Sequestration and Utilization in Cement-Based Materials and Potential Impacts on Durability of Structural Concrete. *Constr. Build. Mater.* **2022**, *361*, 129610. [CrossRef]
10. Sivakrishna, A.; Adesina, A.; Awoyera, P.O.; Rajesh Kumar, K. Green Concrete: A Review of Recent Developments. *Mater. Today Proc.* **2020**, *27*, 54–58. [CrossRef]
11. Habert, G.; Miller, S.A.; John, V.M.; Provis, J.L.; Favier, A.; Horvath, A.; Scrivener, K.L. Environmental Impacts and Decarbonization Strategies in the Cement and Concrete Industries. *Nat. Rev. Earth Environ.* **2020**, *1*, 559–573. [CrossRef]
12. Coppola, L.; Bellezze, T.; Belli, A.; Bignozzi, M.C.; Bolzoni, F.; Brenna, A.; Cabrini, M.; Candamano, S.; Cappai, M.; Caputo, D.; et al. Binders Alternative to Portland Cement and Waste Management for Sustainable Construction—Part 1. *J. Appl. Biomater. Funct. Mater.* **2018**, *16*, 186–202. [CrossRef]
13. Yuan, H. Key Indicators for Assessing the Effectiveness of Waste Management in Construction Projects. *Ecol. Indic.* **2013**, *24*, 476–484. [CrossRef]
14. Wu, H.; Gao, J.; Liu, C.; Guo, Z.; Luo, X. Reusing Waste Clay Brick Powder for Low-Carbon Cement Concrete and Alkali-Activated Concrete: A Critical Review. *J. Clean. Prod.* **2024**, *449*, 141755. [CrossRef]
15. Ma, Z.; Shen, J.; Wang, C.; Wu, H. Characterization of Sustainable Mortar Containing High-Quality Recycled Manufactured Sand Crushed from Recycled Coarse Aggregate. *Cem. Concr. Compos.* **2022**, *132*, 104629. [CrossRef]
16. IEA Global Cement Demand for Building Construction, 2000–2020, and in the Net Zero Scenario, 2025–2030. Available online: https://www.iea.org/data-and-statistics/charts/global-cement-demand-for-building-construction-2000-2020-and-in-the-net-zero-scenario-2025-2030 (accessed on 8 January 2024).
17. Kazemian, M.; Shafei, B. Carbon Sequestration and Storage in Concrete: A State-of-the-Art Review of Compositions, Methods, and Developments. *J. CO2 Util.* **2023**, *70*, 102443. [CrossRef]
18. Bosoaga, A.; Masek, O.; Oakey, J.E. CO_2 Capture Technologies for Cement Industry. *Energy Procedia* **2009**, *1*, 133–140. [CrossRef]

19. Li, J.; Tharakan, P.; Macdonald, D.; Liang, X. Technological, Economic and Financial Prospects of Carbon Dioxide Capture in the Cement Industry. *Energy Policy* **2013**, *61*, 1377–1387. [CrossRef]
20. Kivi, I.R.; Makhnenko, R.Y.; Oldenburg, C.M.; Rutqvist, J.; Vilarrasa, V. Multi-Layered Systems for Permanent Geologic Storage of CO_2 at the Gigatonne Scale. *Geophys. Res. Lett.* **2022**, *49*, e2022GL100443. [CrossRef]
21. Khoo, H.H.; Bu, J.; Wong, R.L.; Kuan, S.Y.; Sharratt, P.N. Carbon Capture and Utilization: Preliminary Life Cycle CO_2, Energy, and Cost Results of Potential Mineral Carbonation. *Energy Procedia* **2011**, *4*, 2494–2501. [CrossRef]
22. Styring, P.; Quadrelli, E.A.; Armstrong, K. *Carbon Dioxide Utilisation: Closing the Carbon Cycle*; Elsevier: Amsterdam, The Netherlands, 2014; ISBN 0-444-62748-0.
23. Cuéllar-Franca, R.M.; Azapagic, A. Carbon Capture, Storage and Utilisation Technologies: A Critical Analysis and Comparison of Their Life Cycle Environmental Impacts. *J. CO2 Util.* **2015**, *9*, 82–102. [CrossRef]
24. Hamada, H.M.; Abdulhaleem, K.N.; Majdi, A.; Al Jawahery, M.S.; Skariah Thomas, B.; Yousif, S.T. The Durability of Concrete Produced from Pozzolan Materials as a Partially Cement Replacement: A Comprehensive Review. *Mater. Today: Proc.* **2023**, *in press*. [CrossRef]
25. Shi, C.; Jiménez, A.F.; Palomo, A. New Cements for the 21st Century: The Pursuit of an Alternative to Portland Cement. *Cem. Concr. Res.* **2011**, *41*, 750–763. [CrossRef]
26. Palmero, P.; Formia, A.; Tulliani, J.-M.; Antonaci, P. Valorisation of Alumino-Silicate Stone Muds: From Wastes to Source Materials for Innovative Alkali-Activated Materials. *Cem. Concr. Compos.* **2017**, *83*, 251–262. [CrossRef]
27. Mijarsh, M.J.A.; Megat Johari, M.A.; Ahmad, Z.A. Synthesis of Geopolymer from Large Amounts of Treated Palm Oil Fuel Ash: Application of the Taguchi Method in Investigating the Main Parameters Affecting Compressive Strength. *Constr. Build. Mater.* **2014**, *52*, 473–481. [CrossRef]
28. Collins, F.; Sanjayan, J.G. Microcracking and Strength Development of Alkali Activated Slag Concrete. *Cem. Concr. Compos.* **2001**, *23*, 345–352. [CrossRef]
29. Provis, J.L. Geopolymers and Other Alkali Activated Materials: Why, How, and What? *Mater. Struct.* **2014**, *47*, 11–25. [CrossRef]
30. Valente, M.; Sambucci, M.; Chougan, M.; Ghaffar, S.H. Reducing the Emission of Climate-Altering Substances in Cementitious Materials: A Comparison between Alkali-Activated Materials and Portland Cement-Based Composites Incorporating Recycled Tire Rubber. *J. Clean. Prod.* **2022**, *333*, 130013. [CrossRef]
31. Bernal, S.A.; Provis, J.L. Durability of Alkali-Activated Materials: Progress and Perspectives. *J. Am. Ceram. Soc.* **2014**, *97*, 997–1008. [CrossRef]
32. Almutairi, A.L.; Tayeh, B.A.; Adesina, A.; Isleem, H.F.; Zeyad, A.M. Potential Applications of Geopolymer Concrete in Construction: A Review. *Case Stud. Constr. Mater.* **2021**, *15*, e00733. [CrossRef]
33. Juenger, M.C.G.; Winnefeld, F.; Provis, J.L.; Ideker, J.H. Advances in Alternative Cementitious Binders. *Cem. Concr. Res.* **2011**, *41*, 1232–1243. [CrossRef]
34. Pace, M.L.; Telesca, A.; Marroccoli, M.; Valenti, G.L. Use of Industrial Byproducts as Alumina Sources for the Synthesis of Calcium Sulfoaluminate Cements. *Environ. Sci. Technol.* **2011**, *45*, 6124–6128. [CrossRef]
35. Klein, A. Calcium Aluminosulfate and Expansive Cements Containing Same. U.S. Patent 3155526A, 3 November 1964.
36. Tao, Y.; Rahul, A.V.; Mohan, M.K.; De Schutter, G.; Van Tittelboom, K. Recent Progress and Technical Challenges in Using Calcium Sulfoaluminate (CSA) Cement. *Cem. Concr. Compos.* **2023**, *137*, 104908. [CrossRef]
37. Aranda, M.A.G.; De la Torre, A.G. 18–Sulfoaluminate Cement. In *Eco-Efficient Concrete*; Pacheco-Torgal, F., Jalali, S., Labrincha, J., John, V.M., Eds.; Woodhead Publishing: Sawston, UK, 2013; pp. 488–522. ISBN 978-0-85709-424-7.
38. Ren, C.; Wang, W.; Mao, Y.; Yuan, X.; Song, Z.; Sun, J.; Zhao, X. Comparative Life Cycle Assessment of Sulfoaluminate Clinker Production Derived from Industrial Solid Wastes and Conventional Raw Materials. *J. Clean. Prod.* **2017**, *167*, 1314–1324. [CrossRef]
39. Kurtis, K.E. Innovations in Cement-Based Materials: Addressing Sustainability in Structural and Infrastructure Applications. *MRS Bull.* **2015**, *40*, 1102–1109. [CrossRef]
40. Mobili, A.; Belli, A.; Giosuè, C.; Telesca, A.; Marroccoli, M.; Tittarelli, F. Calcium Sulfoaluminate, Geopolymeric, and Cementitious Mortars for Structural Applications. *Environments* **2017**, *4*, 64. [CrossRef]
41. Rumman, R.; Bediwy, A.; Alam, M.S. Revolutionizing Concrete Durability: Case Studies on Encapsulation- Based Chemical (Autonomous) Self-Healing Techniques and Future Directions—A Critical Review. *Case Stud. Constr. Mater.* **2024**, *20*, e03216. [CrossRef]
42. Zhang, W.; Zheng, Q.; Ashour, A.; Han, B. Self-Healing Cement Concrete Composites for Resilient Infrastructures: A Review. *Compos. Part B Eng.* **2020**, *189*, 107892. [CrossRef]
43. De Belie, N.; Gruyaert, E.; Al-Tabbaa, A.; Antonaci, P.; Baera, C.; Bajare, D.; Darquennes, A.; Davies, R.; Ferrara, L.; Jefferson, T.; et al. A Review of Self-Healing Concrete for Damage Management of Structures. *Adv Mater. Inter* **2018**, *5*, 1800074. [CrossRef]
44. Cappellesso, V.; Di Summa, D.; Pourhaji, P.; Prabhu Kannikachalam, N.; Dabral, K.; Ferrara, L.; Cruz Alonso, M.; Camacho, E.; Gruyaert, E.; De Belie, N. A Review of the Efficiency of Self-Healing Concrete Technologies for Durable and Sustainable Concrete under Realistic Conditions. *Int. Mater. Rev.* **2023**, *68*, 556–603. [CrossRef]
45. Wang, J.Y.; Soens, H.; Verstraete, W.; De Belie, N. Self-Healing Concrete by Use of Microencapsulated Bacterial Spores. *Cem. Concr. Res.* **2014**, *56*, 139–152. [CrossRef]
46. Dong, B.; Fang, G.; Ding, W.; Liu, Y.; Zhang, J.; Han, N.; Xing, F. Self-Healing Features in Cementitious Material with Urea–Formaldehyde/Epoxy Microcapsules. *Constr. Build. Mater.* **2016**, *106*, 608–617. [CrossRef]

47. Lv, L.; Yang, Z.; Chen, G.; Zhu, G.; Han, N.; Schlangen, E.; Xing, F. Synthesis and Characterization of a New Polymeric Microcapsule and Feasibility Investigation in Self-Healing Cementitious Materials. *Constr. Build. Mater.* **2016**, *105*, 487–495. [CrossRef]

48. Hilloulin, B.; Hilloulin, D.; Grondin, F.; Loukili, A.; De Belie, N. Mechanical Regains Due to Self-Healing in Cementitious Materials: Experimental Measurements and Micro-Mechanical Model. *Cem. Concr. Res.* **2016**, *80*, 21–32. [CrossRef]

49. Li, W.; Jiang, Z.; Yang, Z.; Zhao, N.; Yuan, W. Self-Healing Efficiency of Cementitious Materials Containing Microcapsules Filled with Healing Adhesive: Mechanical Restoration and Healing Process Monitored by Water Absorption. *PLoS ONE* **2013**, *8*, e81616. [CrossRef] [PubMed]

50. Lee, Y.-S.; Ryou, J.-S. Self Healing Behavior for Crack Closing of Expansive Agent via Granulation/Film Coating Method. *Constr. Build. Mater.* **2014**, *71*, 188–193. [CrossRef]

51. Yang, Z.; Hollar, J.; He, X.; Shi, X. A Self-Healing Cementitious Composite Using Oil Core/Silica Gel Shell Microcapsules. *Cem. Concr. Compos.* **2011**, *33*, 506–512. [CrossRef]

52. Kanellopoulos, A.; Giannaros, P.; Al-Tabbaa, A. The Effect of Varying Volume Fraction of Microcapsules on Fresh, Mechanical and Self-Healing Properties of Mortars. *Constr. Build. Mater.* **2016**, *122*, 577–593. [CrossRef]

53. Li, V.C.; Lim, Y.M.; Chan, Y.-W. Feasibility Study of a Passive Smart Self-Healing Cementitious Composite. *Compos. Part B Eng.* **1998**, *29*, 819–827. [CrossRef]

54. Mihashi, H.; Kaneko, Y.; Nishiwaki, T.; Otsuka, K. Fundamental Study on Development of Intelligent Concrete Characterized by Self-Healing Capability for Strength. *Concr. Res. Technol.* **2000**, *11*, 21–28. [CrossRef]

55. Riordan, C.; Anglani, G.; Inserra, B.; Palmer, D.; Al-Tabbaa, A.; Tulliani, J.-M.; Antonaci, P. Novel Production of Macrocapsules for Self-Sealing Mortar Specimens Using Stereolithographic 3D Printers. *Cem. Concr. Compos.* **2023**, *142*, 105216. [CrossRef]

56. Nishiwaki, T.; Mihashi, H.; Jang, B.-K.; Miura, K. Development of Self-Healing System for Concrete with Selective Heating around Crack. *ACT* **2006**, *4*, 267–275. [CrossRef]

57. Šavija, B.; Feiteira, J.; Araújo, M.; Chatrabhuti, S.; Raquez, J.-M.; Van Tittelboom, K.; Gruyaert, E.; De Belie, N.; Schlangen, E. Simulation-Aided Design of Tubular Polymeric Capsules for Self-Healing Concrete. *Materials* **2016**, *10*, 10. [CrossRef] [PubMed]

58. Formia, A.; Terranova, S.; Antonaci, P.; Pugno, N.; Tulliani, J. Setup of Extruded Cementitious Hollow Tubes as Containing/Releasing Devices in Self-Healing Systems. *Materials* **2015**, *8*, 1897–1923. [CrossRef] [PubMed]

59. Formia, A.; Irico, S.; Bertola, F.; Canonico, F.; Antonaci, P.; Pugno, N.M.; Tulliani, J.-M. Experimental Analysis of Self-Healing Cement-Based Materials Incorporating Extruded Cementitious Hollow Tubes. *J. Intell. Mater. Syst. Struct.* **2016**, *27*, 2633–2652. [CrossRef]

60. Dry, C. Matrix Cracking Repair and Filling Using Active and Passive Modes for Smart Timed Release of Chemicals from Fibers into Cement Matrices. *Smart Mater. Struct.* **1994**, *3*, 118–123. [CrossRef]

61. Shields, Y.; Tsangouri, E.; Riordan, C.; De Nardi, C.; Godinho, J.R.A.; Ooms, T.; Antonaci, P.; Palmer, D.; Al-Tabbaa, A.; Jefferson, T.; et al. Non-Destructive Evaluation of Ductile-Porous versus Brittle 3D Printed Vascular Networks in Self-Healing Concrete. *Cem. Concr. Compos.* **2024**, *145*, 105333. [CrossRef]

62. Zhu, X.; Wang, J.; De Belie, N.; Boon, N. Complementing Urea Hydrolysis and Nitrate Reduction for Improved Microbially Induced Calcium Carbonate Precipitation. *Appl. Microbiol. Biotechnol.* **2019**, *103*, 8825–8838. [CrossRef] [PubMed]

63. Erşan, Y.Ç.; Hernandez-Sanabria, E.; Boon, N.; De Belie, N. Enhanced Crack Closure Performance of Microbial Mortar through Nitrate Reduction. *Cem. Concr. Compos.* **2016**, *70*, 159–170. [CrossRef]

64. Basilisk. Basilisk Self-Healing Concrete. Available online: https://basiliskconcrete.com/en/ (accessed on 3 April 2024).

65. Mahmood, A.; Noman, M.T.; Pechočiaková, M.; Amor, N.; Petrů, M.; Abdelkader, M.; Militký, J.; Sozcu, S.; Hassan, S.Z. Geopolymers and Fiber-Reinforced Concrete Composites in Civil Engineering. *Polymers* **2021**, *13*, 2099. [CrossRef]

66. Li, W.; Shumuye, E.D.; Shiying, T.; Wang, Z.; Zerfu, K. Eco-Friendly Fibre Reinforced Geopolymer Concrete: A Critical Review on the Microstructure and Long-Term Durability Properties. *Case Stud. Constr. Mater.* **2022**, *16*, e00894. [CrossRef]

67. Al-Kharabsheh, B.N.; Arbili, M.M.; Majdi, A.; Alogla, S.M.; Hakamy, A.; Ahmad, J.; Deifalla, A.F. Basalt Fiber Reinforced Concrete: A Compressive Review on Durability Aspects. *Materials* **2023**, *16*, 429. [CrossRef]

68. Al-Rousan, E.T.; Khalid, H.R.; Rahman, M.K. Fresh, Mechanical, and Durability Properties of Basalt Fiber-Reinforced Concrete (BFRC): A Review. *Dev. Built Environ.* **2023**, *14*, 100155. [CrossRef]

69. Maduta, C.; Melica, G.; D'Agostino, D.; Bertoldi, P. Towards a Decarbonised Building Stock by 2050: The Meaning and the Role of Zero Emission Buildings (ZEBs) in Europe. *Energy Strategy Rev.* **2022**, *44*, 101009. [CrossRef]

70. Ginga, C.P.; Ongpeng, J.M.C.; Daly, M.K.M. Circular Economy on Construction and Demolition Waste: A Literature Review on Material Recovery and Production. *Materials* **2020**, *13*, 2970. [CrossRef]

71. Huang, B.; Wang, X.; Kua, H.; Geng, Y.; Bleischwitz, R.; Ren, J. Construction and Demolition Waste Management in China through the 3R Principle. *Resour. Conserv. Recycl.* **2018**, *129*, 36–44. [CrossRef]

72. U.S. Environmental Protection Agency. Construction and Demolition Debris: Material-Specific Data. Available online: https://www.epa.gov/facts-and-figures-about-materials-waste-and-recycling/construction-and-demolition-debris-material (accessed on 13 February 2024).

73. Wang, Z.; Xie, W.; Liu, J. Regional Differences and Driving Factors of Construction and Demolition Waste Generation in China. *ECAM* **2022**, *29*, 2300–2327. [CrossRef]

74. Wang, Z.; Zhang, Z.; Jin, X. A Study on the Spatial Network Characteristics and Effects of CDW Generation in China. *Waste Manag.* **2021**, *128*, 179–188. [CrossRef] [PubMed]
75. Jin, R.; Yuan, H.; Chen, Q. Science Mapping Approach to Assisting the Review of Construction and Demolition Waste Management Research Published between 2009 and 2018. *Resour. Conserv. Recycl.* **2019**, *140*, 175–188. [CrossRef]
76. Nasir, M.H.A.; Genovese, A.; Acquaye, A.A.; Koh, S.C.L.; Yamoah, F. Comparing Linear and Circular Supply Chains: A Case Study from the Construction Industry. *Int. J. Prod. Econ.* **2017**, *183*, 443–457. [CrossRef]
77. Ferronato, N.; Fuentes Sirpa, R.C.; Guisbert Lizarazu, E.G.; Conti, F.; Torretta, V. Construction and Demolition Waste Recycling in Developing Cities: Management and Cost Analysis. *Environ. Sci. Pollut. Res.* **2023**, *30*, 24377–24397. [CrossRef]
78. Papastamoulis, V.; London, K.; Feng, Y.; Zhang, P.; Crocker, R.; Patias, P. Conceptualising the Circular Economy Potential of Construction and Demolition Waste: An Integrative Literature Review. *Recycling* **2021**, *6*, 61. [CrossRef]
79. Villoria Sáez, P.; Osmani, M. A Diagnosis of Construction and Demolition Waste Generation and Recovery Practice in the European Union. *J. Clean. Prod.* **2019**, *241*, 118400. [CrossRef]
80. Monsù Scolaro, A.; De Medici, S. Downcycling and Upcycling in Rehabilitation and Adaptive Reuse of Pre-Existing Buildings: Re-Designing Technological Performances in an Environmental Perspective. *Energies* **2021**, *14*, 6863. [CrossRef]
81. Kenai, S. 3–Recycled Aggregates. In *Waste and Supplementary Cementitious Materials in Concrete*; Siddique, R., Cachim, P., Eds.; Woodhead Publishing: Sawston, UK, 2018; pp. 79–120. ISBN 978-0-08-102156-9.
82. Blengini, G.A.; Garbarino, E. Resources and Waste Management in Turin (Italy): The Role of Recycled Aggregates in the Sustainable Supply Mix. *J. Clean. Prod.* **2010**, *18*, 1021–1030. [CrossRef]
83. Martinez-Echevarria, M.J.; Lopez-Alonso, M.; Garach, L.; Alegre, J.; Poon, C.S.; Agrela, F.; Cabrera, M. Crushing Treatment on Recycled Aggregates to Improve Their Mechanical Behaviour for Use in Unbound Road Layers. *Constr. Build. Mater.* **2020**, *263*, 120517. [CrossRef]
84. Panizza, M.; Natali, M.; Garbin, E.; Ducman, V.; Tamburini, S. Optimization and Mechanical-Physical Characterization of Geopolymers with Construction and Demolition Waste (CDW) Aggregates for Construction Products. *Constr. Build. Mater.* **2020**, *264*, 120158. [CrossRef]
85. Vincent, T.; Guy, M.; Louis-César, P.; Jean-François, B.; Richard, M. Physical Process to Sort Construction and Demolition Waste (C&DW) Fines Components Using Process Water. *Waste Manag.* **2022**, *143*, 125–134. [CrossRef] [PubMed]
86. Khatib, J.M. Properties of Concrete Incorporating Fine Recycled Aggregate. *Cem. Concr. Res.* **2005**, *35*, 763–769. [CrossRef]
87. Kou, S.-C.; Poon, C.-S. Properties of Concrete Prepared with Crushed Fine Stone, Furnace Bottom Ash and Fine Recycled Aggregate as Fine Aggregates. *Constr. Build. Mater.* **2009**, *23*, 2877–2886. [CrossRef]
88. Evangelista, L.; De Brito, J. Mechanical Behaviour of Concrete Made with Fine Recycled Concrete Aggregates. *Cem. Concr. Compos.* **2007**, *29*, 397–401. [CrossRef]
89. Singh, R.; Nayak, D.; Pandey, A.; Kumar, R.; Kumar, V. Effects of Recycled Fine Aggregates on Properties of Concrete Containing Natural or Recycled Coarse Aggregates: A Comparative Study. *J. Build. Eng.* **2022**, *45*, 103442. [CrossRef]
90. Restuccia, L. Fracture Properties of Green Mortars with Recycled Sand. *Frat. Integrità Strutt.* **2019**, *13*, 676–689. [CrossRef]
91. Miranda, L.F.R.; Selmo, S.M.S. CDW Recycled Aggregate Renderings: Part I–Analysis of the Effect of Materials Finer than 75 Mm on Mortar Properties. *Constr. Build. Mater.* **2006**, *20*, 615–624. [CrossRef]
92. Restuccia, L.; Spoto, C.; Ferro, G.A.; Tulliani, J.-M. Recycled Mortars with C&D Waste. *Procedia Struct. Integr.* **2016**, *2*, 2896–2904. [CrossRef]
93. Stefanidou, M.; Anastasiou, E.; Georgiadis Filikas, K. Recycled Sand in Lime-Based Mortars. *Waste Manag.* **2014**, *34*, 2595–2602. [CrossRef] [PubMed]
94. Raeis Samiei, R.; Daniotti, B.; Pelosato, R.; Dotelli, G. Properties of Cement–Lime Mortars vs. Cement Mortars Containing Recycled Concrete Aggregates. *Constr. Build. Mater.* **2015**, *84*, 84–94. [CrossRef]
95. Ledesma, E.F.; Jiménez, J.R.; Ayuso, J.; Fernández, J.M.; De Brito, J. Maximum Feasible Use of Recycled Sand from Construction and Demolition Waste for Eco-Mortar Production–Part-I: Ceramic Masonry Waste. *J. Clean. Prod.* **2015**, *87*, 692–706. [CrossRef]
96. Zhao, Z.; Remond, S.; Damidot, D.; Xu, W. Influence of Fine Recycled Concrete Aggregates on the Properties of Mortars. *Constr. Build. Mater.* **2015**, *81*, 179–186. [CrossRef]
97. Ollivier, J.P.; Maso, J.C.; Bourdette, B. Interfacial Transition Zone in Concrete. *Adv. Cem. Based Mater.* **1995**, *2*, 30–38. [CrossRef]
98. Prokopski, G.; Halbiniak, J. Interfacial Transition Zone in Cementitious Materials. *Cem. Concr. Res.* **2000**, *30*, 579–583. [CrossRef]
99. Collepardi, M. A State-of-the-Art Review on Delayed Ettringite Attack on Concrete. *Cem. Concr. Compos.* **2003**, *25*, 401–407. [CrossRef]
100. He, R.; Zheng, S.; Gan, V.J.L.; Wang, Z.; Fang, J.; Shao, Y. Damage Mechanism and Interfacial Transition Zone Characteristics of Concrete under Sulfate Erosion and Dry-Wet Cycles. *Constr. Build. Mater.* **2020**, *255*, 119340. [CrossRef]
101. Zhao, G.; Li, J.; Shi, M.; Fan, H.; Cui, J.; Xie, F. Degradation Mechanisms of Cast-in-Situ Concrete Subjected to Internal-External Combined Sulfate Attack. *Constr. Build. Mater.* **2020**, *248*, 118683. [CrossRef]
102. Castellote, M.; Fernandez, L.; Andrade, C.; Alonso, C. Chemical Changes and Phase Analysis of OPC Pastes Carbonated at Different CO_2 Concentrations. *Mater. Struct.* **2009**, *42*, 515–525. [CrossRef]
103. Shi, C.; Wu, Z.; Cao, Z.; Ling, T.C.; Zheng, J. Performance of Mortar Prepared with Recycled Concrete Aggregate Enhanced by CO_2 and Pozzolan Slurry. *Cem. Concr. Compos.* **2018**, *86*, 130–138. [CrossRef]

104. Wang, L.; Wang, J.; Xu, Y.; Cui, L.; Qian, X.; Chen, P.; Fang, Y. Consolidating Recycled Concrete Aggregates Using Phosphate Solution. *Constr. Build. Mater.* **2019**, *200*, 703–712. [CrossRef]
105. Ismail, S.; Ramli, M. Engineering Properties of Treated Recycled Concrete Aggregate (RCA) for Structural Applications. *Constr. Build. Mater.* **2013**, *44*, 464–476. [CrossRef]
106. Tam, V.W.Y.; Tam, C.M.; Le, K.N. Removal of Cement Mortar Remains from Recycled Aggregate Using Pre-Soaking Approaches. *Resour. Conserv. Recycl.* **2007**, *50*, 82–101. [CrossRef]
107. Montgomery, D.G. Workability and Compressive Strength Properties of Concrete Containing Recycled Concrete Aggregate. In *Sustainable Construction: Use of Recycled Concrete Aggregate*; Thomas Telford Publishing: London, UK, 1998; pp. 287–296.
108. Zhu, Y.-G.; Kou, S.-C.; Poon, C.-S.; Dai, J.-G.; Li, Q.-Y. Influence of Silane-Based Water Repellent on the Durability Properties of Recycled Aggregate Concrete. *Cem. Concr. Compos.* **2013**, *35*, 32–38. [CrossRef]
109. Zhan, B.; Poon, C.S.; Liu, Q.; Kou, S.; Shi, C. Experimental Study on CO_2 Curing for Enhancement of Recycled Aggregate Properties. *Constr. Build. Mater.* **2014**, *67*, 3–7. [CrossRef]
110. Grabiec, A.M.; Klama, J.; Zawal, D.; Krupa, D. Modification of Recycled Concrete Aggregate by Calcium Carbonate Biodeposition. *Constr. Build. Mater.* **2012**, *34*, 145–150. [CrossRef]
111. Tam, V.W.Y.; Wattage, H.; Le, K.N.; Buteraa, A.; Soomro, M. Methods to Improve Microstructural Properties of Recycled Concrete Aggregate: A Critical Review. *Constr. Build. Mater.* **2021**, *270*, 121490. [CrossRef]
112. Al-Kheetan, M.J.; Rahman, M.M.; Chamberlain, D.A. A Novel Approach of Introducing Crystalline Protection Material and Curing Agent in Fresh Concrete for Enhancing Hydrophobicity. *Constr. Build. Mater.* **2018**, *160*, 644–652. [CrossRef]
113. Teng, L.-W.; Huang, R.; Chen, J.; Cheng, A.; Hsu, H.-M. A Study of Crystalline Mechanism of Penetration Sealer Materials. *Materials* **2014**, *7*, 399–412. [CrossRef]
114. Al-Kheetan, M.J.; Rahman, M.M.; Chamberlain, D.A. Optimum Mix Design for Internally Integrated Concrete with Crystallizing Protective Material. *J. Mater. Civ. Eng.* **2019**, *31*, 04019101. [CrossRef]
115. Al-Kheetan, M.J.; Rahman, M.M.; Chamberlain, D.A. Development of Hydrophobic Concrete by Adding Dual-Crystalline Admixture at Mixing Stage. *Struct. Concr.* **2018**, *19*, 1504–1511. [CrossRef]
116. Al-Kheetan, M.J.; Rahman, M.M.; Chamberlain, D.A. Influence of Early Water Exposure on Modified Cementitious Coating. *Constr. Build. Mater.* **2017**, *141*, 64–71. [CrossRef]
117. Reiterman, P.; Pazderka, J. Crystalline Coating and Its Influence on the Water Transport in Concrete. *Adv. Civ. Eng.* **2016**, *2016*, 1–8. [CrossRef]
118. Gojević, A.; Ducman, V.; Netinger Grubeša, I.; Baričević, A.; Banjad Pečur, I. The Effect of Crystalline Waterproofing Admixtures on the Self-Healing and Permeability of Concrete. *Materials* **2021**, *14*, 1860. [CrossRef] [PubMed]
119. Zhang, Y.; Wang, R.; Ding, Z. Influence of Crystalline Admixtures and Their Synergetic Combinations with Other Constituents on Autonomous Healing in Cracked Concrete—A Review. *Materials* **2022**, *15*, 440. [CrossRef]
120. de Souza Oliveira, A.; Dweck, J.; de Moraes Rego Fairbairn, E.; da Fonseca Martins Gomes, O.; Toledo Filho, R.D. Crystalline Admixture Effects on Crystal Formation Phenomena during Cement Pastes' Hydration. *J. Therm. Anal. Calorim.* **2020**, *139*, 3361–3375. [CrossRef]

 *infrastructures*

Article

The Impact of Production Techniques on Pore Size Distribution in High-Strength Foam Concrete

Slava Markin, Genadijs Sahmenko, Aleksandrs Korjakins and Viktor Mechtcherine

1   Sika Deutschland GmbH, TM Concrete, Peter-Schuhmacher-Strasse 8, 69181 Leimen, Germany
2   Institute of Sustainable Building Materials and Engineering Systems, Faculty of Civil and Mechanical Engineering, Riga Technical University (RTU), Kipsala str 6A, LV-1048 Riga, Latvia; genadijs.sahmenko@rtu.lv (G.S.); aleksandrs.korjakins@rtu.lv (A.K.)
3   Institute for Construction Materials, Dresden University of Technology, Georg-Schumann-Straße 7, 01602 Dresden, Germany; viktor.mechtcherine@tu-dresden.de
*   Correspondence: markin.slava@de.sika.com

Abstract: This study examined the impact of various foam concrete production techniques on pore size distribution and its water absorption properties. Techniques such as the use of a cavitation disintegrator and a turbulent mixer were employed to produce foam concrete. Six foam concrete compositions, with dry densities ranging from 820 to 1480 kg/m^3 and compressive strength up to 47 MPa, were prepared. A novel method for digital image correlation was applied to analyse the pore size distribution within the foam concrete specimens. The manufactured foam concrete specimens' porosity and water absorption indices were determined. The experimental results, including compression strength and water absorption, indicated that the production technique significantly affects the pore size distribution in foam concrete, impacting its mechanical and durability properties. Compressive strength was assessed at curing intervals of 7, 28, and 180 days. Cavitation technology was found to promote the formation of a finer porous structure in foam concrete, resulting in enhanced strength properties.

Keywords: foam concrete; pore size distribution; water absorption; production technique

Academic Editors: Chris Goodier and Hong Wong

Received: 14 October 2024
Revised: 22 December 2024
Accepted: 30 December 2024
Published: 9 January 2025

Citation: Slava Markin, Genadijs Sahmenko, Aleksandrs Korjakins and Viktor Mechtcherine. The Impact of Production Techniques on Pore Size Distribution in High-Strength Foam Concrete. *Infrastructures* **2025**, *10*, 14. https://doi.org/10.3390/infrastructures10010014

1. Introduction

Foam concrete (FC) is a lightweight cementitious material with a cellular structure produced by incorporating the air voids into the cement-based matrix. It can be designed to have any density within the range of 200–1900 kg/m^3. FC is characterised by high flowability, low cement and aggregate content, relatively low strength, excellent thermal insulation, high fire resistance, and good sound insulation properties [1–7]. It should be noted that FC is a universal material suitable for application as a heat-insulating material, as well as for self-supporting and load-bearing elements [8].

The properties of the FC are influenced by the mixture components, their proportions, and the mixing method. Key factors include the type and content of cement, the water-to-cement or water-to-binder ratio, the grading distribution, the filler-to-cement ratio, and the type of foaming agent (FA), all of which play a significant role. The strength of FC is closely associated with its density, with lower densities typically leading to reduced strength. For example, decreasing the water-to-cement ratio from 0.5 to 0.38 has been found to result in an 18.1% increase in compressive strength [9].

The strength of FC is generally significantly lower than that of normal-weight concrete due to its lower density and higher porosity [10]. However, when higher densities are achieved—up to approximately 1900 kg/m^3—foam concrete can attain strengths exceeding 50 MPa [11]. This performance enhancement at higher densities is primarily attributed to the reduction in pore size and volume, along with a denser matrix structure. Such high-strength foam concrete finds applications in structural elements where weight reduction and adequate load-bearing capacity are both critical requirements.

FC can be produced using various techniques, with two primary methods being widely utilised: (1) the pre-foaming method and (2) the mixed foaming method, also referred to as the inline method or high-speed mixing method.

In the pre-foaming method (1), the cement-based matrix and preformed foam are prepared separately. First, all ingredients, except for the foam, i.e., constituents of the cement-based matrix, are mixed until a homogeneous matrix with a uniform consistency is achieved. Once the cement-based matrix is ready, the required preformed foam is added. In general, a mixing time of 2–3 min is sufficient to integrate the foam into the matrix; however, the mixing duration can vary depending on the output of the foam generator and the desired concrete density [12]. A low mixing speed is recommended to avoid bubble breakage, especially when dealing with a paste with high viscosity.

In contrast, the mixed foaming method (2) involves producing foam in concrete by mixing the foaming agent directly with the matrix constituents. Usually, the foaming agent is added along with water and the base mix ingredients during the mixing process [3,13]. After forming the foam, the other constituents are introduced into the same mixer and combined until the desired density is achieved. This method is standardised, widely used, and simple to follow [3]. A key advantage of the mixed foaming method is that it simplifies the production process by eliminating the need for separate foam production.

In both production techniques—(1) the pre-foaming method and (2) the mixed foaming method—the foam must remain stable to withstand the pressure from the mortar until the initial set. This stability is crucial for forming a strong concrete structure around the air-filled voids, ensuring the integrity of the foam concrete [7]. The methodology for the assessment of the foam stability was presented earlier in [14]. Notably, the pre-foaming method is more expensive than the mixed foaming method, but it offers better foaming efficiency without compromising the quality of the air bubbles in the mixture, i.e., destroying air bubbles [15]. Furthermore, the pre-foaming method is preferred for producing low-density FCs. At the same time, the amount of air bubbles in the mixture is easier to control by adding a specific volume of preformed foam. In contrast, in the mixed foaming method, the air content depends on various factors that are difficult to control; for instance, the quantity of foaming agent needs to be determined for a particular mixer.

Pore size distribution is one of the critical parameters that significantly influences the physical, mechanical, and durability properties of FC. In particular, the mixture consistency and the production technique highly affect the pore size distribution. Larger pores (macro-pores) weaken the material by serving as stress concentration points that can trigger cracking under load. In contrast, a more significant presence of smaller pores (micro-pores) promotes a more uniform stress distribution, enhancing the material's strength. A previous study [16] found that a finer pore distribution and more concentrated and uniform porosity play a key role in the development of high-strength foam concrete (FC). This emphasises the importance of controlling pore size distribution during FC production to achieve an optimal balance between density and strength. An effective FC mix design should focus on minimising large pores while maximising the presence of well-distributed micro-pores, enhancing the structural performance of the material.

Amran et al. [7] indicated that an ideal pore size distribution, with a higher proportion of small, evenly distributed pores, can significantly improve the compressive strength of FC. However, an overly fine pore structure may lead to shrinkage during curing, resulting in micro-cracks that could weaken the material's structural integrity [17]. Ramamurthy et al. [3] also highlighted the critical role of pore structure in influencing the strength of FC. The authors concluded that FC exhibiting a narrower air-void size distribution demonstrates greater strength. Specifically, a finer distribution of smaller air voids can lead to a denser microstructure and improved strength, even if the overall porosity remains constant. Generally, there is a fundamental inverse relationship between porosity and the FC strength [15]. This relationship indicates that as porosity increases, the strength of the material typically decreases.

Pore size distribution also affects the water absorption characteristics and, consequently, the durability of FC [18,19]. Larger pores allow for more water ingress, which can lead to freeze–thaw damage, efflorescence, and other durability issues. On the other hand, FC with a higher proportion of micro-pores shows lower water permeability, making it more resistant to environmental degradation. Moreover, the connectivity of pores is important. Well-connected pores in FC can negatively affect its durability by allowing capillary water to penetrate more easily into the material. This water can lead to several issues, such as freeze–thaw damage and corrosion of reinforcement. Higher water absorption can lead to greater susceptibility to chemical attacks, such as sulphate or acid reactions, which degrade the concrete. Therefore, controlling the distribution and connectivity of pore sizes is key to enhancing the long-term durability of FC, especially in environments exposed to moisture.

The production technique is crucial in determining the size, shape, and distribution of pores—whether macro- or micro-pores—in FC. As such, this study focused on examining how different production techniques influence the pore size distribution and, in turn, the water absorption characteristics of FC. The primary aim was to develop a highly efficient, high-strength FC by employing innovative production methods. Additionally, the study aims to identify correlations between the qualitative aspects of pore distribution and the physical and mechanical properties of the hardened material. Understanding these relationships is vital for optimising both the performance and durability of foam concrete in various applications.

2. Materials and Methods

2.1. Overview of the Mixers Used

The FC mixers used in this study are presented in Figure 1. The specifications of the mixers used are outlined below.

(a)   (b)

Figure 1. Schemes of used mixers: (**a**) cavitation disintegrator (CD): 1. body; 2. conical lid; 3. toothed disks; 4. rectangular recess; 5. impeller; 6. shaft; 7. inlet branch pipe; 8. outlet branch pipe; 9. electric motor; 10. ferrule; 11. plug; 12. leakage opening; (**b**) laboratory turbulence mixer (TM): 1. electrical engine; 2. bearing; 3. pressure compensated coupling; 4. vertical shaft.

(a) Cavitation Disintegrator (CD): Originally, the CD was developed for producing stable fuel mixtures and water–fuel emulsions and improving heavy fuel oil by dispersing asphalt–resin components to enhance the efficiency of added substances. It can also be applied to prepare and activate other emulsion and dispersion systems [20]. Mironovs et al. [21] patented a novel hydrodynamic dispersion method for fine particles, achieved through mechanical activation driven by cavitation, which occurs during the high-speed rotation of toothed disks (up to 7000 rpm). The use of CD for FC production is an innovative approach. Since CD is not widely known in the construction industry, its technical features are worth noting. In CD, a set of toothed disks and an impeller, mounted on a motorised shaft, form a rotor with a frustoconical shape. The fixed body consists of a conical lid and rectangular grooves matching the rotor's toothed disks (see Figure 1a). The body contains an inlet and outlet branch pipe. The electric motor is shielded from liquid by a ferrule installed in an intermediate disk, which is fastened to the motor flange. A plug is installed in the upper part of the body to discharge air. The cement-based matrix, along with the foaming agent, is first placed in the supply reservoir. During operation, the mixture circulates repeatedly from the supply reservoir through a tube into the disintegrator. The CD used in this study had an output capacity of 10 m^3/h, a power rating of 5.5 kW, and was equipped with a frequency converter to adjust the working shaft's speed from 50 rpm to 7000 rpm.

(b) Turbulence Mixer (TM): The TM used in this study enables the production of FC using both the mixed-foaming and pre-foaming methods. The TM can also be used to produce cement-based slurries and pump them using air pressure. The mixer consists of a conical mixing tank, electric motor, bearing, pressure-compensated coupling, and a vertical shaft, as depicted in Figure 1b. The mixer allows mixing under pressure up to 0.7 bar, enabling the production of low-density FC and the pumping of the mixture through excess pressure. It is equipped with a frequency converter to control the shaft speed from 20 rpm to 1000 rpm. As studied earlier in [6], a key advantage of this mixer, especially when used in conjunction with 3D printing concrete (3DPC), is its ability to pump FC directly, eliminating the need for a separate screw pump. According to the TM mixer's technical data, the FC produced under applied air pressure can be pumped up to a distance of 100 m.

2.2. Raw Materials and Mixtures Under Investigation

Type II Portland composite cement CEM II/A-M (S-LL) 52.5 R (OPTERRA Zement GmbH, Werk Karsdorf, Germany) was used in the production of various mixtures. Hard coal fly ash Steament H4 (STEAG Power Minerals GmbH, Dinslaken, Germany) was chosen as a secondary cementitious material. The chemical composition of cementitious materials is presented in Table 1. A polycarboxylate ether (PCE)-based superplasticiser (SP) (MasterGlenium SKY 593, BASF Construction Solutions GmbH, Trostberg, Germany) was used in the cement-based matrix to adjust the workability at reduced water contents. The SP is characterised by a density of 1050 kg/m^3 and a water content of 77% by mass. Two different foaming agents (FAs) were used for the production of the FC: a tenside-based foaming agent (Centripor SK155, MC-Bauchemie GmbH & Co. KG, Bottrop, Germany) and a protein-based foaming agent (Oxal PLB6, MC-Bauchemie GmbH & Co. KG, Bottrop, Germany).

The particle size distribution of the cementitious materials used in the current study is summarised in Figure 2. According to the results, fly ash has slightly larger particle sizes than Portland cement.

Table 1. Chemical composition of cement and fly ash.

Material	Density (kg/m³)	Chemical Composition (%)											
		Residue	SiO$_2$	Al$_2$O$_3$	Fe$_2$O$_3$	CaO	MgO	SO$_3$	Ka$_2$O	Na$_2$O	Loss on Ign.	CO$_2$	CL
CEM II/A-M(S-LL) 52.5 R	3.120	0.74	20.63	5.35	2.82	60.94	2.14	3.52	1.05	0.22	3.47	2.87	0.07
Fly ash H4	2.220					3.6		0.6		2.9	1.8		<0.01

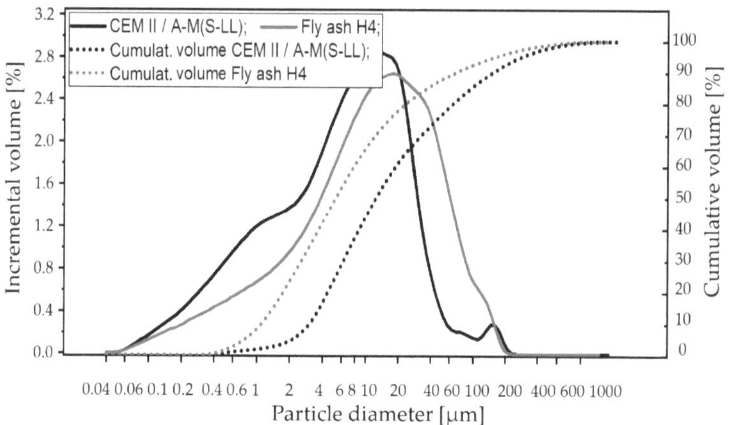

Figure 2. Particle size distribution of used cementitious materials.

The designed FC mix compositions are presented in Table 2. The ratio of components is given in mass proportions. Compositions from M1 to M4 were prepared in a cavitation disintegrator, and compositions M5 and M6 were prepared in a turbulent mixer.

Table 2. Compositions of the foam concretes.

Constituents		Designed Composition of Mixture (Per m³)					
		M-1	M-2	M-3	M-4	M-5	M-6
Cement	(kg)	405	405	405	405	405	405
Fly ash H4	(kg)	192	192	192	192	192	192
Tap water	(kg)	189	189	243	189	189	192
SP SKY 593	(%) *	0.74	1.03	0.15	0.74	0.74	0.88
FA SK-155	(%) *	1.2	1.2	1.2	-	0.7	0.7
FA Oxal PLB6	(%) *	-	-	-	1.2	-	-
w/b	-	0.39	0.39	0.5	0.39	0.39	0.4
Used Mixer		CD	CD	CD	CD	TM	TM

* In the mass percentage of the cement.

The differences in the physical and mechanical properties of compositions M5 and M6 can be explained by the high sensitivity of the mixed foaming method. Even minor changes in the amount of water and plasticising additives may lead to the disruption of the foaming process and changes in the pore structure. As shown in Figure 4f in Section 2.4, the M6 composition has more coarse-sized cells that could have been formed by merging previously formed smaller cells. This could be caused by increased consumption of the plasticiser.

2.3. Experimental Procedure

Dry constituents in proportions according to the mix design in Table 2 were initially mixed with the addition of water and SP for 2 min using a brick trowel. Following the premixing, the foaming agent was added and mixed for an additional 15 s. Finally, the mixture was then transferred to either the CD or TM mixer, depending on the technique used. Considering the technical characteristics of these mixers, the mixing protocols for FC production differed; details are provided in Table 3. After mixing, the FC was cast into moulds. In the case of the CD mixer, the FC circulated numerous times from the supply reservoir through the tube to the disintegrator. At the end of the mixing procedure, the CD's pumping ability allowed the FC to be conveyed directly to the prepared moulds rather than returning to the supply reservoir. For the TM mixer, a pressure of 1.0 bar was settled to transport the FC to the moulds, which were then covered with polyethylene foil to prevent water evaporation. After 24 h, the specimens were unmoulded, wrapped into a polyethylene foil, and stored under constant temperature and humidity conditions (+20 $\pm$ 2 °C, 50%) until testing. The water absorption of the designed FC compositions was measured following EN 772-11 [22]. The dry density of the FC was determined using prism specimens with dimensions of approx. $150 \times 40 \times 40$ mm^3, which were dried in a drying chamber at a temperature of +105 $\pm$ 5 °C until a constant mass was achieved. The compressive strength was determined following EN 12390-3 [23].

Table 3. Foam concrete mixing procedure.

Cavitation Disintegrator (CD)	Turbulence Mixer (TM)
0 min–2.0 min: 2100 rpm	0 min–2.0 min: 1500 rpm
2.0 min–4.0 min: 2400 rpm	2.0 min–4.5 min: 3000 rpm
4.0 min–6.0 min: 3600 rpm	4.5 min–6.0 min: 4800 rpm
6.0 min: Conveying of the FC with settled ca. 2100 rpm	6.0 min: Setting of 1 bar air pressure for conveying of the FC

2.4. Specimen Preparation and Image Processing

There is currently no standardised method for determining the pore size distribution in FC. In addition, the variety of pore classes and surface properties of different FCs complicates establishing a universal measurement method suitable for all types of FC. This study used an automated, software-supported image analysis method to measure pore size distribution. Figure 3 shows the optical digital microscope VHX 600 (Keyence Deutschland GmbH, Neu-Isenburg, Germany), equipped with a high-resolution image analysis tool used for digital photography and image processing. The determination of the pore size distribution and the measurement of porosity were performed on samples with dimensions of $150 \times 100 \times 40$ mm according to EN 480-11 [24]. These samples were extracted from casted specimens with dimensions $150 \times 150 \times 500$ mm. Three samples were cut from each specimen and taken from the middle third of the produced specimens. The analysed surface area corresponded to the cross-section of the prism specimens, ranging between 148 and 150 mm^2. To determine the relevant pore size parameters, the following preparation steps were necessary:

- Polishing of the surface with sandpaper of different grain sizes in uprate from the grain size of 300 µm to 1000 µm;
- Dyeing of the polished surface with a black felt-tip pen;
- Filling of the pores with a contrasting colour powder (white BaSO$_4$).

The black-and-white contrast enabled the simple binarisation of the images. In this process, black areas correspond to the non-porous surfaces of the sample, whereas the white colour depicts the pores. In the binary image, the number of pixels of each colour is counted and referenced to a unit area, e.g., the number of pores per square centimetre. Figure 4 shows a typical binary image of the FCs, with the pictures representing the analysed mix compositions in the current study.

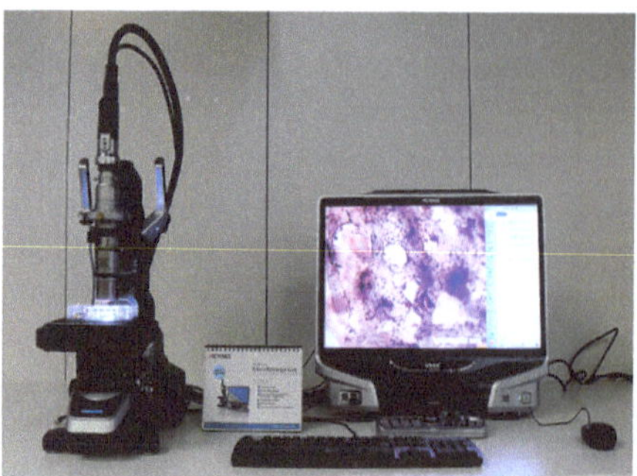

Figure 3. Digital image microscope VHX 600.

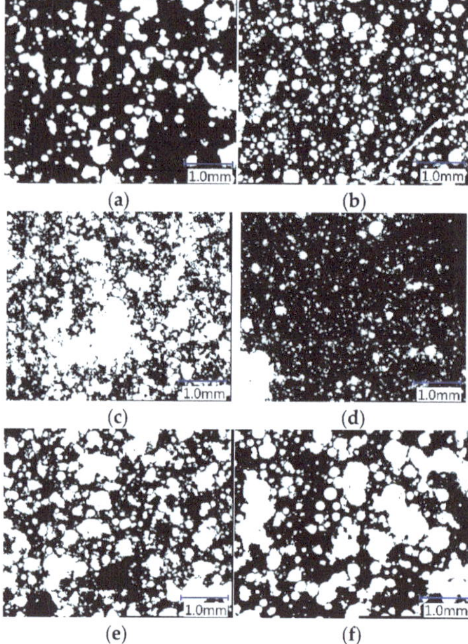

Figure 4. Typical binary images of the pore distribution in FC: (**a**) composition M-1-CD; (**b**) composition M-2-CD; (**c**) composition M-3-CD; (**d**) composition M-4-CD; (**e**) composition M-5-TM; (**f**) composition M-6-TM.

3. Results and Discussion

3.1. Porosity Measurements

Figure 5 shows the measured porosity of six different FC compositions. Compositions M-3-CD and M-5-CD exhibit highest porosities at 53.9% and 48.9%, respectively. Conversely, the M-4-CD composition shows the lowest porosity at 14.3%, which can be explained by the ineffectiveness of the protein-based foaming agent when used with the CD mixer. It is anticipated that a protein-based foaming agent would also be inefficient with the mixed-foaming technique using the TM mixer. This assumption was confirmed in subsequent tests with the TM mixer during the compilation of this manuscript. When comparing the performance of the two different mixers, it is evident that the TM mixer can achieve higher foaming of the cement-based matrix even with a lower dosage of foaming agent than the CD mixer. This can be clearly seen by comparing the M-1-CD composition, which has a porosity of 30.4%, with the M-5-TM composition, which has a porosity of 48.9%. The amount of foaming agent used in the M-1-CD composition was higher than that in the M-5-TM composition. These observations are further supported by the results of dry density presented in Figure 6.

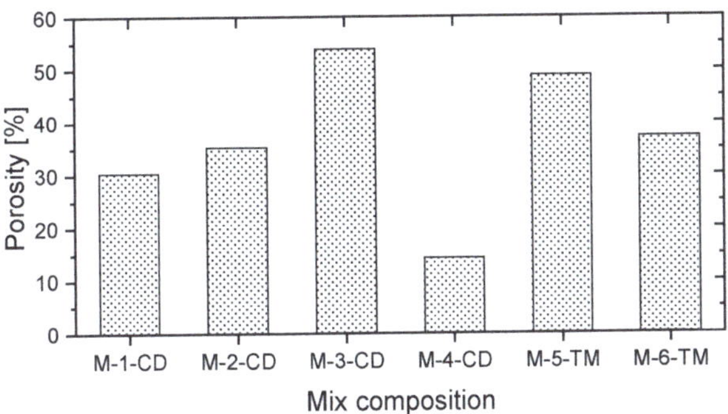

Figure 5. The porosity of the FC samples was measured using digital image analysis.

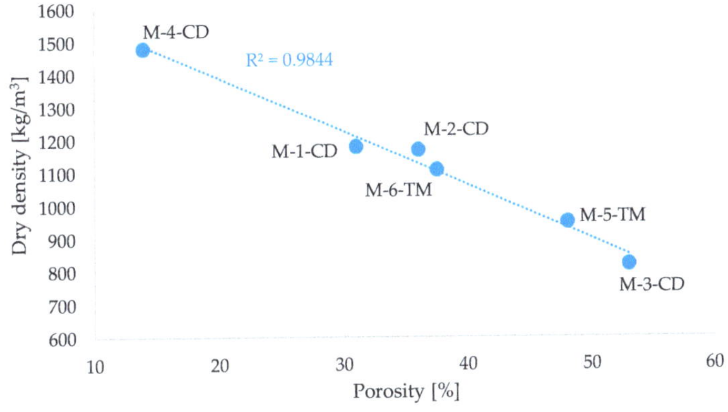

Figure 6. Dry density versus measured porosity with digital image analysis.

The results of the measured porosity of the samples correlate with the measured dry density, as shown in Figure 6. As a result of the study, a direct relationship was found between the porosity obtained by digital image analysis methods and the dry-state density of the material. This correlation shows that digital image analysis can be considered an alternative method for quantification of material porosity. Since porosity and the pore structure influence the mechanical properties of the FC, the measured porosity can be interlinked with mechanical properties. However, to establish a reliable relationship between sample porosity and mechanical properties, a larger database on material properties is needed. In this study, the decision to focus on a single cross-section per sample was made based on the goal of providing a preliminary insight into the influence of production techniques on pore size distribution in high-strength foam concrete. Given the controlled conditions of the production process, the selected cross-section was deemed representative of the overall pore structure for each production technique. While additional cross-sectional data could provide further statistical depth, the primary aim was to compare the qualitative differences in pore formation. Expanding the study to include more cross-sections could be valuable in future research, but for the scope of this work, the approach used allowed for sufficient analysis of the observed trends.

3.2. Air-Void Size Distribution

The pore diameters relative to the number of pores are shown in Table 4. It is noteworthy that the number of counted pores is most variable within the 0–0.1 mm range, with this range showing the highest count compared to other pore size groups. The M-5-TM composition, with a porosity of 48.9% and a dry density of 946 kg/m^3, exhibits a lower total number of pores. This means that the larger pores, with diameters exceeding approximately 0.4 mm, are primarily responsible for the high porosity of the sample; see Figure 6. In contrast, the M-3-CD composition, which has a porosity of 53.9%, shows a significantly higher total pore count than the M-5-TM composition. The sample of composition M-4-CD with the lowest porosity and the highest dry density has the highest total number of pores compared to all compositions. This observation leads to the conclusion that the total number of counted pores or the number of pores in certain groups does not necessarily provide a reliable basis for comparison of the porosity of the sample across different mix compositions.

Table 4. Pore size distribution.

Max. Diameter in [mm]	M-1-CD	M-2-CD	M-3-CD	M-4-CD	M-5-TM	M-6-TM
0–0.1	4349	7919	6607	16,731	4282	7053
0.1–0.2	1114	1969	1207	2153	1289	939
0.2–0.3	344	505	179	191	258	279
0.3–0.4	129	229	62	43	109	99
0.4–0.5	72	109	35	6	74	56
0.5–0.6	35	62	20	6	39	36
0.6–0.7	36	34	13	1	43	30
0.7–0.8	12	16	10	1	22	12
0.8–0.9	12	11	8	0	14	8
0.9–1	9	9	10	0	28	22
>1	13	11	22	3	43	26
Total	6125	10,874	8173	19,135	6201	8560

Figure 7 presents the cumulative frequency distribution of the pores based on the occupied area. Analysing the relationship between porosity and pore size distribution, it can be concluded that specimen M-4-CD, which has the lowest porosity, contains the smallest number of pores with a size greater than 0.4 mm. However, these larger pores are responsible for the total porosity of the specimen.

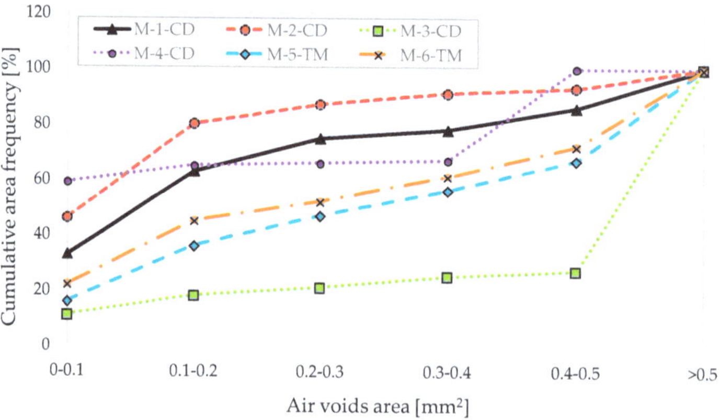

Figure 7. Cumulative frequency distribution of pore areas.

Analysing the obtained results, it can be seen that the lowest values of cumulative area of frequency in the pore range of 0–0.5 mm are found for the M-4-CD composition, which also has the highest density and strength indicators. The highest values of the cumulative area frequency are observed for M-1-CD, M-2-CD, and M-3-CD compositions, which were prepared in a cavitation disintegrator. These compositions have high-frequency values in the range of 0–0.5 mm. The compositions that were prepared in a turbulent mixer (M-5-TM and M-6-TM) have a smoother pore distribution curve, with a lower content of small pores in the range of 0–0.2 mm.

3.3. Compressive Strength

Figure 8 shows the results of determining the compressive strength at different curing times of 7, 28, and 180 days. The obtained curves show a well-confirmed relationship between the strength of FC and its density.

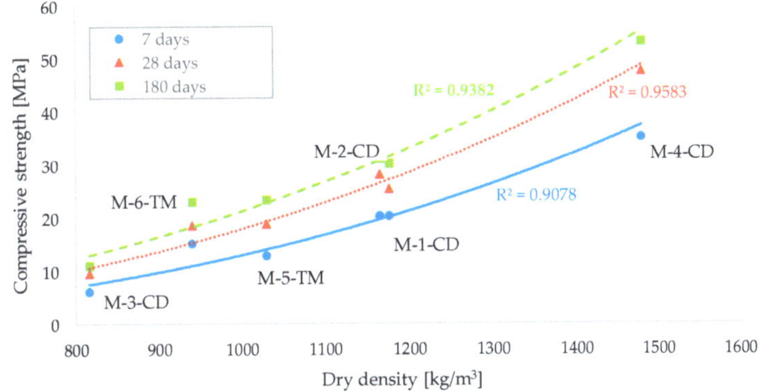

Figure 8. The relationship between density and compressive strength in FC.

The obtained results confirmed a fundamental inverse relationship between porosity and strength, as mentioned in [15]. Summarising the strength results, it should be noted that despite the fact that manufacturing technology has some effect on strength, the determining factors are the material's density and porosity.

Figure 9 presents the relationship between the area of small air voids (0–0.1 mm) and compressive strength at 28 days.

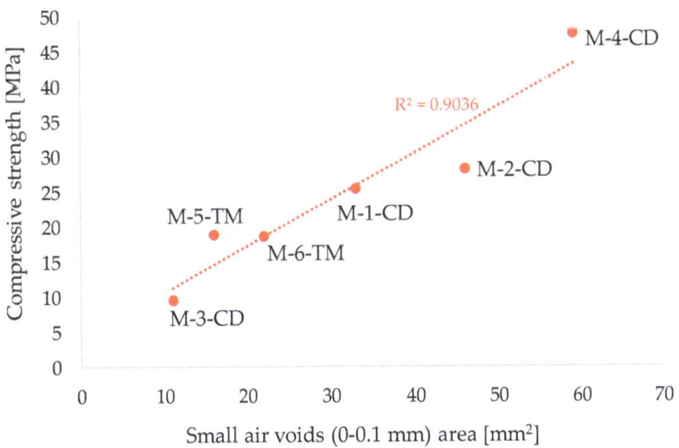

Figure 9. The relationship between the area of small pores and compressive strength.

It can be concluded that there is a direct relationship (correlation factor $R^2 = 0.904$) between the area of small pores and the 28-day strength of FC. The M-4-CD composition has the highest strength and the highest number of small pores. In contrast, the M-3-CD composition, which has a low value of small pores, has the lowest compressive strength of 9.4 MPa.

The results of the relationship between density, strength, and pore distribution are consistent with those of other researchers. For example, in [25], it is shown that high-strength FC has a fine-sized pore structure and a lower total porosity.

3.4. Water Absorption

Analysing the relationship between water absorption and porosity (see Figure 10), it can be concluded that, with an increase in porosity, there is an accelerated increase in water absorption, which can be explained by the more open nature of the pores. The M-4-CD composition with the lowest porosity is also characterised by the lowest water absorption rate (14.4%). The highest porosity and water absorption value is shown by composition M-3-CD, which also has the lowest density. Additionally, from the previous graphs, it can also be concluded that water absorption decreases with an increase in the area of small pores (0–0.1 mm).

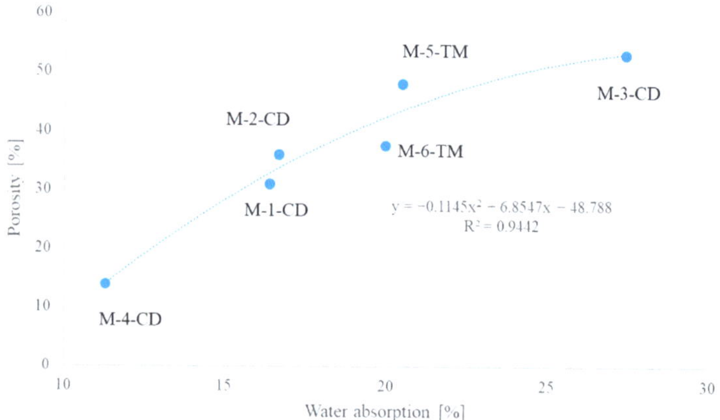

Figure 10. The relationship between porosity and water absorption.

4. Conclusions

This research study has indicated how different production techniques can significantly impact the material properties of foam concrete. The pore structure of foam concrete significantly influences its thermal insulation properties, compressive strength, and durability [3]. For example, controlling the pore size distribution, as shown in our study, can improve the material's insulation properties, which is essential for energy-efficient building construction. Furthermore, finer and more uniformly distributed pores may increase the overall strength of the concrete, as noted by [3,10,25], who reported that pore size has a direct correlation with mechanical performance. Our study emphasises the importance of carefully selecting production techniques to optimise these properties for use in sustainable and high-performance construction materials.

In particular, this study highlights how varying methods of FC production, such as the cavitation disintegrator (CD) and turbulence mixer (TM), affect not only the porosity but also the overall durability, strength, and performance of the material. By comparing these techniques and obtained properties, the study provides insights into how production choices influence the FC's vulnerability to environmental factors and its long-term structural integrity. The outcome of the study can be concluded as follows:

- Both the TM and CD mixers demonstrated superior performance in FC preparation using the mixed-foaming method. With the CD mixer, a wide range of densities from 820 to 1480 kg/m^3 was achieved. However, the TM mixer allowed FC to be obtained with a lower dosage of foaming agent.
- Porosity measurements were correlated with dry density, suggesting that digital image analysis is a viable method for quantifying material properties. However, a larger dataset is needed to link porosity with mechanical properties reliably.
- The number of pores was most variable in the 0–0.1 mm range, and the larger pores over 0.4 mm significantly influenced the overall porosity.
- Water absorption increased with porosity, but it was also influenced by pore distribution and shape. Therefore, it was affected by both, not just the total porosity.
- A 28-day compressive strength ranging from 9.4 to 47.4 MPa was achieved. Moreover, the highest compressive strength was exhibited by the M-4-CD composition, characterised by the lowest porosity and the highest frequency of small pores in the range of 0–0.1 mm.
- Similarly, the finest porous composition, M-4-CD, had the lowest water absorption rate, which can be mainly explained by the closed nature of the pores.
- By ensuring the hardening process, the strength of the specimens increased with age. At 180 days, the compressive strength increased by 10–15% compared to the 28-day results, with the composition M-4-CD reaching 53 MPa at a density of 1480 kg/m^3.

These findings highlight the importance of carefully selecting production techniques to optimise FC's structural and durability properties. Future studies should expand the dataset to better establish the relationship between porosity and mechanical strength and refine pore characterisation techniques using advanced imaging methods. Also, future research could include a more extensive statistical analysis involving multiple cross-sections of each sample. This would provide a broader understanding of how pore size distribution varies across different parts of a foam concrete sample and could help validate the representativeness of a single cross-section. Additionally, the use of advanced imaging techniques such as micro-computed tomography (micro-CT) scanning could offer a more detailed three-dimensional view of the pore structure and allow for more precise quantification of pore size and distribution. Such approaches would improve the reliability and depth of analysis and provide a more comprehensive understanding of the factors influencing foam concrete's structural properties.

Author Contributions: The experimental work, data analysis, and the initial draft of the manuscript were performed by S.M. and G.S. The research was conceived by all the authors and performed under the supervision of V.M. and A.K. All authors have read and agreed to the published version of the manuscript.

Funding: The authors express their sincere gratitude to the German Federal Ministry for the Environment, Nature Conservation, Building, and Nuclear Safety (BMUB) for funding this project in the framework of the research initiative Zukunft Bau of the Federal Institute for Research on Building, Urban Affairs and Spatial Development (BBSR), grant number SWD-10.08.18.7-17.07.

Data Availability Statement: The data supporting the findings of this study will be made available upon reasonable request.

Acknowledgments: We also thank our industrial partners OPTERRA Zement GmbH, MC-Bauchemie Müller GmbH & Co. KG, Kniele GmbH, and BAM Deutschland AG.

Conflicts of Interest: Slava Markin is employed by Sika Deutschland GmbH, TM Concrete. The authors declare that they have no known competing financial interests or personal relationships that could have appeared to influence the work reported in this paper.

References

1. Brady, K.C.; Watts, G.R.A.; Jones, M.R. Specification for Foamed Concrete: Project Record: TF 3/31 The Use of Foamed Concrete as Backfill. 2001. Available online: https://trid.trb.org/View/715810 (accessed on 7 September 2024).
2. Jones, M.R.; McCarthy, A. Preliminary views on the potential of foamed concrete as a structural material. *Mag. Concr. Res.* **2005**, *57*, 21–31. [CrossRef]
3. Ramamurthy, K.; Kunhanandan Nambiar, E.K.; Indu Siva Ranjani, G. A classification of studies on properties of foam concrete. *Cem. Concr. Compos.* **2009**, *31*, 388–396. [CrossRef]
4. Jones, M.R.; McCarthy, A. Behaviour and assessment of foamed concrete for construction applications. In *Use of Foamed Concrete in Construction*; ICE Publishing: Ashton-Under-Lyne, UK, 2015; pp. 61–88.
5. Kara, P.; Shishkins, A.; Korjakins, A. Eco cellular concrete. In Proceedings of the Environmentally Friendly Concrete Eco-Crete conference, Reykjavik, Iceland, 13–15 August 2014; Prentmet-Green Environmental Printing: Los Angeles, CA, USA, 2014; pp. 37–42.
6. Markin, V.; Sahmenko, G.; Nerella, V.N.; Näther, M.; Mechtcherine, V. Investigations on the foam concrete production techniques suitable for 3D-printing with foam concrete. *IOP Conf. Ser. Mater. Sci. Eng.* **2019**, *660*, 012039. [CrossRef]
7. Amran, Y.H.M.; Farzadnia, N.; Abang Ali, A.A. Properties and applications of foamed concrete: A review. *Constr. Build. Mater.* **2015**, *101*, 990–1005. [CrossRef]
8. Eltayeb, E.; Ma, X.; Zhuge, Y.; Youssf, O.; Mills, J.E.; Xiao, J.; Singh, A. Structural performance of composite panels made of profiled steel skins and foam rubberised concrete under axial compressive loads. *Eng. Structr.* **2020**, *211*, 110448. [CrossRef]
9. Eltayeb, E.; Ma, X.; Zhuge, Y.; Youssf, O.; Mills, J.E. Influence of rubber particles on the properties of foam concrete. *J. Build. Eng.* **2020**, *30*, 101217. [CrossRef]
10. Khan, M.Y.; Baqi, A.; Sadique, M.R.; Khan, R.A. Development of High Strength Lightweight Foamed Concrete with Low Cement Content. Preprint (Version 1). 2022. Available online: https://doi.org/10.21203/rs.3.rs-1721278/v1 (accessed on 20 November 2024).
11. Lv, X.-S.; Cao, W.-X.; Yio, M.; Ji, W.-Y.; Lu, J.-X.; She, W.; Poon, C.S. High-performance lightweight foam concrete enabled by compositing ultra-stable hydrophobic aqueous foam. *Cem. Concr. Compos.* **2024**, *152*, 105675. [CrossRef]
12. Sobolev, K.; Bindiganavile, V.S.; Babbitt, F.; Barnett, R.E.; Chan, C.; Cornelius, M.L.; Dye, B.T.; Fouad, F.H.; Glysson, M.; Gomez, M.R., Jr.; et al. ACI 523.3R-14: Guide for Cellular Concretes above 50 lb/ft3 (800 kg/m3). ACI Committee 523, American Concrete Institute. Available online: https://www.concrete.org/Portals/0/Files/PDF/Previews/523.3R-14_PREVIEW.pdf (accessed on 7 September 2024).
13. Saint-Jalmes, A.; Peugeot, M.L.; Ferraz, H.; Langevin, D. Differences between protein and surfactant foams: Microscopic properties, stability and coarsening. *Colloids Surf. A: Physicochem. Eng. Asp.* **2005**, *263*, 219–225. [CrossRef]
14. Mechtcheirine, V.; Otto, J.; Will, F.; Markin, V.; Schröfl, C.; Nerella, V.N.; Krause, M.; Dorn, C.; Näther, M. CONPrint3D Ultralight–Herstellung monolithischer, tragender, wärmedämmender Wandkonstruktionen durch additive Fertigung mit Schaumbeton/Production of monolithic, load-bearing, heat-insulating wall structures by additive manufacturing with foam concrete. *Bauingenieur* **2019**, *94*, 405–415. [CrossRef]
15. Kumar Mehta, P.; Monteiro, P.J.M. *Concrete: Microstructure, Properties, and Materials*, 4th ed.; McGraw-Hill Education: New York, NY, USA, 2014; p. 675. Available online: https://www.accessengineeringlibrary.com/content/book/9780071797870 (accessed on 7 September 2024).
16. Bian, P.; Zhang, M.; Yu, Q.; Zhan, B.; Gao, P.; Guo, B.; Chen, Y. Prediction model of compressive strength of foamed concrete considering pore size distribution. *Constr. Build. Mater.* **2023**, *409*, 133705. [CrossRef]
17. Roslan, A.F.; Awang, H.; Mydin, M.A.O. Effects of Various Additives on Drying Shrinkage, Compressive and Flexural Strength of Lightweight Foamed Concrete (LFC). *Adv. Mater. Res.* **2013**, *626*, 594–604. [CrossRef]
18. Panesar, D.K. Cellular concrete properties and the effect of synthetic and protein foaming agents. *Constr. Build. Mater.* **2013**, *44*, 575–584. [CrossRef]
19. Kashani, A.; Ngo, T.D.; Nguyen, T.N.; Hajimohammadi, A.; Sinaie, S.; Mendis, P. The effects of surfactants on properties of lightweight concrete foam. *Mag. Concr. Res.* **2018**, *72*, 163–172. [CrossRef]
20. Justs, J.; Shakhmenko, G.; Mironovs, V.; Kara, P. Cavitation treatment of nano and micro filler and its effect on the properties of UHPC. In Proceedings of the HiPerMat 2012, 3rd International Symposium on UHPC and Nanotechnology for High Performance Construction Materials, Kassel, Germany, 7–9 March 2012; Kassel University Press: Kassel, Germany, 2012; pp. 87–92.
21. Mironovs, V.; Polakovs, A.; Korjakins, A. A Method and apparatus for suspension preparation. Patent LV14364B Int.cl. C01B33/00 RTU, 20 October 2011.

22. *EN 772-11:2011*; Methods of Test for Masonry Units—Part 11: Determination of Water Absorption of Aggregate Concrete, Autoclaved Aerated Concrete, Manufactured Stone and Natural Stone Masonry Units due to Capillary Action and the Initial Rate of Water Absorption of Clay Masonry Units. CEN, European Committee for Standartization: Brussels, Belgium, 2011.
23. *EN 12390-3*; Testing Hardened Concrete—Part 3: Compressive Strength of Test Specimens. European Committee for Standardization (CEN): Brussels, Belgium, 2019.
24. *EN 480-11:2005*; Admixtures for Concrete, Mortar and Grout—Test Methods—Part 11: Determination of Air Void Characteristics in Hardened Concrete. CEN, European Committee for Standartization: Brussels, Belgium, 2005.
25. Just, A.; Middendorf, B. Microstructure of high-strength foam concrete. *Mater. Charact.* **2009**, *60*, 741–748. [CrossRef]

Article

A Path towards SDGs: Investigation of the Challenges in Adopting 3D Concrete Printing in India

Bandoorvaragerahalli Thammannagowda Shivendra, Shahaji, Sathvik Sharath Chandra, Atul Kumar Singh, Rakesh Kumar and Nitin Kumar et al.

[1] Department of Civil Engineering, Dayananda Sagar College of Engineering, Bengaluru 560111, India; shivendra-cvl@dayanandasagar.edu (B.T.S.); sathvik-cvl@dayanandasagar.edu (S.S.C.); atulkumar-cvl@dayanandasagar.edu (A.K.S.); rakesh-cvl@dayanandasagar.edu (R.K.); nitin-cvl@dayanandasagar.edu (N.K.)

[2] Department of Civil Engineering, Manipal Institute of Technology Bengaluru, Manipal Academy of Higher Education, Manipal 576104, India; sujay.n@manipal.edu

[*] Correspondence: shahaji-cvl@dayanandasagar.edu (S.); tantri.adithya@manipal.edu (A.T.)

Abstract: In recent years, three dimensional concrete printing (3DCP) has gained traction as a promising technology to mitigate the carbon footprint associated with construction industry. However, despite its environmental benefits, studies frequently overlook its impact on social sustainability and its overall influence on project success. This research investigates how strategic decisions by firms shape the tradeoffs between economic, environmental, and social sustainability in the context of 3DCP adoption. Through interviews with 20 Indian industry leaders, it was found that companies primarily invest in 3DCP for automation and skilled workforce development, rather than solely for environmental reasons. The lack of incentives for sustainable practices in government procurement regulations emerges as a significant barrier to the widespread adoption of 3DCP. Our study identifies five key strategies firms employ to promote sustainability through 3DCP and proposes actionable measures for government intervention to stimulate its advancement. Addressing these issues is crucial for realizing the full societal and environmental benefits of 3DCP technology.

Keywords: 3D concrete printing (3DCP); social–sustainability tradeoffs; carbon footprint; environmental impacts; government intervention; sustainable development goals

Citation: Bandoorvaragerahalli Thammannagowda Shivendra, Shahaji, Sathvik Sharath Chandra, Atul Kumar Singh, Rakesh Kumar and Nitin Kumar et al. . A Path towards SDGs: Investigation of the Challenges in Adopting 3D Concrete Printing in India. *Infrastructures* **2024**, *9*, 166. https://doi.org/10.3390/infrastructures9090166

Academic Editor: Patricia Kara De Maeijer

Received: 29 July 2024
Revised: 3 September 2024
Accepted: 20 September 2024
Published: 23 September 2024

1. Introduction

The building sector is a pivotal driver of global economic growth and energy consumption but is also a significant contributor to greenhouse gas emissions [1,2]. Globally, the construction industry accounts for 39% of the CO_2 emissions and uses 36% of the energy. Concrete, a cornerstone material in the construction industry, is particularly noteworthy due to its extensive use and environmental impact [3]. Concrete's advantages include low cost, excellent fire resistance, and high compressive strength, which make it a preferred material in construction. However, its production is a primary environmental concern: cement alone contributes approximately 6% of global CO_2 emissions, and about 9% of all industrial water withdrawals are used in concrete production [4].

Despite its widespread application and benefits, the environmental footprint of concrete continues to worsen as urbanization and construction activities expand. Innovations like 3D concrete printing (3DCP) is being explored to address these challenges. Proponents of 3DCP argue that it can enhance sustainability by reducing material usage, minimizing waste, improving productivity, and mitigating the skilled labor shortage in the construction industry [5,6]. Although case studies and lifecycle assessments have quantified the economic and environmental benefits of 3DCP, they have often neglected the impact on social sustainability [7–9].

Previous research in 3DCP has focused on the technological advancements and potential benefits of 3DCP in the construction industry. An early study explored the potential for 3DCP to revolutionize construction by reducing material waste, labor costs, and construction time, thus offering significant environmental and economic advantages [1]. Examining the technical aspects of 3DCP highlights its ability to produce complex geometries and customized structures, which are difficult to achieve with traditional construction methods [2].

However, the social sustainability implications of 3DCP have received less attention. The environmental benefits of 3DCP, such as reduced carbon emissions and resource efficiency, are studied, but the social aspects, such as job displacement or the impact on local communities, are not deeply investigated [3]. More recent studies have explored the broader implications of 3DCP adoption, including economic and environmental factors, but again, social sustainability remains underexplored [4–6].

Moreover, studies have emphasized the role of governmental policies and institutional frameworks in promoting sustainable construction practices [5,6]. They argue that while 3DCP has the potential to contribute significantly to sustainability, the lack of comprehensive policies and standards hinders its widespread adoption [5]. Analyzing the regulatory challenges that 3DCP faces and recommending that sustainability criteria be incorporated into government tenders encourages the adoption of innovative technologies [6].

Moreover, previous analyses have not sufficiently addressed how managerial decisions affect the adoption of 3DCP and its implications for social sustainability. The institutional framework, which includes government commissioning large construction projects, is crucial in influencing companies' sustainability outcomes [10–12]. To fill this gap, this study aims to investigate how management decisions regarding the application of 3D concrete printing (3DCP) affect the balance between environmental, economic, and social sustainability. The specific objectives are:

(1) To examine how management decisions regarding 3DCP application impact sustainability across environmental, economic, and social dimensions.
(2) To explore how current institutional conditions, such as government commissioning and tendering processes, influence management incentives to invest in 3DCP.
(3) To identify key challenges in implementing 3DCP and propose policy changes that could facilitate greater adoption of this technology, enhancing overall sustainability.

The study uses a qualitative approach involving 20 interviews with 3DCP pioneers in India to analyze data across three dimensions: people, planet, and profit. The findings reveal that government tenders often lack sufficient sustainability criteria to promote widespread 3DCP adoption, highlighting the need to incorporate social sustainability considerations into implementing new technologies. The study's significance lies in its focus on the intersection of management decisions, institutional frameworks, and 3DCP technology adoption. It addresses gaps in existing research by emphasizing social sustainability and the influence of institutional conditions, offering practical recommendations for policymakers and industry leaders to promote a more balanced approach to sustainability in the construction industry.

2. Background

2.1. Challenges and Promises of 3DCP in Construction

Various technologies are used to construct items by layering concrete until the final geometry is achieved. As with 3D printers that print metals or polymers, 3DCP machines work similarly [13–15]. In contrast to conventional printing methods, 3DCP allows structures, such as walls and even entire floors, to be printed in much larger sizes than those obtained from traditional printing methods. The larger the printed component, the more challenging it is to implement effective quality control, as minute faults are more likely to arise during printing.

3DCP technology most appeals to the construction industry because it allows complex geometric shapes to be produced using no formwork. A formwork project can cost over 10% of the total project cost, depending on the location and the type. As formwork can only be reused a limited number of times, it is typically made from wood and is a significant source of waste [16,17]. In conjunction with 3DCP, the printing procedure can be set up so that the least amount of material is used. Additionally, 3DCP reduces the environmental impact of concrete manufacturing and construction operations. Architects can also experiment with innovative building shapes using 3DCP to improve energy efficiency or airflow, thereby minimizing the environmental impact.

In addition to material savings, 3DCP is expected to generate financial benefits as well. The lack of technological sophistication and high demand for manual labor have contributed to a stagnation in construction productivity in recent years [18,19]. Construction in western countries is also experiencing a labor shortage. Employees from less developed countries may temporarily fill the vacancy, but there are more chances they will face social and professional difficulties. Human trafficking and forced labor are the most common ways criminal networks use undocumented and illegal workers.

Governments and businesses have suggested increasing prefabrication and off-site construction and digitizing the entire supply chain of construction activities to boost productivity. The construction process could be automated to eliminate structural defects and consistency issues with manual labor. Construction project planning is made more difficult by rework expenses related to these flaws, which could comprise 5–15% of the total budget [20–22]. Moreover, 3DCP equipment may theoretically continue working while manual building activities cease at night or during difficult weather conditions [22,23].

Even though 3DCP has the potential to contribute to sustainability significantly, it also has some drawbacks. One of the significant limitations of 3DCP is that it offers constructors few material options [24]. For instance, alternative binders like geopolymers, fly ash, and limestone are being incorporated into 3DCP to enhance the sustainability of 3D-printed concrete. Their rheological properties are different from traditional mortars. 3DCP materials must be produced, distributed, and recycled sustainably [25–28].

Furthermore, printed components are far less mechanically strong than cast concrete and susceptible to even the smallest manufacturing process modifications. This makes 3DCP not quite relevant. To build strong quality control and lower manufacturing variability, pioneers should extensively invest in R&D and go through protracted trial-and-error. To reap the benefits of 3DCP and save on materials, designers will also need retraining and time [29]. New technical standards and long-term durability tests for 3DCP materials will need to be developed.

The sustainability of 3DCP has been studied, but these studies focus on its benefits rather than its limitations. Studies that examine its limitations concentrate more on technical aspects than sustainability [30,31]. Usually measuring environmental and economic effects, 3DCP's lifecycle assessment ignores elements of social sustainability and is limited to single case studies without accounting for regulatory influences [32,33]. A technology's sustainability benefits, as well as its business model sustainability and the context in which it is adopted, can all influence how a technology is adopted, thus affecting the extent of its sustainability benefits. Although these interactions are crucial for assessing sustainability, the literature does not explain how 3DCP interacts with the triple bottom line [34,35]. Studies enhance the literature by incorporating social sustainability into evaluating 3DCP's overall sustainability. It also examines how managerial and design decisions affect the trade-offs in environmental, social, and economic sustainability [36–39].

Certain features of the building sector hinder the application of 3DCP: the lack of specialized training programs and the unattractiveness of the construction sector as a career choice for young people; low levels of R&D spending; and a lack of skilled labor-led managers to adopt new technologies cautiously with low and unstable profit margins [37–41]. As a result, construction R&D focuses on incremental innovations, and market forces often drive sustainability. To help businesses overcome the sensitivity to risk that the industry has, governments have created laws and policies that provide incentives for companies to increase their sustainability to help them improve their efficiency [42,43].

3D concrete printing (3DCP) faces challenges such as rigid procurement rules, high initial costs, and the need for established quality standards, alongside potential resistance from traditional industry stakeholders and a shortage of skilled professionals. However, it also promises significant benefits, including innovation, increased efficiency, material savings, and reduced construction waste [44]. Governments can drive adoption by using public procurement to set technology mandates and waste management requirements, fostering long-term cost savings, job creation, and enhanced sustainability. By developing new standards and regulations, public procurement can effectively support the integration of 3DCP into mainstream construction practices [45]. Standardized indicators usually quantify the highest permitted degree of environmental harm caused by a project. Such performance-based indicators have the advantage that businesses can choose the solution that best fits their environment to meet the required threshold instead of being forced to employ a specific technology [46]. Regulations may have a technology-forcing effect if they cannot be met with current technology, causing businesses to develop new technologies [47].

Technical requirements and architectural codes are frequently tied to procurement regulations. To promote innovation throughout the supply chain, the government may adopt more stringent technical criteria [48–50]. How well new technologies follow established standards has dramatically affected their acceptance in the building sector. While standardization has substantially increased construction safety and homogenized procedures across enterprises, an overreliance on standards risks preventing the commercialization of technology advancements compatible with those standards [51,52]. With the advent of 3DCP, its design flexibility may clash with current standards favoring prefabricated assemblies with specific shapes where manufacturing quality can be more carefully monitored. The technology of 3D printing has also been under dispute with technical standards, resulting in a conflict between the two technologies [53,54].

A firm's institutional framework will likely affect its decision to use 3DCP and, therefore, whether sustainability benefits can be realized. This institutional setting may have significant regional variation. The impact of the institutional context on the sustainability of 3DCP has not been considered in previous analyses [55–59]. Our paper contributes to the literature by offering insights into how managers in India perceive institutional incentives for adopting 3D concrete printing (3DCP). We anticipate that our findings will also be relevant to other countries in the Indian subcontinent, as they face similar challenges related to sustainability goals and construction standards, despite varying regulatory frameworks.

2.2. Research Gap

In the Indian context, while 3D concrete printing (3DCP) presents significant potential for enhancing construction sustainability, there remains a notable research gap. Existing studies often focus primarily on the environmental and economic benefits of 3DCP, with limited exploration of its social sustainability aspects and the influence of institutional frameworks on technology adoption. Specifically, there is insufficient data and analysis of how managerial decisions and government policies in India impact the integration of 3DCP and its effectiveness in achieving sustainability goals. The lack of specialized training programs, high initial costs, and rigid procurement rules further complicate its adoption. Moreover, the regional variations in regulatory frameworks and institutional incentives have not been thoroughly examined. This gap underscores the need for research investigating how these factors affect the implementation of 3DCP in India, including its impact on social sustainability and the broader implications for construction practices across the Indian subcontinent.

3. Methodology

This study comprehensively addresses each step of the research process, from the initial stages of research design and data collection to data analysis and the interpretation of findings, as illustrated in Figure 1.

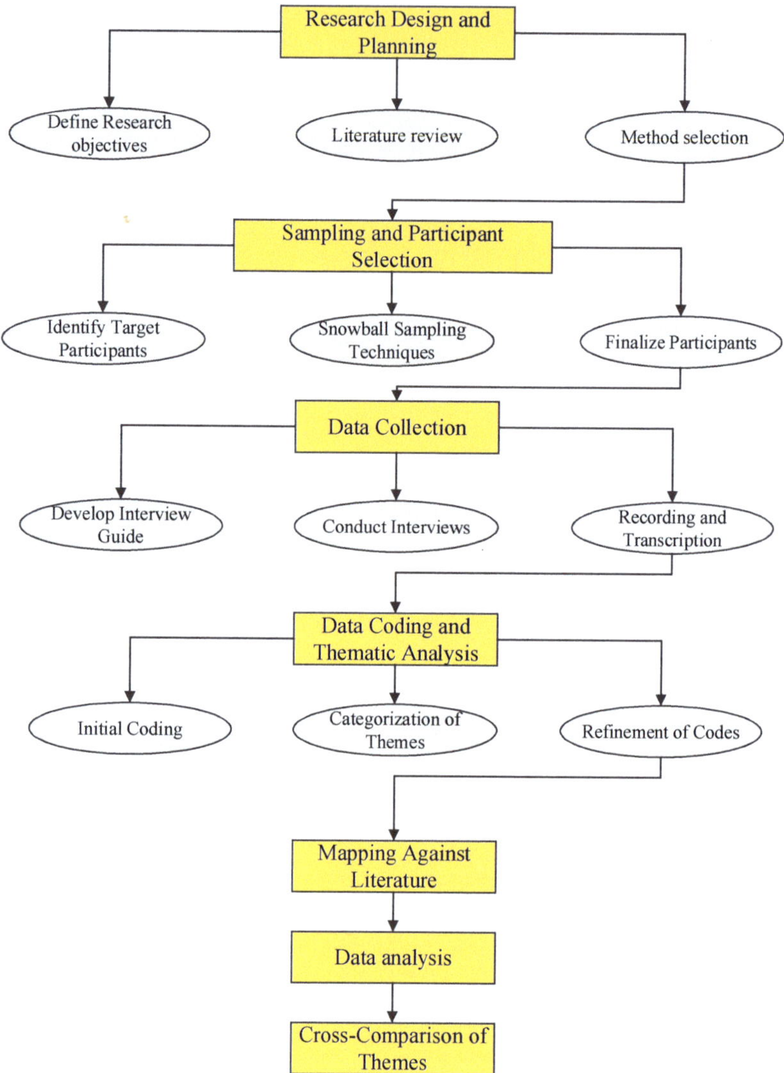

Figure 1. Research methodology flowchart.

3.1. Research Approach

This study utilizes a qualitative research approach to understand how management decisions and institutional conditions impact the adoption of 3D concrete printing (3DCP) in the construction industry, particularly concerning sustainability outcomes. A qualitative approach is appropriate for exploring complex, context-dependent phenomena that require rich, detailed insights [13]. To collect data, this study employed open-ended interview questions, which allowed participants to freely express their views and experiences regarding the adoption of 3DCP. These questions were designed to explore the three dimensions of sustainability—people, planet, and profit—and how they intersect with management and policy decisions. Preparing the interview questions involved a thorough literature review and consultations with subject matter experts to ensure that the questions were relevant and capable of eliciting comprehensive responses [32]. The selection of experts for the interviews followed a purposeful sampling strategy, focusing on individuals with extensive experience and expertise in 3DCP. Criteria for selection included their involvement in pioneering 3DCP projects, their role in decision-making processes related to technology adoption, and their understanding of the sustainability implications of 3DCP. This approach ensured that the insights gathered were informed by knowledgeable and experienced professionals, contributing to the study's credibility and depth [33].

3.2. Data Collection

The data collection for this study involved a comprehensive qualitative approach, utilizing semi-structured interviews to explore the sustainability trade-offs and institutional challenges associated with the adoption of 3D concrete printing (3DCP). The interviews were conducted with a diverse group of 20 experts [60,61], including managers, structural designers, innovation managers, and sustainability consultants, and all had significant experience with 3DCP and its application in the construction industry (see Table 1). The interview process was designed to ensure a broad representation of perspectives from various stakeholders involved in 3DCP adoption. Interviews were conducted in English and local languages, including Kannada, Tamil, Telugu, and Hindi, to accommodate the participants' preferences and ensure clear communication. Each interview lasted between 30 and 90 min, allowing for an in-depth exploration of the topics.

The interviews were structured into four main sections. The first section focused on understanding the characteristics of 3DCP and the challenges associated with its implementation. Subsequent sections explored how the adoption of 3DCP influenced the triple bottom line dimensions—people, planet, and profit—alongside the technical and financial hurdles of bringing the technology from lab to market. The final round of interviews was designed to identify sustainability trade-offs related to 3DCP and to compare it with other technologies available to businesses. A snowball sampling technique was employed, where initial interviewees recommended other potential participants, ensuring the sample was comprehensive and relevant to the study's objectives. The iterative coding process involved three phases of hand-coding, focusing on technological sustainability features and their impact on the triple bottom line. The authors reviewed and refined the coding results collaboratively, ensuring consistency and reducing bias. This methodical approach provided a robust framework for analyzing the data and deriving insights into the sustainability implications of 3DCP adoption.

3.3. Data Analysis

The data analysis in this study was conducted through a rigorous and systematic process, following a qualitative approach to uncover patterns and insights related to the adoption of 3D concrete printing (3DCP) in the construction industry. The analysis focused on understanding the sustainability trade-offs and institutional influences on 3DCP implementation. Initially, the interview data were transcribed and subjected to a thematic analysis involving identifying, analyzing, and reporting patterns within the data [41]. The first phase of the analysis involved coding the interview transcripts line by line and categorizing responses based on key themes related to technological sustainability features, social sustainability considerations, and economic impacts. These themes were identified based on the literature and the data, ensuring that the analysis was grounded in the participants' experiences and the study's theoretical framework [44].

The coding process was carried out in three stages. In the first stage, technological sustainability features were coded, emphasizing how these features affected the triple bottom line—people, planet, and profit. The codes were then mapped against existing technical literature on 3DCP to assess whether the identified features positively or negatively impacted sustainability outcomes [47–49]. The second stage involved linking these codes to the strategic decisions made by organizations regarding 3DCP adoption and identifying trade-offs between sustainability and other operational priorities. The final stage of the coding process involved comparing the findings across different sectors and organizations, revealing commonalities and differences in how 3DCP is perceived and implemented. This comparative analysis highlighted the varied challenges and opportunities faced by different stakeholders in adopting 3DCP and the role of institutional frameworks in shaping these outcomes [50,51].

Table 1. A chronological list of the interviewees.

Expert	Qualification	Experience	Area of Expertise	Interview Duration (mins)
Expert 1	Ph.D. in Structural Engineering	15 years	3D Concrete Printing, Structural Design	54
Expert 2	M.Sc. in Construction Management.	12 years	Project Management, Sustainable Construction	95
Expert 3	B.Eng. in Civil Engineering	20 years	Concrete Technology, Building Materials	74
Expert 4	M.Sc. in Environmental Engineering	10 years	Environmental Impact Assessment, Green Building Technologies	86
Expert 5	Ph.D. in Architecture	18 years	Innovative Building Design, 3D Printing Applications	46
Expert 6	B.Arch. in Architecture	14 years	Sustainable Design, Building Information Modeling	46
Expert 7	M.Sc. in Construction Technology	8 years	Construction Innovation, 3DCP Implementation	76
Expert 8	Ph.D. in Mechanical Engineering	22 years	Robotics in Construction, Automation Technologies	46
Expert 9	M.Sc. in Structural Engineering	16 years	Structural Analysis, Material Science	44
Expert 10	B.Sc. in Civil Engineering	13 years	Infrastructure Projects, Concrete Durability	46
Expert 11	M.Sc. in Environmental Design	11 years	Environmental Sustainability, Circular Economy	45
Expert 12	Ph.D. in Building Science	17 years	Building Physics, Thermal Efficiency	60
Expert 13	B.Eng. in Construction Engineering	15 years	Construction Technology, Project Management	58
Expert 14	M.Sc. in Architectural Engineering	9 years	Sustainable Architecture, 3D Printing	49
Expert 15	Ph.D. in Civil Engineering	19 years	Structural Integrity, Advanced Construction Materials	33
Expert 16	M.Sc. in Sustainability	12 years	Sustainable Building Practices, Environmental Policy	59
Expert 17	B.Eng. in Mechanical Engineering	20 years	Construction Robotics, Automation Systems	71
Expert 18	M.Sc. in Project Management	14 years	Construction Projects, Resource Management	43
Expert 19	Ph.D. in Environmental Science	16 years	Eco-friendly Materials, Life Cycle Assessment	54
Expert 20	M.Sc. in Urban Planning	13 years	Urban Development, Sustainable Design	87

Throughout the analysis, the data were first independently coded, and then an additional perspective ensured the reliability and validity of the findings (see Table 2). There was a high level of agreement between the co-coders [52–55]. The iterative nature of the coding process, coupled with the triangulation of data from multiple sources, ensured that the analysis was thorough and reflective of the complexities involved in 3DCP adoption. The findings from this analysis contribute to the academic understanding of 3DCP's sustainability implications and offer practical insights for policymakers and industry leaders.

Table 2. Trade-off in sustainability over three-dimensional CP acceptance parameters. Positive impacts as (+) and negative effects as (−) [62].

3DCP Trait	Effects on Earth	Impact on Individuals	Effect on Profit
Reduction in material use	+ Production of concrete and transportation help to lessen environmental effects. + Formwork is not needed, which is limited in reuse. − Still, concrete is a big and difficult substance. − There is limited material availability for 3DCP, hence longer distance travel could be necessary.	− Formwork not needed.	+ Potential cost and savings in formwork. − 3DCP materials are more expensive than traditional concrete. − Costlier alternatives to concrete.
Geometry freedom for complex designs	+ A holistic design approach can improve energy efficiency. − Holistic design could be less modular and contradict circularity.	+ The designs allow one to more readily fit the requirements of particular users. + Promotes high-skilled employment.	− Since integrated designs are difficult to evaluate, 3DCP's cost-competitiveness cannot be readily matched with conventional designs.
Automation in 3DCP	+ A higher level of quality control results in less waste and fewer errors.	+ Automation will be the only choice should future construction personnel be lacking. − With 3DCP, concrete pouring calls for less work.	+ Significant savings connected to expenses related to failures. Low variability helps project managers to lessen their uncertainty. High equipment, R&D, and quality control expenses.

4. Findings

4.1. Environmental Sustainability: Good Promise, but There Are Questions Regarding Complete Circularity

The careful balance between the advantages they can get and the costs they pay determines construction managers' application of environmentally friendly technologies. Usually, major actors in big infrastructure projects are governments. Therefore, the government tendering process has great impact ("Decision-making is often at the project or tender level, so the tender manager or project manager must convince them of the added value."). Usually, the candidate proposing the most economically advantageous tender (MEAT) gets the project. Tenders could contain environmental factors like predicted effects on acidification, water eutrophication, and climate change. MEAT requirements are decided upon in India project by project. It is necessary to evaluate applications twice. Initial evaluations are conducted by independent experts who are not aware of the final price of the tender.

The procurer uses these qualitative factors to calculate the price based on evaluating these factors. A MEAT price (ECI) is calculated based on environmental cost indications associated with each material in the project. Shadow costs per kilogram of material consumed are computed using an ECI that accounts for a material's production process, transportation distance, and disposal. ECI materials must be purchased from a qualified source or certified by the contractor to determine the total cost. 3DCP materials do not contain these certificates. A company is not encouraged to undergo the certification process, as the estimated amount of 3DCP would represent only a tiny portion of the total material used in the project. "For concrete, we first have to get all kinds of certificates, and it takes much time, so you cannot easily implement it … it does not always have to be cheaper, it can also have a better environmental score … [but] we are not there yet". In addition to its rapidly expanding composition, 3DCP blends are exclusive to a select number of material sources that consider these blends a significant source of competitive advantage, making certification even more challenging.

ECI ratings and MEAT allowances must reflect values that promote innovation and technological advancement to be effective. Still, our interviewees felt that "the [ECI] values now set as the maximum are so high that, in practice, you are almost always below these numbers". Consequently, there is no actual urgency to lower environmental effects. Some interviewees identified a 3DCP opportunity in tenders where MEAT was not the primary criterion. They underlined "a tender in which 50% of your plan is rated on image/quality (aesthetics), 35% on flora/fauna, environmental nuisance, road safety, and 15% on price, ranging from 0 to 100". To win, then, you must make investments in image and quality. An applicant making such a bid should be advised to use 3DCP to create complex geometries that are more sensitive to local ecosystems and restrict the visual impact instead of using it to reduce the ECI score.

Though they are now lacking, ECI scores should become tighter going forward. The building sector claims that if the E.U. is to reach climate neutrality by then, all building components must be circular by 2025. All significant Bengaluru construction firms have made their strategic plans for circular buildings public to comply with future regulations.

There are tensions between opportunities and challenges with circular concrete, according to 3DCP. Due to the high amounts of trash and the requirement for formwork, "the demand for material is more than what becomes available. Secondary [recycling] is a significant problem in constructing concrete structures". By radically reducing material usage, 3DCP eliminates formwork. Reducing material use will not help if it cannot be used to print 3D again. One way to solve the recycling issue is creating new material combinations with less or no cement and simpler treatment at the end of their lifetime. The results of such mixes are still relatively unknown despite the efforts of academic and industrial studies. Many interviewees mentioned they were considering replacing concrete structures with alternative materials like wood or composites. While acknowledging that "we must continue to use [concrete] because, for certain applications, there is no fully-fledged alternative", *one manager we spoke with said: "it is better to invest time and energy in these kinds of concepts than in a technique/material that by definition can never be sustainable".* Today, concrete is admired for its affordability and durability, and substituting concrete is challenging. "In civil concrete construction, production is fully functional, and all concrete is there because of its strength".

Furthermore, while 3D concrete printing (3DCP) offers significant design freedom, it is important to consider how this potential aligns with the actual needs of the construction industry. 3DCP also allows for the creation of intricate and aesthetic shapes that would be difficult or impossible to achieve with traditional construction methods. This opens up new possibilities for architectural design and expression [63]. A notable advantage is the ability to create modular constructions that can be easily deconstructed and replaced. However, this modularity contrasts with the benefits of building entire structures in one go with minimal assembly. The technology can better address practical needs by integrating 3DCP's design flexibility with industry requirements for modularity and efficiency while enhancing overall construction practices. Modular construction offers significant productivity benefits by reducing construction time and increasing efficiency. With prefabricated modules that can be easily assembled, construction projects can be completed in a fraction of the time compared to traditional construction methods. It saves time and money and allows for faster occupancy and quicker return on investment. Building a demountable is also essential to your design, as you must consider how the connection between your design and the monolith will differ. Modular designs become more circular as everything is baked together in 3D printing. By contrast, consolidated designs offer higher performance, reduce weight, and extend component life. Modular designs are easily circularized. To protect the integrity, additional material is needed whenever two components are joined.

The technology appears to represent a significant advancement in reducing the building sector's environmental impact over the current situation. Although 3DCP might eventually satisfy circularity requirements, there are several unanswered questions. In some cases, alternative materials can be used instead of concrete, but it is unclear whether concrete can be replaced entirely.

4.2. Social Sustainability: Lower Reliance on Seasonal Labor and Higher Levels of Satisfaction among Building Occupants

For its operations today, the Bengaluru building sector mostly relies on seasonal foreign labor. "If you look at civil concrete construction, such as iron braids, workers are mostly UP and Bihar, and you hardly see any Bengaluru". Importing labor is less expensive economically than hiring workers from India. Managers claim that over time, this condition is not sustainable. Labor prices will probably increase as the economies of the nations where seasonal workers come from project growth in the next years.

Furthermore, the labor shortage is anticipated to worsen over the next several decades. Automation is driven primarily by these two trends: "In 20, 30 to 40 years, you will have few people who still understand the profession, so you will have to switch production ... you can see that the number of people learning a trade like carpenter is decreasing, there are only a few". From this vantage point, technology would not drive workers out of the market; instead, a lack of labor would drive technology into industry. 3DCP would aid in removing the social issues brought on by seasonal workers' unstable employment, their challenges assimilating into new societies, and the possibility of immigrant labor exploitation by diminishing the demand for seasonal employees.

By replacing conventional concrete pouring duties with its components, 3DCP will impact various jobs, depending on its use. Whether 3DCP should be done on-site or off-site—prefabricated in a factory—is a significant question with much ambiguity. If prefabricated modular components were constructed by a team of workers at the building site, 3DCP would become merely another off-site tool. Labor consequences would be negligible and only affect some professions, such carpenters in countries like India, where a considerable amount of the construction is prefabricated (since wooden formworks for pouring concrete would no longer be needed). The impact might be more important in other countries where prefabrication is rather less widespread. Using 3DCP on-site might theoretically allow a huge robot to print the whole building structure. In such a case, a bigger workforce would be affected since labor would only be needed for operation management and the building and calibration of robotic equipment. Human effort would still be needed for installation of plumbing and electrical equipment that is not yet automated.

Moreover, 3DCP should increase worker safety ("Creating formwork is relatively dangerous; more than one carpenter lost a finger; sawing or cutting wood"). Furthermore, even in cases of an accident involving 3DCP technology, the employee's injuries will be far less severe ("A printer may break, but no one will be injured or killed"). As with current prefabrication methods, 3DCP has risks similar to those of factory workers since it can be used off-site. Using 3DCP technology for construction can significantly improve worker safety by reducing the risk of accidents and injuries. With 3DCP, carpenters do not need to handle dangerous tools and materials, such as in cutting wood or sawing, which are often associated with accidents. Furthermore, 3DCP is automated, drastically reducing the possibility of equipment-related mishaps and guaranteeing staff safety. Off-site buildings offer fewer work-at-height chores and less total control over the manufacturing environment than on-site buildings.

As is sometimes observed on building sites, *"around 150 individuals stroll about during busy times. A lot of people are still on site. I wonder whether 3DCP printing could be useful there."* 3DCP could stop mishaps by easing worker interaction and congestion.

Developing training plans is crucial if one is to enjoy these advantages. One of the main obstacles to extensive acceptance of 3DCP (*"If you want to do it [3DCP] yourself, you also have to train your people"*). Although there are only a few professional training courses, universities have begun instructing their students in 3DCP construction. Apart from the staff, 3DCP could improve the usability of a construction for its final users. Raising the thermal efficiency of a building has great possibilities. While concrete provides inadequate insulation, 3DCP allows the building of hollow constructions that may be filled with better insulators much more easily.

Additionally, depending on their location and intended usage, buildings might be constructed to maximize aspects like sunlight, ventilation, or acoustics. A comprehensive approach to building design is necessary to *"add this integration without adding the extra costs"* 3DCP promises. In theory, achieving this full integration would enhance a building's occupants' comfort and quality of life while lowering energy usage and, consequently, the energy bill.

However, according to two people we spoke with, moisture control is a significant barrier to using 3DCP structures. Concrete does not "breathe", unlike other wall surface materials like gypsum. Maintaining consistent moisture throughout the day is crucial in places with high humidity, and this can be a serious problem to deal with. During freezing weather, where water can freeze, no water drops must enter the concrete structure. These issues are resolved using specialized surface treatments. Another option is to build structural features into the walls, like rain screens, at the building time. The building's location will inevitably affect the chosen solution because it must also adhere to local regulations.

4.3. Sustainability Economic: High R&D Expenditures Due to the Immaturity of Technologies

The current state of 3DCP, while innovative, presents challenges compared to traditional concrete pouring methods, particularly in cost and long-term performance. One interviewee noted, "All these companies advertise that you can print a house in 24 or 48 h. However, one key drawback of 3DCP is the potential compromise on structural integrity". While it may be possible to print a house quickly, concerns about the long-term durability and strength of the printed structure remain significant. Rigorous research and testing are essential to ensure that 3DCP can meet the same safety and sustainability standards as traditional methods.

Moreover, 3DCP employs fundamentally different design principles, requires operators with specialized skills, and involves distinct quality control and maintenance protocols. While the initial costs of 3DCP structures may be higher, their advantages in comfort, thermal efficiency, and potential energy savings can offset these expenses over time. However, comprehensive cost–benefit analyses are complex and require advanced tools. Furthermore, the long-term performance and maintenance of 3DCP structures are critical factors in assessing their overall sustainability. Construction firms may hesitate to invest heavily in this still-maturing technology without additional institutional support, especially when direct cost reductions are not immediately evident.

According to interviewees, two main factors contribute most to the cost of using 3DCP, and these factors could be reduced if they were considered: materials and quality control. There is a challenge in that the average cost of a large construction varies from a low cost per kilogram to a high cost per ton. In general, a house costs one euro a kilogram. 3DCP materials are much more expensive than traditional concrete mixtures because there is not enough demand to scale up production and achieve economies of scale, and there are not many suppliers. Without competition, material suppliers do not have to lower their prices. Furthermore, because there are only a few suppliers, materials must be transported over longer distances, increasing transportation costs and environmental impact.

A significant portion of the final cost of 3DCP structures is devoted to quality control, as there are no standardized procedures for designing 3DCP structures. "What Eurocodes has now is Design by Testing . . . we have to make a 1:1 prototype, test it, and based on that test, we can 3D print it and get it certified as safe . . . This process is costly because you have to make a bridge or house twice". Once the final product has been built, non-destructive diagnostics testing is carried out. For instance, a bridge might be tested by putting the maximum load the bridge design allows on it and checking that the structure holds. As researchers produce digital twins to replicate the behavior of 3DCP structures, "the more testing you do, the more efficient your digital twin becomes as you feed it more data in the future, these digital twins will help to create building codes". Testing costs should thus drop.

Further, there are concerns regarding the long-term behavior of materials. In some tenders, contractors must ensure that a bridge lasts 100 years. However, there is considerable uncertainty about whether 3DCP structures will fail in the future in unknown ways. "It will cause great anxiety; should we have a bad experience, the guidelines [construction codes] will once more be tightened. People want [innovation], but if at some point it gets project-specific and you discuss it at the project level, their overwhelming worry is the 100-year lifespan". Tightening the building codes probably would make 3DCP less appealing.

3DCP presents two appealing benefits despite the economic challenges: better consistency and lower formwork costs. "Formwork costs almost a third of the total costs; cement and labor makes up the rest proportionate; that's right, formwork accounts for about 50% of the material costs". Although 3DCP materials may be more expensive, the time and money saved on constructing formwork and hiring labor to pour the concrete may offset the cost. The interviewee said the most interesting uses are "where people are in danger, where you print highly complex objects, somewhere in between, under the ground, or in tough places".

There is unpredictability in 3DCP compared to manual labor, which is like any other automated process. "People can do things right today but wrong tomorrow; you are not in control and the 3D printer either prints something right or wrong". Companies who can employ 3DCP to attain structural characteristics can save much money. The extra cost of 3DCP could be traded off in terms of the reduced chance of rebuilding a given component. Failure expenses are great, and the profit margin ranges from 2% to 3%. Our actions either reduce or convert earnings into losses. If we are talking about risk, it would be better to purchase one more costly prefabricated pile than three inferior piles, for which I will have to build two afterwards. Reducing the demand for rework will also help to lessen environmental effects.

4.4. Preserving Harmony between the Surroundings, People, and Business Interests

Our study especially illustrates the industry's opinions on 3DCP possibilities and difficulties. Although 3DCP has the potential to increase construction sustainability significantly, early adopters will have to make some tradeoffs along the triple bottom line. We summarize our findings in Table 2, which emphasizes 3DCP's tensions between people, planet, and profit. In Table 2, 3DCP proponents summarize its main benefits in horizontal rows. Each column represents the triple bottom line.

5. Discussion

5.1. Decisions That Affect the Sustainability of the 3DCP

Our research suggests that how a company uses 3DCP will significantly impact the sustainability benefits it offers. There is no established adoption plan due to the current level of technological uncertainty [56–59]. As a result, businesses seem to be approaching technology with caution. We examine five crucial options for construction organizations regarding 3DCP, together with their consequences for sustainability, to aid managers in their decision-making (see Table 1) [64–67]. Every company's short- and long-term strategic goals will determine the optimal course of action; hence, these decisions are interrelated.

First, whether to invest in 3D printing technology and, if so, whether to employ concrete as a base material is fundamental to the second choice [68,69]. Using 3DCP could help to lower the environmental effect of building activities and enhance structural performance. Modern robotic equipment requires significant research and development investments to acquire the knowledge to design and construct sturdy structures [68–70]. Concrete may not be sustainable in the long run. 3DCP can simplify the process of designing safe concrete constructions since businesses are already familiar with the properties of concrete [71]. One of the downsides of concrete is that it is no longer recyclable or reusable. Although numerous research initiatives are underway to create more ecologically friendly material mixtures suitable for 3DCP, the details of the materials and the timing of their commercialization are unclear [72–75]. 3DCP may delay the shift to a circular economy; concrete must be discontinued as soon as possible, as some interviewees assert [76,77].

On the other hand, a 3DCP implementation can help companies become acquainted with digital infrastructures and material design ideas. Installed in Amsterdam in 2018, MX3D is a 12-m-long 3D-printed stainless-steel bridge [78–82]. Since concrete would have had detrimental environmental effects, this bridge benefited from 3D printing.

When a corporation uses 3DCP to design a product, three options govern the design process [83]. One of the decisions that needs to be made is creating integrated or modular structures [84]. Modular structures are considered more sustainable than integrated ones to reduce waste and boost circularity in some parts of the finished building that can be replaced without altering the entire structure [85,86]. However, it is essential to note that modular constructions require additional materials due to the stress concentration areas created by the connections between the different components [87]. The authors recommend that, to figure out whether a design should be modular or integrated, three factors should be taken into consideration: (1) the possibility of achieving light weight, (2) the capability of improving performance, and (3) the expected longevity of the system [88–90].

Additionally, businesses must decide whether to "produce or acquire" 3DCP components [91]. If a company lacks the expertise to make specific components or there is little technological opportunity, they will often outsource some of their production. On the other hand, companies typically vertically integrate component manufacture if the final product's complexity is notable or if they consider it as a crucial expertise and competitive advantage [92,93]. Companies now contract with specialized suppliers to produce 3DCP components [94]. Businesses could find it beneficial to insource 3DCP activities, nevertheless, if technology develops to the point of replacing specialist building technologies as the standard [95,96]. General Electric acquired the 3D printing machinery manufacturers Arcam and Concept Las to perform metal 3D printing [97,98].

The final consideration for 3DCP adopters is whether it is preferable to make components locally or elsewhere [98]. On the one hand, off-site component production could enhance process control and part quality, two significant technological constraints in developing manufacturing technologies [99,100]. Still, off-site components must be transported from the manufacturer to the building site. The demand for transportation limits the biggest component size that can be produced [101,102]. On the other hand, building contractors may quickly erect larger structures without the trouble of transportation when components are produced on-site. Printing is also less reliable and possibly more prone to error on a construction site because of the constantly changing environmental conditions [103–105].

Location, outsourcing, and modularization considerations are all closely related [106]. Modular designs make a more adaptable supply chain possible, which may encourage supplier rivalry [107]. Outsourcing could be less expensive than making components internally if supplier competition lowers prices [108]. Modular design can be the most financially realistic choice when off-site manufacturing is desired, and quality control is a top priority [109]. In contrast, companies that wish to take advantage of the performance advantages of 3DCP, such as the reduction in manual assembly, may consider using on-site 3DCP for large, consolidated constructions [110–113]. Under such circumstances, 3DCP would become essential for their company operations, and companies would choose to internalize 3DCP responsibilities [114–116]. It is necessary to consider each organization's characteristics and the anticipated technological advancement rate.

5.2. Policy Implications

A shortage of skilled labor and low tender requirements currently hinder the widespread adoption of 3D concrete printing (3DCP) technology. To address these barriers, targeted government intervention is needed in three key areas: experimentation and data sharing, workforce development, and establishing stringent tender requirements. The government can significantly influence the adoption of 3DCP by promoting regulation and standardization, garnering public support for funding pilot projects and forming public–private partnerships for data collection and analysis. Reducing the uncertainty around 3DCP's structural behavior will aid industry acceptance; thus, public committees could facilitate this by standardizing and characterizing 3DCP mixtures to overcome suppliers' reluctance to share data.

The skilled labor shortage, exacerbated by the introduction of 3DCP, underscores the need for robust training programs tailored to this technology. Expanding existing 3D printing training initiatives to include materials like metals and polymers can support this goal. However, given the significant differences between 3DCP and other 3D printing applications in mechanical properties and component sizes, specialized construction-oriented training programs are essential. Strengthening these socio-economic aspects will provide a more comprehensive understanding of the changes 3DCP may bring to the labor market and social structure.

6. Conclusions and Future Scope

Three-dimensional concrete printing (3DCP) presents significant potential in reducing construction's carbon footprint; however, many challenges remain, particularly concerning the environmental impact of raw materials. Conducting a comprehensive environmental impact assessment of the raw materials used in the 3DCP process is essential to identify whether alternative material combinations could further reduce environmental impact. Additionally, a thorough study of how design decisions, such as the level of modularity and the thickness of 3DCP structures, influence both circularity and overall performance is necessary to realize this technology's sustainability benefits fully.

Social sustainability can be significantly influenced by 3D concrete printing (3DCP) through its impact on the construction industry and the availability of low-skilled labor. While 3DCP can reduce reliance on seasonal labor, it may also affect employment opportunities for low-skilled workers. Construction automation could decrease the demand for these workers, leading to significant challenges for the industry and its workforce. To fully understand the implications of this technological change, further investigation is needed to analyze the full impact on the quantity and quality of construction jobs that could be affected by a higher level of automation.

3DCP's cost-competitiveness, compared to other options, is currently unavailable to construction managers as a decision-making tool. Technoeconomic models should consider cost, quality, labor, and maintenance factors. It is difficult to predict the effects of technological inflation on costs due to rapid technological advancements and limited market strategies for 3DCP materials and equipment. Analyzing historical cost evolution in prefabricated construction or other industries can help forecast expenses better.

Author Contributions: Conceptualization, B.T.S., S., S.S.C. and A.K.S.; Data curation, S.S.C. and A.T.; Formal analysis, B.T.S., S., S.S.C., A.K.S., N.K. and A.T.; Funding acquisition, S.R.N.; Investigation, S., R.K. and N.K.; Methodology, R.K.; Project administration, S.R.N.; Resources, A.K.S., N.K. and A.T.; Supervision, S.R.N.; Validation, S. and R.K.; Writing—original draft, B.T.S., S.S.C. and A.K.S.; Writing—review and editing, S., R.K., N.K., A.T. and S.R.N. All authors have read and agreed to the published version of the manuscript.

Funding: This research received no funding.

Data Availability Statement: The data presented in this study are available on request from the corresponding author (The data are not publicly available due to privacy restrictions).

Acknowledgments: The authors would like to extend their appreciation to all participants who completed the survey questionnaire. Their feedback has been invaluable to this research.

Conflicts of Interest: The authors declare no conflicts of interest.

References

1. Aramburu, A.; Calderon-Uriszar-Aldaca, I.; Puente, I.; Castano-Alvarez, R. Effects of 3D-printing on the tensile splitting strength of concrete structures. *Case Stud. Constr. Mater.* **2024**, *20*, e03090. [CrossRef]
2. Dananjaya, V.; Marimuthu, S.; Yang, R.C.; Grace, A.N.; Abeykoon, C. Synthesis, properties, applications, 3D printing and machine learning of graphene quantum dots in polymer nanocomposites. *Prog. Mater. Sci.* **2024**, *144*, 101282. [CrossRef]
3. Tinoco, M.P.; de Mendonça, É.M.; Fernandez, L.I.C.; Caldas, L.R.; Reales, O.A.M.; Filho, R.D.T. Life cycle assessment (LCA) and environmental sustainability of cementitious materials for 3D concrete printing: A systematic literature review. *J. Build. Eng.* **2022**, *52*, 104456. [CrossRef]
4. Rajeev, P.; Ramesh, A.; Navaratnam, S.; Sanjayan, J. Using Fibre recovered from face mask waste to improve printability in 3D concrete printing. *Cem. Concr. Compos.* **2023**, *139*, 105047. [CrossRef]
5. Cuevas, K.; Weinhold, J.; Stephan, D.; Kim, J.S. Effect of printing patterns on pore-related microstructural characteristics and properties of materials for 3D concrete printing using in situ and ex situ imaging techniques. *Constr. Build. Mater.* **2023**, *405*, 133220. [CrossRef]
6. Hojati, M.; Memari, A.M.; Zahabi, M.; Wu, Z.; Li, Z.; Park, K.; Nazarian, S.; Duarte, J.P. Barbed-wire reinforcement for 3D concrete printing. *Autom. Constr.* **2022**, *141*, 104438. [CrossRef]
7. Haar, B.T.; Kruger, J.; van Zijl, G. Off-site construction with 3D concrete printing. *Autom. Constr.* **2023**, *152*, 104906. [CrossRef]
8. Castro, B.M.; Elbadawi, M.; Ong, J.J.; Pollard, T.; Song, Z.; Gaisford, S.; Pérez, G.; Basit, A.W.; Cabalar, P.; Goyanes, A. Machine learning predicts 3D printing performance of over 900 drug delivery systems. *J. Control. Release* **2021**, *337*, 530–545. [CrossRef]
9. Chang, Z.; Zhang, H.; Liang, M.; Schlangen, E.; Šavija, B. Numerical simulation of elastic buckling in 3D concrete printing using the lattice model with geometric nonlinearity. *Autom. Constr.* **2022**, *142*, 104485. [CrossRef]
10. Su, Z.; Zhao, K.; Ye, Z.; Cao, W.; Wang, X.; Liu, K.; Wang, Y.; Yang, L.; Dai, B.; Zhu, J. Overcoming the penetration-saturation trade-off in binder jet additive manufacturing via rapid in situ curing. *Addit. Manuf.* **2022**, *59*, 103157. [CrossRef]
11. Christ, J.; Perrot, A.; Ottosen, L.M.; Koss, H. Rheological characterization of temperature-sensitive biopolymer-bound 3D printing concrete. *Constr. Build. Mater.* **2024**, *411*, 134337. [CrossRef]
12. Khan, S.A.; Koç, M. Numerical modelling and simulation for extrusion-based 3D concrete printing: The underlying physics, potential, and challenges. *Results Mater.* **2022**, *16*, 100337. [CrossRef]
13. Liu, J.; Li, S.; Fox, K.; Tran, P. 3D concrete printing of bioinspired Bouligand structure: A study on impact resistance. *Addit. Manuf.* **2022**, *50*, 102544. [CrossRef]
14. Geng, S.Y.; Luo, Q.L.; Cheng, B.Y.; Li, L.X.; Wen, D.C.; Long, W.J. Intelligent multi-objective optimization of 3D printing low-carbon concrete for multi-scenario requirements. *J. Clean. Prod.* **2024**, *445*, 141361. [CrossRef]
15. Ahmed, G.H. A review of "3D concrete printing": Materials and process characterization, economic considerations and environmental sustainability. *J. Build. Eng.* **2023**, *66*, 105863. [CrossRef]
16. Dey, D.; Srinivas, D.; Panda, B.; Suraneni, P.; Sitharam, T.G. Use of industrial waste materials for 3D printing of sustainable concrete: A review. *J. Clean. Prod.* **2022**, *340*, 130749. [CrossRef]
17. Nasiri, H.; Dadashi, A.; Azadi, M. Machine learning for fatigue lifetime predictions in 3D-printed polylactic acid biomaterials based on interpretable extreme gradient boosting model. *Mater. Today Commun.* **2024**, *39*, 109054. [CrossRef]
18. Alabbasi, M.; Agkathidis, A.; Chen, H. Robotic 3D printing of concrete building components for residential buildings in Saudi Arabia. *Autom. Constr.* **2023**, *148*, 104751. [CrossRef]
19. Rehman, A.U.; Kim, I.G.; Kim, J.H. Towards full automation in 3D concrete printing construction: Development of an automated and inline sensor-printer integrated instrument for in-situ assessment of structural build-up and quality of concrete. *Dev. Built Environ.* **2024**, *17*, 100344. [CrossRef]
20. Lyu, Q.; Dai, P.; Chen, A. Sandwich-structured porous concrete manufactured by mortar-extrusion and aggregate-bed 3D printing. *Constr. Build. Mater.* **2023**, *392*, 131909. [CrossRef]
21. Wang, Y.; Qiu, L.C.; Hu, Y.Y.; Chen, S.G.; Liu, Y. Influential factors on mechanical properties and microscopic characteristics of underwater 3D printing concrete. *J. Build. Eng.* **2023**, *77*, 107571. [CrossRef]
22. Rizzieri, G.; Cremonesi, M.; Ferrara, L. A 2D numerical model of 3D concrete printing including thixotropy. *Mater. Today Proc.* **2023**, *20*, 21–29. [CrossRef]
23. Yang, H.; Fang, L.; Yuan, Z.; Teng, X.; Qin, H.; He, Z.; Wan, Y.; Wu, X.; Zhang, Y.; Guan, L.; et al. Machine learning guided 3D printing of carbon microlattices with customized performance for supercapacitive energy storage. *Carbon* **2023**, *201*, 408–414. [CrossRef]
24. Ma, G.; Buswell, R.; da Silva, W.R.L.; Wang, L.; Xu, J.; Jones, S.Z. Technology readiness: A global snapshot of 3D concrete printing and the frontiers for development. *Cem. Concr. Res.* **2022**, *156*, 106774. [CrossRef]
25. Yu, H.; Zhang, W.; Yin, B.; Sun, W.; Akbar, A.; Zhang, Y.; Liew, K.M. Modeling extrusion process and layer deformation in 3D concrete printing via smoothed particle hydrodynamics. *Comput. Methods Appl. Mech. Eng.* **2024**, *420*, 116761. [CrossRef]
26. Geng, S.Y.; Mei, L.; Cheng, B.Y.; Luo, Q.L.; Xiong, C.; Long, W.J. Revolutionizing 3D concrete printing: Leveraging RF model for precise printability and rheological prediction. *J. Build. Eng.* **2024**, *88*, 109127. [CrossRef]
27. Voydie, D.; Goupil, L.; Chanthery, E.; Travé-Massuyès, L.; Delautier, S. Machine Learning Based Fault Anticipation for 3D Printing. *IFAC PapersOnLine* **2023**, *56*, 2927–2932. [CrossRef]

28. Li, S.; Nguyen-Xuan, H.; Tran, P. Digital design and parametric study of 3D concrete printing on non-planar surfaces. *Autom. Constr.* **2023**, *145*, 104624. [CrossRef]
29. Lu, B.; Li, M.; Wong, T.N.; Qian, S. Spray-based 3D concrete printing with calcium and polymeric additives: A feasibility study. *Mater. Today Proc.* **2022**, *70*, 252–257. [CrossRef]
30. Rosseau, L.R.S.; Jansen, J.T.A.; Roghair, I.; van Sint Annaland, M. Favorable trade-off between heat transfer and pressure drop in 3D printed baffled logpile catalyst structures. *Chem. Eng. Res. Des.* **2023**, *196*, 214–234. [CrossRef]
31. Wan, Q.; Wang, L.; Ma, G. Continuous and adaptable printing path based on transfinite mapping for 3D concrete printing. *Autom. Constr.* **2022**, *142*, 104471. [CrossRef]
32. Zhang, N.; Sanjayan, J. Mechanisms of rheological modifiers for quick mixing method in 3D concrete printing. *Cem. Concr. Compos.* **2023**, *142*, 105218. [CrossRef]
33. Huang, X.; Yang, W.; Song, F.; Zou, J. Study on the mechanical properties of 3D printing concrete layers and the mechanism of influence of printing parameters. *Constr. Build. Mater.* **2022**, *335*, 127496. [CrossRef]
34. Gebhard, L.; Esposito, L.; Menna, C.; Mata-Falcón, J. Inter-laboratory study on the influence of 3D concrete printing set-ups on the bond behaviour of various reinforcements. *Cem. Concr. Compos.* **2022**, *133*, 104660. [CrossRef]
35. Liu, D.; Zhang, Z.; Zhang, X.; Chen, Z. 3D printing concrete structures: State of the art, challenges, and opportunities. *Constr. Build. Mater.* **2023**, *405*, 133364. [CrossRef]
36. Li, Z.; Liu, H.; Nie, P.; Cheng, X.; Zheng, G.; Jin, W.; Xiong, B. Mechanical properties of concrete reinforced with high-performance microparticles for 3D concrete printing. *Constr. Build. Mater.* **2024**, *411*, 134676. [CrossRef]
37. Cipollone, D.; Yang, H.; Yang, F.; Bright, J.; Liu, B.; Winch, N.; Wu, N.; Sierros, K.A. 3D printing of an anode scaffold for lithium batteries guided by mixture design-based sequential learning. *J. Mater. Process. Technol.* **2021**, *295*, 117159. [CrossRef]
38. Lu, Y.; Xiao, J.; Li, Y. 3D printing recycled concrete incorporating plant fibres: A comprehensive review. *Constr. Build. Mater.* **2024**, *425*, 135951. [CrossRef]
39. Rehman, A.U.; Perrot, A.; Birru, B.M.; Kim, J.H. Recommendations for quality control in industrial 3D concrete printing construction with mono-component concrete: A critical evaluation of ten test methods and the introduction of the performance index. *Dev. Built Environ.* **2023**, *16*, 100232. [CrossRef]
40. Zhao, B.; Zhang, M.; Dong, L.; Wang, D. Design of grayscale digital light processing 3D printing block by machine learning and evolutionary algorithm. *Compos. Commun.* **2022**, *36*, 101395. [CrossRef]
41. Barjuei, E.S.; Courteille, E.; Rangeard, D.; Marie, F.; Perrot, A. Real-time vision-based control of industrial manipulators for layer-width setting in concrete 3D printing applications. *Adv. Ind. Manuf. Eng.* **2022**, *5*, 100094. [CrossRef]
42. Ramesh, S.; Deep, A.; Tamayol, A.; Kamaraj, A.; Mahajan, C.; Madihally, S. Advancing 3D bioprinting through machine learning and artificial intelligence. *Bioprinting* **2024**, *38*, e00331. [CrossRef]
43. Breseghello, L.; Hajikarimian, H.; Naboni, R. 3DLightSlab. Design to 3D concrete printing workflow for stress-driven ribbed slabs. *J. Build. Eng.* **2024**, *91*, 109573. [CrossRef]
44. Pan, Z.; Si, D.; Tao, J.; Xiao, J. Compressive behavior of 3D printed concrete with different printing paths and concrete ages. *Case Stud. Constr. Mater.* **2023**, *18*, e01949. [CrossRef]
45. Liu, Z.; Li, M.; Quah, T.K.N.; Wong, T.N.; Tan, M.J. Comprehensive investigations on the relationship between the 3D concrete printing failure criterion and properties of fresh-state cementitious materials. *Addit. Manuf.* **2023**, *76*, 103787. [CrossRef]
46. Lowke, D.; Vandenberg, A.; Pierre, A.; Thomas, A.; Kloft, H.; Hack, N. Injection 3D concrete printing in a carrier liquid—Underlying physics and applications to lightweight space frame structures. *Cem. Concr. Compos.* **2021**, *124*, 104169. [CrossRef]
47. Yuan, P.F.; Zhan, Q.; Wu, H.; Beh, H.S.; Zhang, L. Real-time toolpath planning and extrusion control (RTPEC) method for variable-width 3D concrete printing. *J. Build. Eng.* **2022**, *46*, 103716. [CrossRef]
48. Rubin, A.P.; Quintanilha, L.C.; Repette, W.L. Influence of structuration rate, with hydration accelerating admixture, on the physical and mechanical properties of concrete for 3D printing. *Constr. Build. Mater.* **2023**, *363*, 129826. [CrossRef]
49. Lee, K.W.; Lee, H.J.; Choi, M.S. Correlation between thixotropic behavior and buildability for 3D concrete printing. *Constr. Build. Mater.* **2022**, *347*, 128498. [CrossRef]
50. Sun, L.; Hua, G.; Cheng, T.C.E.; Teunter, R.H.; Dong, J.; Wang, Y. Purchase or rent? Optimal pricing for 3D printing capacity sharing platforms. *Eur. J. Oper. Res.* **2023**, *307*, 1192–1205. [CrossRef]
51. Jaji, M.B.; van Zijl, G.P.A.G.; Babafemi, A.J. Slag-modified metakaolin-based geopolymer for 3D concrete printing application: Evaluating fresh and hardened properties. *Clean Eng. Technol.* **2023**, *15*, 100665. [CrossRef]
52. Chen, Y.; Zhang, Y.; Zhang, Y.; Pang, B.; Zhang, W.; Liu, C.; Liu, Z.; Wang, D.; Sun, G. Influence of gradation on extrusion-based 3D printing concrete with coarse aggregate. *Constr. Build. Mater.* **2023**, *403*, 133135. [CrossRef]
53. El Abbaoui, K.; Al Korachi, I.; El Jai, M.; Šeta, B.; Mollah, M.T. 3D concrete printing using computational fluid dynamics: Modeling of material extrusion with slip boundaries. *J. Manuf. Process.* **2024**, *118*, 448–459. [CrossRef]
54. Rehman, A.U.; Birru, B.M.; Kim, J.H. Set-on-demand 3D Concrete Printing (3DCP) construction and potential outcome of shotcrete accelerators on its hardened properties. *Case Stud. Constr. Mater.* **2023**, *18*, e01955. [CrossRef]
55. Mollah, M.T.; Comminal, R.; da Silva, W.R.L.; Šeta, B.; Spangenberg, J. Computational fluid dynamics modelling and experimental analysis of reinforcement bar integration in 3D concrete printing. *Cem. Concr. Res.* **2023**, *173*, 107263. [CrossRef]
56. Besklubova, S.; Tan, B.Q.; Zhong, R.Y.; Spicek, N. Logistic cost analysis for 3D printing construction projects using a multi-stage network-based approach. *Autom. Constr.* **2023**, *151*, 104863. [CrossRef]

57. Bai, G.; Wang, L.; Wang, F.; Ma, G. Assessing printing synergism in a dual 3D printing system for ultra-high performance concrete in-process reinforced cementitious composite. *Addit. Manuf.* **2023**, *61*, 103338. [CrossRef]
58. Wang, L.; Ye, K.; Wan, Q.; Li, Z.; Ma, G. Inclined 3D concrete printing: Build-up prediction and early-age performance optimization. *Addit. Manuf.* **2023**, *71*, 103595. [CrossRef]
59. Ramakrishnan, S.; Pasupathy, K.; Mechtcherine, V.; Sanjayan, J. Printhead mixing of geopolymer and OPC slurries for hybrid alkali-activated cement in 3D concrete printing. *Constr. Build. Mater.* **2024**, *430*, 136439. [CrossRef]
60. Singh, A.K.; Kumar, V.P.; Dehdasht, G.; Mohandes, S.R.; Manu, P.; Rahimian, F.P. Investigating the barriers to the adoption of blockchain technology in sustainable construction projects. *J. Clean. Prod.* **2023**, *403*, 136840. [CrossRef]
61. Singh, A.K.; Kumar, V.P.; Shoaib, M.; Adebayo, T.S.; Irfan, M. A strategic roadmap to overcome blockchain technology barriers for sustainable construction: A deep learning-based dual-stage SEM-ANN approach. *Technol. Forecast. Soc. Change* **2023**, *194*, 122716. [CrossRef]
62. Adaloudis, M.; Roca, J.B. Sustainability tradeoffs in the adoption of 3D Concrete Printing in the construction industry. *J. Clean Prod.* **2021**, *307*, 127201. [CrossRef]
63. Naganna, S.R.; Ibrahim, H.A.; Yap, S.P.; Tan, C.G.; Mo, K.H.; El-Shafie, A. Insights into the multifaceted applications of architectural concrete: A state-of-the-art review. *Arab. J. Sci. Eng.* **2021**, *46*, 4213–4223. [CrossRef]
64. Lyu, Q.; Dai, P.; Chen, A. Mechanical strengths and optical properties of translucent concrete manufactured by mortar-extrusion 3D printing with polymethyl methacrylate (PMMA) fibers. *Compos. B Eng.* **2024**, *268*, 111079. [CrossRef]
65. Liu, X.; Cai, H.; Ma, G.; Hou, G. Spray-based 3D concrete printing parameter design model: Actionable insight for high printing quality. *Cem. Concr. Compos.* **2024**, *147*, 105446. [CrossRef]
66. Salaimanimagudam, M.P.; Jayaprakash, J. Effect of introducing dummy layers on interlayer bonding and geometrical deformations in concrete 3D printing. *Mater. Lett.* **2024**, *366*, 136575. [CrossRef]
67. Qu, Z.; Yu, Q.; Ong, G.P.; Cardinaels, R.; Ke, L.; Long, Y.; Geng, G. 3D printing concrete containing thermal responsive gelatin: Towards cold environment applications. *Cem. Concr. Compos.* **2023**, *140*, 105029. [CrossRef]
68. Cheng, H.; Radlińska, A.; Hillman, M.; Liu, F.; Wang, J. Modeling concrete deposition via 3D printing using reproducing kernel particle method. *Cem. Concr. Res.* **2024**, *181*, 107526. [CrossRef]
69. Li, H.; Alkahtani, M.E.; Basit, A.W.; Elbadawi, M.; Gaisford, S. Optimizing environmental sustainability in pharmaceutical 3D printing through machine learning. *Int. J. Pharm.* **2023**, *648*, 123561. [CrossRef]
70. Zhu, J.; Ren, X.; Cervera, M. Buildability modeling of 3D-printed concrete including printing deviation: A stochastic analysis. *Constr. Build. Mater.* **2023**, *403*, 133076. [CrossRef]
71. Li, L.; Hao, L.; Li, X.; Xiao, J.; Zhang, S.; Poon, C.S. Development of CO_2-integrated 3D printing concrete. *Constr. Build. Mater.* **2023**, *409*, 134233. [CrossRef]
72. Ma, G.; Hu, T.; Wang, F.; Liu, X.; Li, Z. Magnesium phosphate cement for powder-based 3D concrete printing: Systematic evaluation and optimization of printability and printing quality. *Cem. Concr. Compos.* **2023**, *139*, 105000. [CrossRef]
73. Zhao, Z.; Ji, C.; Xiao, J.; Yao, L.; Lin, C.; Ding, T.; Ye, T. A critical review on reducing the environmental impact of 3D printing concrete: Material preparation, construction process and structure level. *Constr. Build. Mater.* **2023**, *409*, 133887. [CrossRef]
74. Chang, Z.; Chen, Y.; Schlangen, E.; Šavija, B. A review of methods on buildability quantification of extrusion-based 3D concrete printing: From analytical modelling to numerical simulation. *Dev. Built Environ.* **2023**, *16*, 100241. [CrossRef]
75. Lyu, Q.; Wang, Y.; Dai, P. Multilayered plant-growing concrete manufactured by aggregate-bed 3D concrete printing. *Constr. Build. Mater.* **2024**, *430*, 136453. [CrossRef]
76. Zhang, R.C.; Wang, L.; Xue, X.; Ma, G.W. Environmental profile of 3D concrete printing technology in desert areas via life cycle assessment. *J. Clean. Prod.* **2023**, *396*, 136412. [CrossRef]
77. Yang, W.; Wang, L.; Ma, G.; Feng, P. An integrated method of topological optimization and path design for 3D concrete printing. *Eng. Struct.* **2023**, *291*, 116435. [CrossRef]
78. Nguyen-Van, V.; Nguyen-Xuan, H.; Panda, B.; Tran, P. 3D concrete printing modelling of thin-walled structures. *Structures* **2022**, *39*, 496–511. [CrossRef]
79. Yin, Y.; Huang, J.; Wang, T.; Yang, R.; Hu, H.; Manuka, M.; Zhou, F.; Min, J.; Wan, H.; Yuan, D.; et al. Effect of Hydroxypropyl methyl cellulose (HPMC) on rheology and printability of the first printed layer of cement activated slag-based 3D printing concrete. *Constr. Build. Mater.* **2023**, *405*, 133347. [CrossRef]
80. Westphal, E.; Seitz, H. Machine learning for the intelligent analysis of 3D printing conditions using environmental sensor data to support quality assurance. *Addit. Manuf.* **2022**, *50*, 102535. [CrossRef]
81. Bi, M.; Tran, P.; Xia, L.; Ma, G.; Xie, Y.M. Topology optimization for 3D concrete printing with various manufacturing constraints. *Addit. Manuf.* **2022**, *57*, 102982. [CrossRef]
82. Jia, Z.; Kong, L.; Jia, L.; Ma, L.; Chen, Y.; Zhang, Y. Printability and mechanical properties of 3D printing ultra-high performance concrete incorporating limestone powder. *Constr. Build. Mater.* **2024**, *426*, 136195. [CrossRef]
83. Wei, Y.; Han, S.; Chen, Z.; Lu, J.; Li, Z.; Yu, S.; Cheng, W.; An, M.; Yan, P. Numerical simulation of 3D concrete printing derived from printer head and printing process. *J. Build. Eng.* **2024**, *88*, 109241. [CrossRef]
84. An, D.; Zhang, Y.X.; Yang, R.C. Numerical modelling of 3D concrete printing: Material models, boundary conditions and failure identification. *Eng. Struct.* **2024**, *299*, 117104. [CrossRef]

85. Krishna, D.V.; Sankar, M.R. Machine learning-assisted extrusion-based 3D bioprinting for tissue regeneration applications. *Ann. 3D Print. Med.* **2023**, *12*, 100132. [CrossRef]
86. Gao, H.; Chen, Y.; Chen, Q.; Yu, Q. Thermal and mechanical performance of 3D printing functionally graded concrete: The role of SAC on the rheology and phase evolution of 3DPC. *Constr. Build. Mater.* **2023**, *409*, 133830. [CrossRef]
87. Duan, Z.; Deng, Q.; Xiao, J.; Lv, Z.; Hu, B. Experimental realization on stress distribution monitoring during 3D concrete printing. *Mater. Lett.* **2024**, *358*, 135878. [CrossRef]
88. Nguyen-Van, V.; Li, S.; Liu, J.; Nguyen, K.; Tran, P. Modelling of 3D concrete printing process: A perspective on material and structural simulations. *Addit. Manuf.* **2023**, *61*, 103333. [CrossRef]
89. Ovhal, M.M.; Kumar, N.; Lee, H.B.; Tyagi, B.; Ko, K.J.; Boud, S.; Kang, J.W. Roll-to-roll 3D printing of flexible and transparent all-solid-state supercapacitors. *Cell. Rep. Phys. Sci.* **2021**, *2*, 100562. [CrossRef]
90. Liu, T.; Chen, S.; Ruan, K.; Zhang, S.; He, K.; Li, J.; Chen, M.; Yin, J.; Sun, M.; Wang, X.; et al. A handheld multifunctional smartphone platform integrated with 3D printing portable device: On-site evaluation for glutathione and azodicarbonamide with machine learning. *J. Hazard. Mater.* **2022**, *426*, 128091. [CrossRef]
91. Xiong, B.; Nie, P.; Liu, H.; Li, X.; Li, Z.; Jin, W.; Cheng, X.; Zheng, G.; Wang, L. Optimization of fiber reinforced lightweight rubber concrete mix design for 3D printing. *J. Build. Eng.* **2024**, *88*, 109105. [CrossRef]
92. Wei, Y.; Han, S.; Yu, S.; Chen, Z.; Li, Z.; Wang, H.; Cheng, W.; An, M. Parameter impact on 3D concrete printing from single to multi-layer stacking. *Autom. Constr.* **2024**, *164*, 105449. [CrossRef]
93. Wang, Y.; Qiu, L.C.; Chen, S.G.; Liu, Y. 3D concrete printing in air and under water: A comparative study on the buildability and interlayer adhesion. *Constr. Build. Mater.* **2024**, *411*, 134403. [CrossRef]
94. Rollakanti, C.R.; Prasad, C.V.S.R. Applications, performance, challenges and current progress of 3D concrete printing technologies as the future of sustainable construction—A state of the art review. *Mater. Today Proc.* **2022**, *65*, 995–1000. [CrossRef]
95. Liu, S.; Lu, B.; Li, H.; Pan, Z.; Jiang, J.; Qian, S. A comparative study on environmental performance of 3D printing and conventional casting of concrete products with industrial wastes. *Chemosphere* **2022**, *298*, 134310. [CrossRef]
96. Liu, X.; Sun, B. The influence of interface on the structural stability in 3D concrete printing processes. *Addit. Manuf.* **2021**, *48*, 102456. [CrossRef]
97. Khan, S.A.; Ilcan, H.; Imran, R.; Aminipour, E.; Şahin, O.; Al Rashid, A.; Şahmaran, M.; Koç, M. The impact of nozzle diameter and printing speed on geopolymer-based 3D-Printed concrete structures: Numerical modeling and experimental validation. *Results Eng.* **2024**, *21*, 101864. [CrossRef]
98. Zhang, N.; Sanjayan, J. Quick nozzle mixing technology for 3D printing foam concrete. *J. Build. Eng.* **2024**, *83*, 108445. [CrossRef]
99. Dabbagh, S.R.; Ozcan, O.; Tasoglu, S. Machine learning-enabled optimization of extrusion-based 3D printing. *Methods* **2022**, *206*, 27–40. [CrossRef]
100. Zhao, Y.; Gao, Y.; Chen, G.; Li, S.; Singh, A.; Luo, X.; Liu, C.; Gao, J.; Du, H. Development of low-carbon materials from GGBS and clay brick powder for 3D concrete printing. *Constr. Build. Mater.* **2023**, *383*, 131232. [CrossRef]
101. Tu, H.; Wei, Z.; Bahrami, A.; Kahla, N.B.; Ahmad, A.; Özkılıç, Y.O. Recent advancements and future trends in 3D concrete printing using waste materials. *Dev. Built Environ.* **2023**, *16*, 100187. [CrossRef]
102. Zhang, N.; Sanjayan, J. Surfactants to enable quick nozzle mixing in 3D concrete printing. *Cem. Concr. Compos.* **2023**, *142*, 105226. [CrossRef]
103. Salaimanimagudam, M.P.; Jayaprakash, J. Effect of printing parameters on inter-filament voids, bonding, and geometrical deviation in concrete 3D printed structures. *Mater. Lett.* **2023**, *349*, 134815. [CrossRef]
104. Dong, W.; Wang, J.; Hang, M.; Qu, S. Research on printing parameters and salt frost resistance of 3D printing concrete with ferrochrome slag and aeolian sand. *J. Build. Eng.* **2024**, *84*, 108508. [CrossRef]
105. Waqar, A.; Othman, I.; Almujibah, H.R.; Sajjad, M.; Deifalla, A.; Shafiq, N.; Azab, M.; Qureshi, A.H. Overcoming implementation barriers in 3D printing for gaining positive influence considering PEST environment. *Ain Shams Eng. J.* **2024**, *15*, 102517. [CrossRef]
106. Pott, U.; Jakob, C.; Dorn, T.; Stephan, D. Investigation of a shotcrete accelerator for targeted control of material properties for 3D concrete printing injection method. *Cem. Concr. Res.* **2023**, *173*, 107264. [CrossRef]
107. Christ, J.; Leusink, S.; Koss, H. Multi-axial 3D printing of biopolymer-based concrete composites in construction. *Mater. Des.* **2023**, *235*, 112410. [CrossRef]
108. Tao, Y.; Ren, Q.; Vantyghem, G.; Lesage, K.; Van Tittelboom, K.; Yuan, Y.; De Corte, W.; De Schutter, G. Extending 3D concrete printing to hard rock tunnel linings: Adhesion of fresh cementitious materials for different surface inclinations. *Autom. Constr.* **2023**, *149*, 104787. [CrossRef]
109. Kwon, S.W.; Kim, J.S.; Lee, H.M.; Lee, J.S. Physics-added neural networks: An image-based deep learning for material printing system. *Addit. Manuf.* **2023**, *73*, 103668. [CrossRef]
110. Li, H.; Addai-Nimoh, A.; Kreiger, E.; Khayat, K.H. Methodology to design eco-friendly fiber-reinforced concrete for 3D printing. *Cem. Concr. Compos.* **2024**, *147*, 105415. [CrossRef]
111. Wan, Q.; Yang, W.; Wang, L.; Ma, G. Global continuous path planning for 3D concrete printing multi-branched structure. *Addit. Manuf.* **2023**, *71*, 103581. [CrossRef]
112. Zeng, J.J.; Yan, Z.T.; Jiang, Y.Y.; Li, P.L. 3D printing of FRP grid and bar reinforcement for reinforced concrete plates: Development and effectiveness. *Compos. Struct.* **2024**, *335*, 117946. [CrossRef]

113. Wang, X.; Banthia, N.; Yoo, D.Y. Reinforcement bond performance in 3D concrete printing: Explainable ensemble learning augmented by deep generative adversarial networks. *Autom. Constr.* **2024**, *158*, 105164. [CrossRef]
114. Motalebi, A.; Khondoker, M.A.H.; Kabir, G. A systematic review of life cycle assessments of 3D concrete printing. *Sustain. Oper. Comput.* **2024**, *5*, 41–50. [CrossRef]
115. Ahi, O.; Ertunç, Ö.; Bundur, Z.B.; Bebek, Ö. Automated flow rate control of extrusion for 3D concrete printing incorporating rheological parameters. *Autom. Constr.* **2024**, *160*, 105319. [CrossRef]
116. Zhang, N.; Sanjayan, J. Extrusion nozzle design and print parameter selections for 3D concrete printing. *Cem. Concr. Compos.* **2023**, *137*, 104939. [CrossRef]

 *infrastructures*

Article

Collision Milling of Oil Shale Ash as Constituent Pretreatment in Concrete 3D Printing

Lucija Hanžič, Mateja Štefančič, Katarina Šter, Vesna Zalar Serjun, Māris Šinka and Alise Sapata et al.

1 Slovenian National Building and Civil Engineering Institute (ZAG), SI-1000 Ljubljana, Slovenia; mateja.stefancic@zag.si (M.Š.); katarina.ster@zag.si (K.Š.); vesna.zalar@zag.si (V.Z.S.); lidija.korat@zag.si (L.K.B.)

2 Institute of Sustainable Building Materials and Engineering Systems, Riga Technical University (RTU), LV-1048 Riga, Latvia; maris.sinka@rtu.lv (M.Š.); alise.sapata@rtu.lv (A.S.); genadijs.sahmenko@rtu.lv (G.Š.)

3 Faculty of Civil Engineering and Architecture, Kaunas University of Technology (KTU), 51367 Kaunas, Lithuania; evaldas.serelis@ktu.lt

4 Sakret, LV-2121 Rumbula, Latvia; baiba.migliniece@sakret.lv

* Correspondence: lucija.hanzic@zag.si; Tel.: +386-2-330-2422

Abstract: Concrete is an essential construction material, and infrastructures, such as bridges, tunnels, and power plants, consume large quantities of it. Future infrastructure demands and sustainability issues necessitate the adoption of non-conventional supplementary cementitious materials (SCMs). At the same time, global labor shortages are compelling the conservative construction sector to implement autonomous and digital fabrication methods, such as 3D printing. This paper thus investigates the feasibility of using oil shale ash (OSA) as an SCM in concrete suitable for 3D printing, and collision milling is examined as a possible ash pretreatment. OSA from four different sources was collected and analyzed for its physical, chemical, and mineralogical composition. Concrete formulations containing ash were tested for mechanical performance, and the two best-performing formulations were assessed for printability. It was found that ash extracted from flue gases by the novel integrated desulfurizer has the greatest potential as an SCM due to globular particles that contain β-calcium silicate. The 56-day compression strength of concrete containing this type of ash is ~60 MPa, the same as in the reference composition. Overall, collision milling is effective in reducing the size of particles larger than 10 μm but does not seem beneficial for ash extracted from flue gasses. However, milling bottom ash may unlock its potential as an SCM, with the optimal milling frequency being ~100 Hz.

Keywords: digital concrete; 3D printing; oil shale ash; supplementary cementitious material; collision milling

Academic Editor: Patricia Kara De Maeijer

Received: 15 November 2024
Revised: 24 December 2024
Accepted: 30 December 2024
Published: 13 January 2025

Citation: Lucija Hanžič, Mateja Štefančič, Katarina Šter, Vesna Zalar Serjun, Māris Šinka and Alise Sapata et al. . Collision Milling of Oil Shale Ash as Constituent Pretreatment in Concrete 3D Printing. *Infrastructures* 2025, 10, 18. https://doi.org/10.3390/infrastructures10010018

1. Introduction

The construction sector is notorious for a conservative approach to change and, consequently, a low level of innovation compared to other industry sectors. On one hand, its inherent attributes, such as site-based operations, project-based business models, and complex value chains, make it difficult to measure innovation returns using criteria from other industries [1]. On the other hand, it has been asserted that construction firms do not need to innovate to remain successful or viable [2] and that historically, the construction sector has ignored investments in research and development [3]. As a consequence, the level of automation in construction is low. However, workforce issues, such as aging [4], a shortage of young talent due to poor industry image [3], and poor representation of women [5,6], are pressing the industry into adopting less labor-intensive and more attractive fabrication methods. These challenges call for a paradigm shift presented in *Construction 4.0* formwork [3] with digitalization as its core component.

Digitalization in construction became a hot topic in the 1990s due to a pressing need to digitally exchange data already created by digital means. As a result, Building Information Models and Digital Twins emerged [7,8]. However, it took the construction industry almost another 20 years to start considering digital means for translating digital objects into physical ones. One of the most promising technologies for large-scale digital construction is the additive manufacturing of concrete structures, colloquially referred to as 3D concrete printing. Its exponential growth has resulted in a rather unprecedented leapfrog in the Technology Readiness Level (TRL), which is currently estimated to be between 6 and 7 [9]. To achieve full market implementation of TRL9, a number of challenges still need to be addressed [10–14]. In this respect, three areas of expertise have been recognized as crucial [13], namely, (i) digital modeling and printer operation, (ii) engineering of mechanical components, and (iii) material functionalities.

One of the central issues related to the material functionalities is that in terms of the material per-unit consumption, printed structures have a considerably worse carbon footprint than the structures they are replacing [12]. One of the reasons is the high paste content, of which cement is the major component [12]. Therefore, replacing clinker, which is the primary constituent of cement, with supplementary cementitious materials (SCMs) is considered to have huge potential [15,16]. The problem is a limited supply of conventional SCMs, in particular, fly ash (FA) and ground granulated blast furnace slag (GGBS) [16]. Thus, alternative sources of SCMs need to be investigated. One such source is oil shale ash (OSA), a residue of power plants using oil shale as an energy source.

The use of oil shale in power plants is not as widespread as the use of coal; however, in a few countries, such as Estonia, it is an important natural resource primarily consumed by power plants [17]. The annual amount of OSA formed in the combustion process in Estonia alone is estimated to be about 5–7 Mt, and the majority of it is deposited in stockpiles in the vicinity of power plants [18]. While OSA is a considerably different material than FA, its potential for use in construction materials [19] and, in particular, in concrete [20–22] has already been demonstrated, and Estonian standard EVS 927 [23] for specification, performance, and conformity of OSA for building materials was published in 2018.

The characterization of six OSA streams established [19] that the principal constituents of OSA are SiO_2, CaO, and Al_2O_3, forming the following major minerals: calcite, K-feldspar, quartz, and lime. Also present are two cementitious minerals, namely, dicalcium silicate (C_2S) and tetracalcium aluminoferrite (C_4AF). Furthermore, the specific surface area and the mean particle size were found to be ~1–6 $m^2 \, g^{-1}$ and ~20–45 μm, respectively.

The fineness of OSA compares favorably to cement whose specific surface area is ~0.01–1 $m^2 \, g^{-1}$ and whose particle sizes range between ~5 and 80 μm [24]. In concrete, fine powders together with water and admixtures form the paste and, as mentioned above, high paste content is required for printable concrete. Namely, the paste provides the necessary cohesiveness to the mix, with the size and morphology of the powder particles playing a crucial role [25]. In principle, the smaller the particles, the more pronounced the cohesive forces and, therefore, reducing their size by milling or grinding can prove beneficial.

It has been reported [26] that collision milling is an effective method to reduce the particle size of brittle materials. It has been successfully employed in the pretreatment of quartz and dolomite sand used as aggregate in cement-based mortar. Despite the sand being considered an inert component, the compressive strength of the mortar increased if the milled sand was used immediately after grinding [27]. However, if the sand was exposed to environmental conditions for 28 days before being mixed into mortar, the strength decreased compared to mortar with untreated sand [27]. This indicates that collision milling might activate the material.

This paper thus aims to assess the performance of OSA in concrete mixtures for extrusion 3D printing and collision milling as a pretreatment for OSA. It is hypothesized that reducing the particles' size can improve the printability of concrete mixtures and possibly activate the ash, thus improving the mechanical characteristics of concrete. The focus of this study is the characterization of different types of OSA, the identification of the most suitable ash for concrete 3D printing applications, and the proof of concept on a laboratory-scale gantry printer. The formulation, optimization, and detailed assessment of printable concrete compositions, as well as their viability, are out of the scope of this work.

Additionally, it should be stressed that this paper showcases how digital fabrication can address the industry challenges related to workforce shortages and attract women and young people to construction jobs. Namely, seven out of the ten authors of this paper are women, and two of them are early-stage researchers, while another five authors were awarded a PhD within the last ten years.

2. Materials and Methods

In pursuance of the aims identified above, the following investigation methodology was implemented:

1. Collection of OSA from three power plants;
2. Pretreatment of OSA with collision milling in a laboratory-scale disintegrator;
3. Physical, chemical, and mineralogical characterization of OSA before and after pretreatment;
4. Design of printable concrete compositions containing OSA;
5. Proof of concept—testing the strength and printability of a small number of concrete compositions.

The ash samples were collected at selected power plants and were bagged, sealed, and brought to *Riga Technical University* (RTU) laboratories for further processing. They were split into two portions—one was retained in its original condition without pretreatment while the other was collision milled. The original and processed ash samples were sent to the *Slovenian National Building and Civil Engineering Institute* (ZAG) laboratories for characterization, which comprises physical, chemical, and mineralogical analysis. Furthermore, the ash samples were used in concrete mixtures at RTU to assess mechanical properties and printability.

2.1. Oil Shale Ash

The OSA was collected at three power plants, namely, *Eesti Elektrijaam*, *Enefit280 Elektrijaam*, and *Auvere Elektrijaam*. The first two use only oil shale as fuel while the latter combines oil shale with wood. At all plants, the fly ash was collected, while at the *Auvere* plant, the bottom ash (ba) was also collected. The fly ash at the *Eesti* plant is extracted by a novel integrated desulfurizer (nid), while at *Enefit280* and *Auvere*, the extraction is carried out by electrostatic filters (ef). The ash samples were collected from the silos located at the power plant sites where the ash is temporarily stored. The information about ash sources and assigned designations are summarized in Table 1. Photographs of unprocessed ash are collated in Figure 1.

Table 1. Oil shale ash collected for this study.

Ash Designation	Power Plant	Fuel	Ash Extraction Point	Ash Type
OSA-Ees(nid)	Eesti	Oil shale	Novel integrated desulfurizer	Fly ash
OSA-Ene(ef)	Enefit280	Oil shale	Electrostatic filter	Fly ash
OSWA-Auv(ef)	Auvere	Oil shale and wood	Electrostatic filter	Fly ash
OSWA-Auv(ba)	Auvere	Oil shale and wood	Grate	Bottom ash

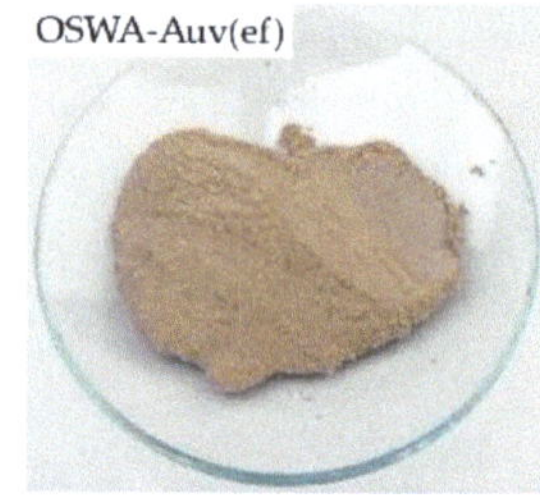

Figure 1. Macrophotographs of collected ash. Notation: Ash: OSA—oil shale ash, OSWA—oil shale + wood ash; power plants: Ees—*Eesti*, Ene—*Enefit280,* and Auv—*Auvere*; ash extraction method: nid—novel integrated desulfurizer, ef—electrostatic filters, ba—bottom ash.

2.2. Collision Milling Pretreatment

The pretreatment of ash by collision milling was executed on a laboratory-scale disintegrator *Desi-11* shown in Figure 2. The central unit of the disintegrator is the milling chamber, which is hermetically sealed during the operation. It houses two vertical rotors driven in opposite directions by asynchronous electric motors, which generate a nominal rotation speed of 3000 rpm. The rotation speed is controlled through the frequency converters and can be adjusted in the range of ±100 % of the nominal speed. The rotors are fitted with grinding discs whose concentrically shaped segments are spaced in such a way that segments of one disk fit between the segments of the other disk.

Before milling, the ash was dried at 80 °C for 24 h and cooled to room temperature. It was fed into the milling chamber via the loading hopper at a flow rate of ~170–200 g min^{-1}. Bottom ash OSWA-Auv(ba), which is the coarsest, as seen in Figure 1, was milled first, and a range of frequencies was tested, namely, 30, 50, 100, or 130 Hz. These frequencies correspond to the tangential speed of ~14, 24, 47, and 61 m s^{-1}, respectively, considering the disk diameter is 150 mm. Based on preliminary research on OSA extracted from the flue gases and the results obtained on OSWA-Auv(ba), the milling frequency of 100 Hz was selected for OSA-Ees(nid), OSA-Ene(ef), and OSWA-Auv(ef). The milled material is discharged into a sealed collection vessel. The time required to mill a batch of ~700 g of ash was 5–10 min. Table 2 summarizes the ash treatment and designations.

Table 2. Summary of ash pretreatment by collision milling and milled sample designations.

Ash Designation	Milling Frequency (Hz)	Milled Ash Designation
OSA-Ees(nid)	100	OSA-Ees(nid)/100
OSA-Ene(ef)	100	OSA-Ene(ef)/100
OSWA-Auv(ef)	100	OSWA-Auv(ef)/100
OSWA-Auv(ba)	30	OSWA-Auv(ba)/30
	50	OSWA-Auv(ba)/50
	100	OSWA-Auv(ba)/100
	130	OSWA-Auv(ba)/130

Figure 2. Laboratory-scale disintegrator *Desi-11* for collision milling. Labeled parts: 1—disintegrator with an open milling chamber showing rotors and grinding disks; 2—vibro feeder with a hopper; 3—air filter; 4—receiving container; and 5—control board.

2.3. Characterization of OSA

2.3.1. Physical Characteristics

The following physical characteristics of the ash specimens were determined:

- Moisture content;
- Particle density;
- Particle size distribution (PSD);
- Particle morphology.

The first two characteristics were measured only on the unprocessed material.

Moisture content was determined on the as-received materials according to standard EN 1097-5 [28]. A test portion of ~20–50 g was scooped from the bag, and its mass was immediately determined. Next, it was dried at 110 °C to a constant mass and cooled in a desiccator to room temperature. Finally, the mass of the dry test portion was determined, and the moisture content (wt%) was calculated against the dry mass.

Particle density was measured by the pycnometer method following standard EN 1097-7 [29]. A test portion of ~20 g was scooped from the bagged material and dried before being loaded into the 1.8 cm³ cell of the helium pycnometer *Quantachrome Ultrapyc 1200e*. The instrument flow purge was set to 1.0 min, and at least three measurements were automatically performed. If there was a larger discrepancy between the results, the instrument automatically took further measurements until the deviation between the results was sufficiently small. Hence, up to five measurements were recorded on some test portions.

The PSD was determined by laser diffraction in compliance with standard ISO 13320 [30] using the *Microtrac SYNC 5001* instrument. One test portion of ~1 g was scooped from the bag, loaded into the test cell, and analyzed using a wet configuration in isopropanol. For comparison purposes, the PSD of cement CEM I was also measured.

The morphology of the ash particles was analyzed by scanning electron microscopy (SEM). Specimens were prepared by scooping material from the bag and tapping it onto ~1 cm^2 of the double-sided conductive carbon tape attached to the standard SEM stand. The specimens were examined with the *JEOL IT500 LV* microscope equipped with a Wolfram filament. The microscope was operated at an accelerating voltage of 15 kV in a low vacuum mode at a working distance of ~10 mm.

2.3.2. Chemical Characteristics

In terms of chemical characteristics of OSA, the following were analyzed:

- Elemental composition;
- Free calcium oxide (CaO);
- Soundness;
- Chemical reactivity.

The first three characteristics were determined only on unprocessed ash.

In the scope of elemental composition, the loss on ignition (LOI) was performed first, according to the method specified by standard EN 196-2 [31]. A change in mass of ~1 g test portion was determined after successive 15 min ignitions at 950 °C. Afterwards, the quantitative elemental analysis was executed with X-ray fluorescence spectroscopy (XRF). The test portion was obtained by scooping the ash from the bag. It was first dried at 110 °C and then ignited at 950 °C. If necessary, the test portion was ground and sieved through a 0.090 mm sieve. The powder thus obtained was mixed with lithium–tetraborate, serving as a flux, in a 1:10 ratio. The mixture was fused at 1100 °C to create beads. These were analyzed with the *Bruker Tiger S8—4 kW* wavelength dispersive X-ray fluorescence spectrometer and the *Geomaj-Quant* program.

The quantity of free CaO was determined with the method specified in standard EN 451-1 [32]. A portion of ~20 g of material was sieved on a 63 μm sieve, and the residue was ground by mortar and pestle until it passed through the sieve. A homogenized portion of ~1.0–1.5 g was placed into a 250 mL flask and mixed with 12 mL of butanoic acid and 80 mL of butan-2-ol. The flask was fitted with the spiral reflux condenser and absorption tube and boiled for 3 h. The warm dispersion was filtered, and the residue was washed with 50 mL propan-2-ol. A few drops of bromophenol blue indicator were added to the filtrate and titrated with hydrochloric acid until the color changed to yellow. The free CaO content was calculated from the volume of titrant and expressed as wt% of the dry portion of tested ash.

In terms of the cement and concrete industry, soundness refers to the reactivity of free lime, magnesia, and excess sulfates in cementitious materials. These reactions are expansive and may cause cracking in young concrete. The soundness is normally determined with the Le Chatelier method, detailed in EN 196-3 [33] standard. The test specimen was prepared according to EN 450-1 [34] by first blending 70 wt% of cement with 30 wt% of OSA and then adding a sufficient amount of water to form a paste of standard consistency. The paste was placed in the Le Chatelier mold and stored at ≥90 % relative humidity at 20 ± 1 °C for 24 h when the initial distance between the needles was measured. The mold was then placed in a water bath and boiled for 3 h. The molds were then removed from the bath and left to cool to room temperature whereupon the final distance between the needles was determined and the difference between the final and initial distance was calculated.

The chemical reactivity was assessed with isothermal calorimetry following the procedure suitable for the SCMs. The procedure specified in the ASTM C1897-20 [35] standard detects hydraulic and pozzolanic reactions, which are exothermic by nature. Thus, the amount of released heat is the measure of reactivity. One specimen of 10.00 g was used for the analysis. It was collected from the bag by scooping. The specimen was combined with 30.00 g of calcium hydroxide, 5.00 g of calcium carbonate, and 54.00 g of potassium solution and mixed for 3 min to obtain a smooth paste. A test portion of 15 ± 0.01 g of the paste was placed into a glass ampoule. The ampule was inserted into the *TA Instruments calorimeter TAM Air 8*, where distilled water was used as the reference material. The measurements were conducted over seven days at a temperature of 40 °C. The cumulative heat of hydration H_{SCM} (J kg^{-1}) stated as heat released per unit mass of SCM in the test portion was calculated.

2.3.3. Mineralogical Characteristics

The qualitative phase analysis, which gives an overall insight into the mineralogical composition, was performed by X-ray Diffraction (XRD). The test portions were obtained by quartering the received samples to a suitable amount, which was dried in an oven at 60 °C to a constant mass. A portion of 100 g of material was ground with a disc mill to a particle size < 500 μm. The quartering of the material was repeated, and a suitable quantity was ground in an agate mortar to a particle size of <63 μm. The test portion was placed into a 27 mm diameter sample holder for analysis.

The analysis was performed on the *Panalytical Malvern Empyrean* diffractometer with Cu–Kα radiation. The measurements were carried out at laboratory temperature. The tube voltage was set to 45 kV and current to 40 mA. Data were collected over the 2θ range from 5 to 70° in increments of 0.013°. The increment measurement time was 150 s. The results were analyzed with *Panalytical Highscore 4.8* diffraction software using the *ICDD Powder Diffraction File PDF-4+* database as a source of references for the crystalline phases.

2.4. Characterization of Printable Concrete

The performance of OSA in concrete formulations for 3D printing was assessed in three stages. In the first stage, the formulations containing OSA were adjusted to the same workability as the reference formulation. Secondly, the mechanical tests were conducted on all formulations, and in the third stage, the reference formulation and two OSA formulations with the best mechanical performance were selected for printability tests.

In terms of dry components, the formulations consisted of 33 wt% of powder and 67 wt% of sand. In the reference formulation, the powder contained only cement, while in OSA formulations, it was split into 10 wt% of ash and 23 wt% of cement [34,36]. A superplasticizer in the amount of 1 wt% of mass of water was also added. The following types of materials were used: cement CEM I 42.5N from *Schwenk Ltd.* (Brocēni, Latvia), sand 0/2 mm from *Sakret Ltd.* (Rumbula, Latvia), and superplasticizer *Floormix* from *Vincents Polyline Ltd.* (Kalngale, Latvia).

The workability of OSA formulations was adjusted according to EN 450-1 [34] to maintain the same consistency for all formulations. The consistency was confirmed with the flow table method specified in EN 1015-3 [37]. A sufficient amount of water containing 1 wt% of superplasticizer was added so that OSA formulations were within the ±10 mm range of the reference, whose spread after 25 jolts was 180 mm. The reference formulation was known to be printable from previous printing tests. Proportions of the components in tested concrete formulations are given in Table 3.

Table 3. Composition of concrete formulations expressed as mass per 1000 g of dry components. Superplasticizer was added in the amount of 1 wt% of mass of water, and water was adjusted to retain the consistency in the range of ±10 mm of the reference formulation measured on the flow table. Notation: ash: OSA—oil shale ash, OSWA—oil shale + wood ash; power plants: Ees—*Eesti*, Ene—*Enefit280*, and Auv—*Auvere*; ash extraction method: nid—novel integrated desulfurizer, ef—electrostatic filters, ba—bottom ash; collision milling frequency on a *Desi-11* disintegrator: 50, 100, 130 Hz.

Ash and Concrete Formulation Designation	Mass of Components per 1000 g of Dry Materials (g)				
	Ash	**Cement**	**Sand**	**Water**	**Superplasticizer**
Ref	---	333.3	666.7	141.7	1.42
OSA-Ees(nid)	100.0	233.3	666.7	146.7	1.47
OSA-Ees(nid)/100	100.0	233.3	666.7	143.3	1.43
OSA-Ene(ef)	100.0	233.3	666.7	166.7	1.67
OSA-Ene(ef)/100	100.0	233.3	666.7	160.0	1.60
OSWA-Auv(ef)	100.0	233.3	666.7	186.7	1.87
OSWA-Auv(ef)/100	100.0	233.3	666.7	173.3	1.73
OSWA-Auv(ba)/<1 mm *	100.0	233.3	666.7	160.6	1.61
OSWA-Auv(ba)/50	100.0	233.3	666.7	156.1	1.56
OSWA-Auv(ba)/100	100.0	233.3	666.7	147.7	1.48
OSWA-Auv(ba)/130	100.0	233.3	666.7	148.1	1.48

* The bottom ash in its original form was too coarse to be used as cement replacement; thus, only the portion passing through the 1 mm sieve was used.

The choice of fixing the spread by adjusting the water and superplasticizer content was made because printability is the principal parameter in this study. As a result, the water–powder ratio is not fixed, which may have a bearing on the mechanical properties. However, printable formulations need to be cohesive, and to achieve this, a high powder content with a sufficient amount of water to wet the particles is required. The water requirement depends on particle size, morphology, texture, and mineralogical composition.

2.4.1. Mechanical Properties

The mechanical characteristics of printable concrete mixtures containing unprocessed and milled ash were assessed with compressive and flexural strength tests 7, 28, and 56 days after casting the specimens. The specimens for compression tests were cubes with nominal dimensions of 20 mm, while specimens for the flexural test were prisms with nominal dimensions $b \times h \times L$ of $40 \times 40 \times 160$ mm^3. The preparation and curing of the specimens were carried out according to EN 1015-11 [38]. The specimens were demolded ~24 h after casting and kept in a storage chamber at a relative humidity of 95 ± 5 % and a temperature of 20 ± 2 °C until the time of the test. The tests were carried out on the *Controls 50-C56Z00* machine in a force-controlled mode at a stress rate of 0.8 MPa s^{-1} for compression tests and in a displacement-controlled mode at 0.5 mm s^{-1} for the flexural tests.

2.4.2. Printability Assessment

Two formulations exhibiting the best mechanical performance, namely, OSA-Ees(nid) and OSA-Ees(nid)/100, were selected for printability assessment, which was carried out with the slug test and direct buildability test. Both tests were performed on the custom-made gantry-type laboratory concrete printer located at RTU and shown in Figure 3. Its print area dimensions are 1500×1000 m^2, while the clearance is 1 m. The printhead comprises a hopper of 15 L from where the material is pushed through the nozzle by means of a helix screw conveyor. The nozzle of a circular cross-section with a diameter of 25 mm was used, and the material flow was regulated to form a 40–45 mm wide filament at a pre-selected layer height of 10 mm.

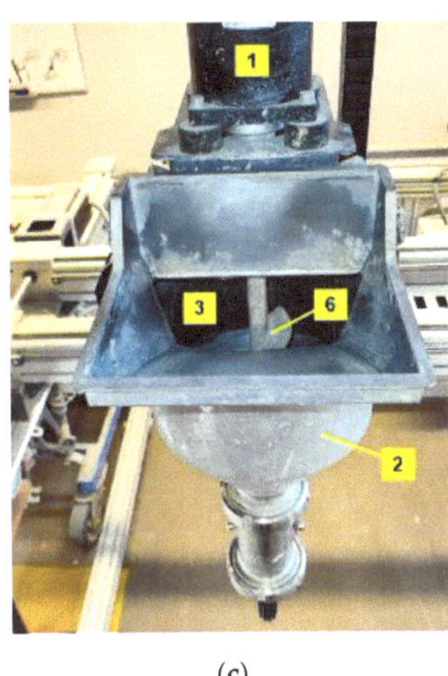

(a) (b) (c)

Figure 3. Custom-made laboratory concrete printer at *Riga Technical University* (RTU): (**a**) the printer set up with the aluminum frame and print area; (**b**) the printhead closeup; and (**c**) the hopper. Labeled parts: 1—motor; 2—hopper, 3—inlet; 4—pipe; 5—nozzle; and 6— helix screw conveyor.

The slug test was performed according to the method proposed by Ducoulombier et al. [39] where the material is extruded under a constant force through the printer's nozzle. The nozzle is positioned vertically ~50 cm above the print area, and the extruded filament breaks as a result of uniaxial yielding, thus forming a droplet or a slug. Once a uniform flow is achieved, 25 slugs are collected in a container, and their cumulative mass is measured. The yield stress τ_y (Pa) is determined as per Equation (1)

$$\tau_y = m_s \frac{g}{S\sqrt{3}} \, , \tag{1}$$

where m_s (kg) is the average mass of a slug, g (m s^{-2}) is gravity, and S (m^2) is the nozzle cross-section.

The direct buildability test was performed immediately after the slugs test by printing a cylinder with a diameter of 250 mm until plastic collapse occurred [40]. The lap time needed for a single layer to print was ~4 s, resulting in a total printing time of ~1 min. Therefore, the mixture was extruded ~16–17 min after water was added to dry components. The compressive stress at plastic collapse σ_p (Pa) and yield stress τ_y (Pa) are calculated according to Equations (2) and (3) [41], respectively

$$\sigma_p = \rho \, n \, h \, g \, , \tag{2}$$

$$\tau_p = \frac{\rho \, n \, h \, g}{\sqrt{3}} \tag{3}$$

where ρ (kg m^{-3}) is density, n (1) is the number of layers at the time of collapse, and h (m) is the height of the layer. Additionally, the surface quality was visually checked, and the printed cylinder was measured to assess the dimensional consistency of the filament width.

Before commencing the printability tests and 15 min after adding water to the dry components, the density of the fresh mixture was determined according to EN 12350-6 [42]. It was measured by filling a 1 L container in two layers, compacting each layer 25 times using a compacting rod, leveling the top surface with a trowel, and then weighing the container.

3. Results

This chapter presents and discusses the results in relation to the test method. The results are coupled in Chapter 4 where the discussion pertaining to the individual type of ash is given.

3.1. Properties of OSA

Moisture content was determined to find if a noteworthy amount of moisture is absorbed during pretreatment, as it may cause hydration reactions that impact the mineralogical composition. It is observed in Table 4 that moisture content is negligible. The particle density values, also collated in Table 4, are somewhat higher than values normally observed on fly ash from coal-burning power plants, which are between 2.2 and 2.5 kg L^{-1}, and somewhat lower than values found in cement, those normally standing at ~3 kg L^{-1}. The particle density values are needed in the concrete mix design process.

Table 4. Moisture content and particle density measured on ash milled at 100 Hz. Notation: ash: OSA—oil shale ash, OSWA—oil shale + wood ash; power plants: Ees—*Eesti*, Ene—*Enefit280*, and Auv—*Auvere*; ash extraction method: nid—novel integrated desulfurizer, ef—electrostatic filters, ba—bottom ash; collision milling frequency on a *Desi-11* disintegrator: 100 Hz.

	Moisture Content (wt%)	Particle Density (kg L^{-1})
OSA-Ees(nid)/100	0.7	2.6130 ± 0.002
OSA-Ene(ef)/100	0.9	2.7504 ± 0.004
OSWA-Auv(ef)/100	0.4	2.6825 ± 0.003
OSWA-Auv(ba)/100	0	2.6862 ± 0.004

The PSDs of OSA before and after pretreatment with collision milling are collated in Figure 4 as differential passing curves and in Figure 5 as 50 and 95 wt% demarcation diameters. The latter is the diameter at which either 50 or 95 wt% of the particles are smaller than the stated value. The results for cement CEM I are added for comparison. Out of the three types of fly ash, namely, OSA-Ees(nid), OSA-Ene(ef), and OSWA-Auv(ef), the first one is the finest with the peak in differential particle size curve at ~6 μm, while the latter two peak between ~30 and 50 μm. When comparing these three types of ash to their milled counterparts, namely, OSA-Ees(nid)/100, OSA-Ene(ef)/100, and OSWA-Auv(ef)/100, the differential curves indicate that collision milling at 100 Hz predominantly reduces particles larger than 10 μm so that in case of the latter two, the peak is shifted to between ~10 and 20 μm. In the case of OSA-Ees(nid)/100, the peak remains at ~6 μm; however, the percentage passing increases from ~6 to ~8 wt%. This is also reflected in the demarcation diameter of OSA-Ees(nid)/100. Namely, the 50 wt% demarcation diameter remains approximately the same at 8–9 μm, whereas the 95 wt% one is reduced from 100 to 40 μm. Comparison to CEM I shows that OSA-Ees(nid) is finer than cement already in its original condition, while in the case of OSA-Ene(ef) and OSWA-Auv(ef), collision milling at 100 Hz reduces the particles closer to the CEM I distribution.

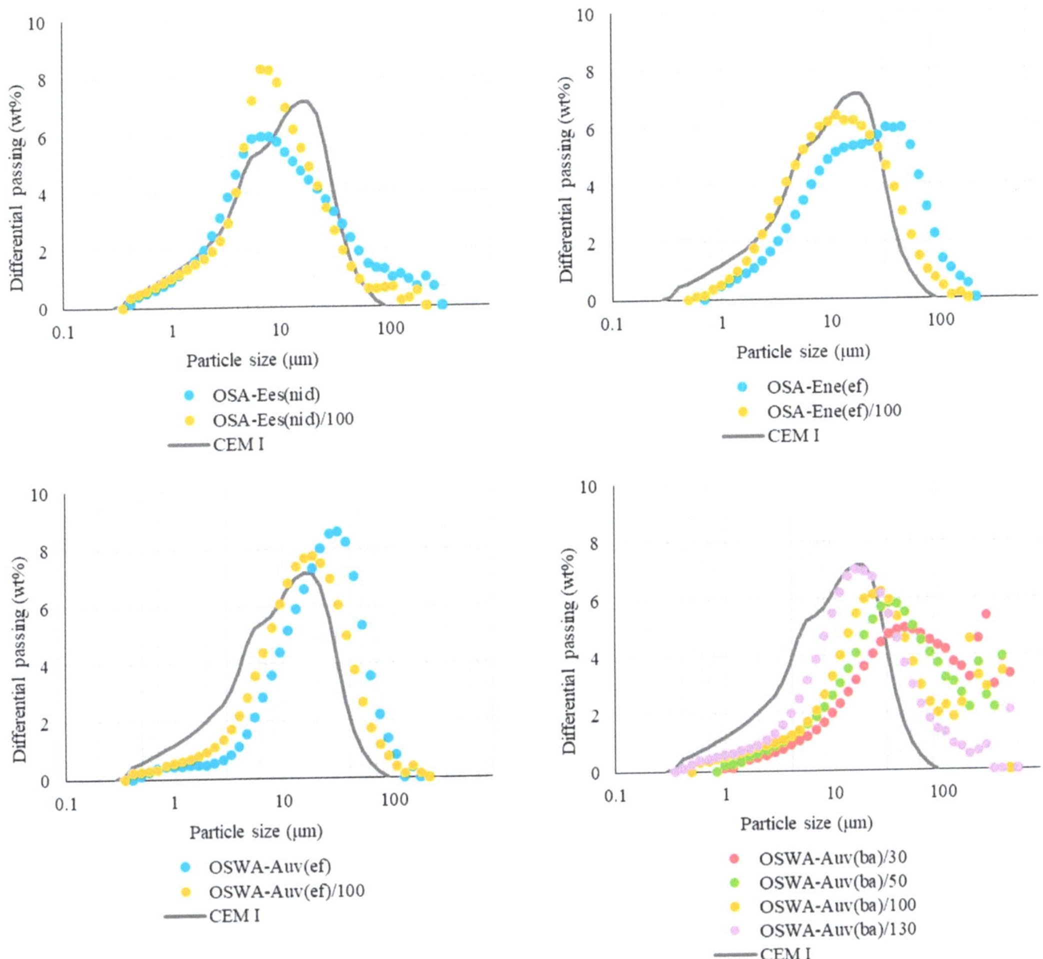

Figure 4. Particle size distribution (PSD) of ash before and after collision milling measured with laser diffraction. For comparison purposes, the PSD of CEM I 42.5N from *Schwenk Ltd.* (Brocēni, Latvia) is also presented. Notation: ash: OSA—oil shale ash, OSWA—oil shale + wood ash; power plants: Ees—*Eesti*, Ene—*Enefit280*, and Auv—*Auvere*; ash extraction method: nid—novel integrated desulfurizer, ef—electrostatic filters, ba—bottom ash; collision milling frequency on a *Desi-11* disintegrator: 30, 50, 100, 130 Hz.

The OSWA-Auv(ba) ash, being collected at the bottom of the furnace, is too coarse for the laser diffraction analysis prior to milling. The curves, therefore, show the ash milled at 30, 50, 100, and 130 Hz. The peak of OSWA-Auv(ba)/30 was at ~50 μm, and with increased frequency, it moved to ~10 μm for OSWA-Auv(ba)/130. At the same time, the curve gradually narrowed so that the peak increased in height from ~5 to ~7 wt%. Similarly, the 50 wt% demarcation diameter decreased from ~50 to ~15 μm and the 95 wt% diameter from ~320 to ~130 μm. Therefore, as a result of collision milling, the PSD curve of OSWA-Auv(ba) gradually shifts closer to the PSD curve of CEM I as the milling frequency increases. Overall, it can, therefore, be concluded that collision milling is efficient for reducing the size of particles larger than 10 μm.

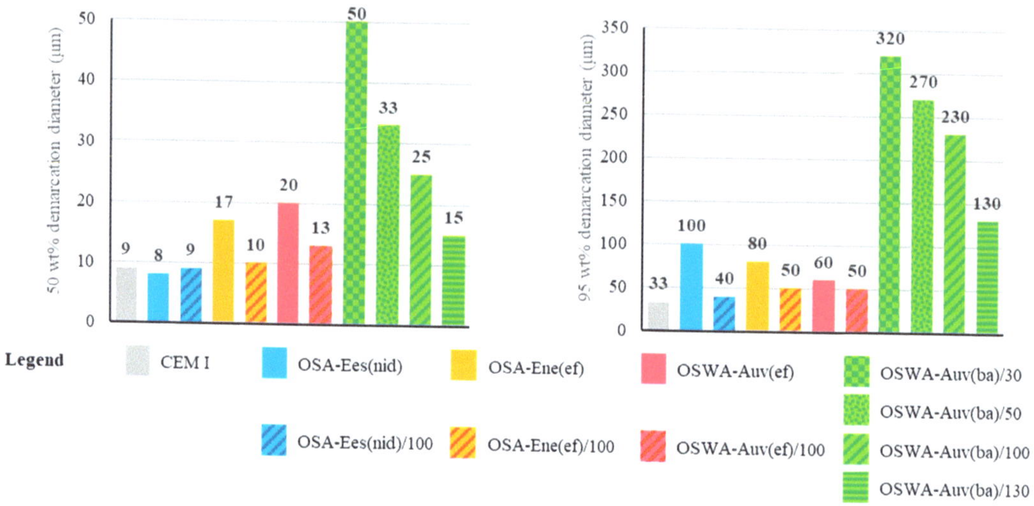

Figure 5. Demarcation diameter at which either 50 or 95 wt% of the particles are smaller than the stated value. For comparison purposes, the results for CEM I 42.5N from *Schwenk Ltd.* (Brocēni, Latvia) are also presented. Notation: ash: OSA—oil shale ash, OSWA—oil shale + wood ash; power plants: Ees—*Eesti*, Ene—*Enefit280*, and Auv—*Auvere*; ash extraction method: nid—novel integrated desulfurizer, ef—electrostatic filters, ba—bottom ash; collision milling frequency on a *Desi-11* disintegrator: 30, 50, 100, 130 Hz.

The results of SEM are summarized in Figures 6 and 7, where a heterogenic distribution of particles is observed in all types of ash. The particles that are similar to cenospheres found in fly ash from coal power plants, having a smooth surface and dimensions up to 30 µm, are abundant in OSA-Ees(nid), while in OSA-Ene(ef), only a few are found. The morphology of the two types of ash collected at the electrostatic filters, namely, OSA-Ene(ef) and OSWA-Auv(ef), is predominantly angular with a rather smooth texture.

The effect of pretreatment with collision milling on particle morphology and texture can be analyzed in Figure 6, where images of ash before and after milling at 100 Hz are lined up. Overall, it is observed that milling does not significantly affect the shape of the particles. Namely, the particles of OSA-Ene(ef) and OSWA-Auv(ef) are angular even before the milling, while in OSA-Ees(nid) it is mostly larger, angular particles that are crushed while milling does not reduce the size of the smaller spherical particles.

The impact of milling frequency can be studied in Figure 7, which lines up images of OSWA-Auv(ba) milled at 30, 50, 100, and 130 Hz. The particles of this type of ash are also angular, and no impact on morphology and texture is observed. However, it is clear that as frequency increases, the number of large particles decreases.

The elemental composition shown in Table 5 indicates that ash filtered from the flue gasses contains ~20–30 wt% of SiO_2 and ~30–40 wt% CaO, while the content of these two oxides in bottom ash is ~5 and ~50 wt%, respectively. Compared to the results of free CaO content collated in Table 6, it can be concluded that in these three types of ash, the majority of CaO is consumed for the formation of minerals, such as dicalcium silicate. Nevertheless, the requirement for the presence of free CaO in fly ash from coal-burning power plants set by EN 450-1 [34] is ≤1.5 wt%. Ash that exceeds this limit must be further tested for soundness, and the expansion measured between the tips of the Le Chatelier ring should be ≤10 mm. The soundness results in Table 6 show that all types of ash meet this requirement. The same standard also sets the limit for the total equivalent alkali content at ≤ 5 wt%. The latter, calculated as (wt% Na_2O + 0.658 × wt% K_2O), is between ~1 and 3 wt%.

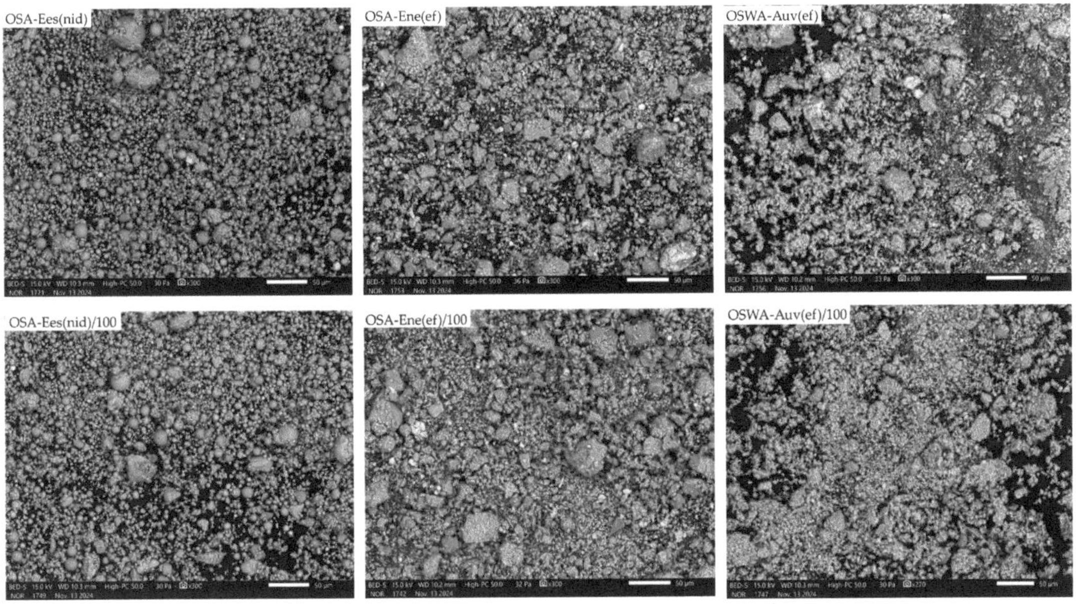

Figure 6. Scanning electron microscope (SEM) images of ash particles before (**top** row) and after collision milling (**bottom** row). Notation: ash: OSA—oil shale ash, OSWA—oil shale + wood ash; power plants: Ees—Eesti, Ene—Enefit280, and Auv—Auvere; ash extraction method: nid—novel integrated desulfurizer, ef—electrostatic filters; collision milling frequency on a Desi-11 disintegrator: 100 Hz.

The LOI, in principle, originates either from the dehydration and decomposition of minerals or from unburnt organic matter, such as hydrocarbons. The LOI limit set by EN 450-1 [34] is ≤ 9 wt%, and in comparison, the values obtained on OSA-Ene(ef), OSWA-Auv(ef), and OSWA-Auv(ba) of ~20–30 wt% are high. This indicates a possibility that they still contain a significant amount of kerogen, a complex mixture of hydrocarbon compounds that is a primary organic component of oil shale. However, decomposition is a complex process that depends on the mineralogy of oil shale and the thermal maturity of kerogen [43], and in order to identify the decomposing constituents, further analysis with thermogravimetry (TG) coupled with Fourier-transform infrared spectroscopy (FTIR) is needed.

The results of isothermal calorimetry are summarized in Figure 8. The chemical reactivity is assessed as the cumulative heat released due to the exothermic reactions. The three types of flue gas ash, namely, OSA-Ees(nid), OSA-Ene(ef), and OSWA-Auv(ef), exhibit low chemical reactivity before milling as the cumulative heat is between ~10 and 20 J g^{-1}. Collision milling at 100 Hz results in a more than seven-fold increase in reactivity, with a cumulative heat release of ~100–150 J g^{-1} found in OSA-Ees(nid)/100, OSA-Ene(ef)/100, and OSWA-Auv(ef)/100. For comparison, fly ash, certified for use in concrete, released ~70 J g^{-1} after 200 h. There are two possible causes for increased reactivity due to milling, firstly, the smaller the particles, the larger the surface area in contact with water, and secondly, reactive phases in the kernel may be coated by a shell of inert phases—as milling breaks the inert shell, the reactive kernel becomes exposed.

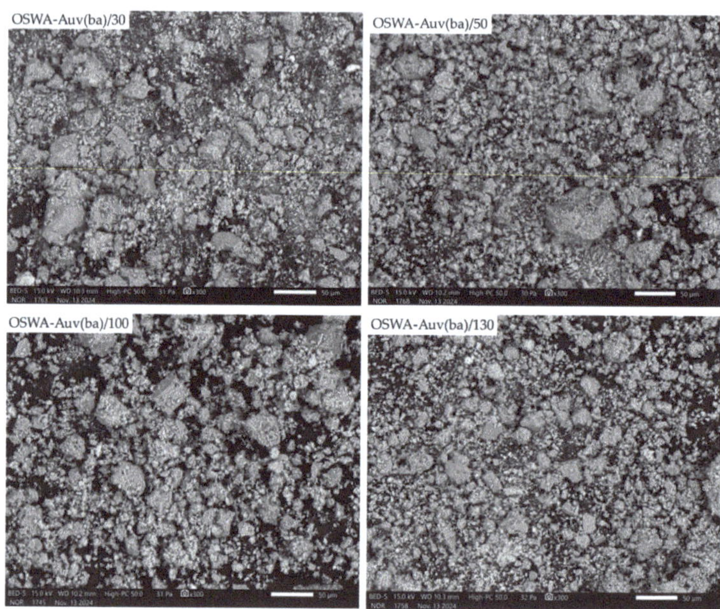

Figure 7. Scanning electron microscope (SEM) images of ash particles. Notation: ash OSWA—oil shale + wood ash; power plant: Auv—*Auvere*; ash extraction method: ba—bottom ash; collision milling frequency on a *Desi-11* disintegrator: 30, 50, 100, 130 Hz.

Table 5. Elemental composition measured with X-ray fluorescence spectroscopy (XRF) and loss on ignition (LOI). Notation: ash: OSA—oil shale ash, OSWA—oil shale + wood ash; power plants: Ees—*Eesti*, Ene—*Enefit280,* and Auv—*Auvere*; ash extraction method: nid—novel integrated desulfurizer, ef—electrostatic filters, ba—bottom ash.

Ash Designation	Composition (wt%)											
	SiO_2	Al_2O_3	Fe_2O_3	CaO	MgO	SO_3	Na_2O	K_2O	Mn_2O_3	TiO_2	P_2O_5	LOI
OSA-Ees(nid)	21.5	5.6	2.9	35.6	3.0	17.3	0.4	2.9	<0.1	0.3	0.1	9.6
OSA-Ene(ef)	28.0	6.9	4.0	29.5	3.0	4.9	0.1	3.4	<0.1	0.4	0.1	19.1
OSWA-Auv(ef)	22.2	5.5	3.7	41.8	3.4	4.9	<0.1	2.8	<0.1	0.3	0.2	14.4
OSWA-Auv(ba)	6.0	1.4	2.5	50.8	2.3	3.1	0.1	0.6	<0.1	0.1	0.1	32.9

Table 6. Free lime content measured with hydrochloric acid titration and the soundness of ash determined with the Le Chatelier method. Notation: ash: OSA—oil shale ash, OSWA—oil shale + wood ash; power plants: Ees—*Eesti*, Ene—*Enefit280,* and Auv—*Auvere*; ash extraction method: nid—novel integrated desulfurizer, ef—electrostatic filters, ba—bottom ash.

Ash Designation	Free CaO (wt%)	Soundness (mm)
OSA-Ees(nid)	4.5	3.6
OSA-Ene(ef)	<0.1	2.3
OSWA-Auv(ef)	21.8	2.0
OSWA-Auv(ba)	8.1	0.9

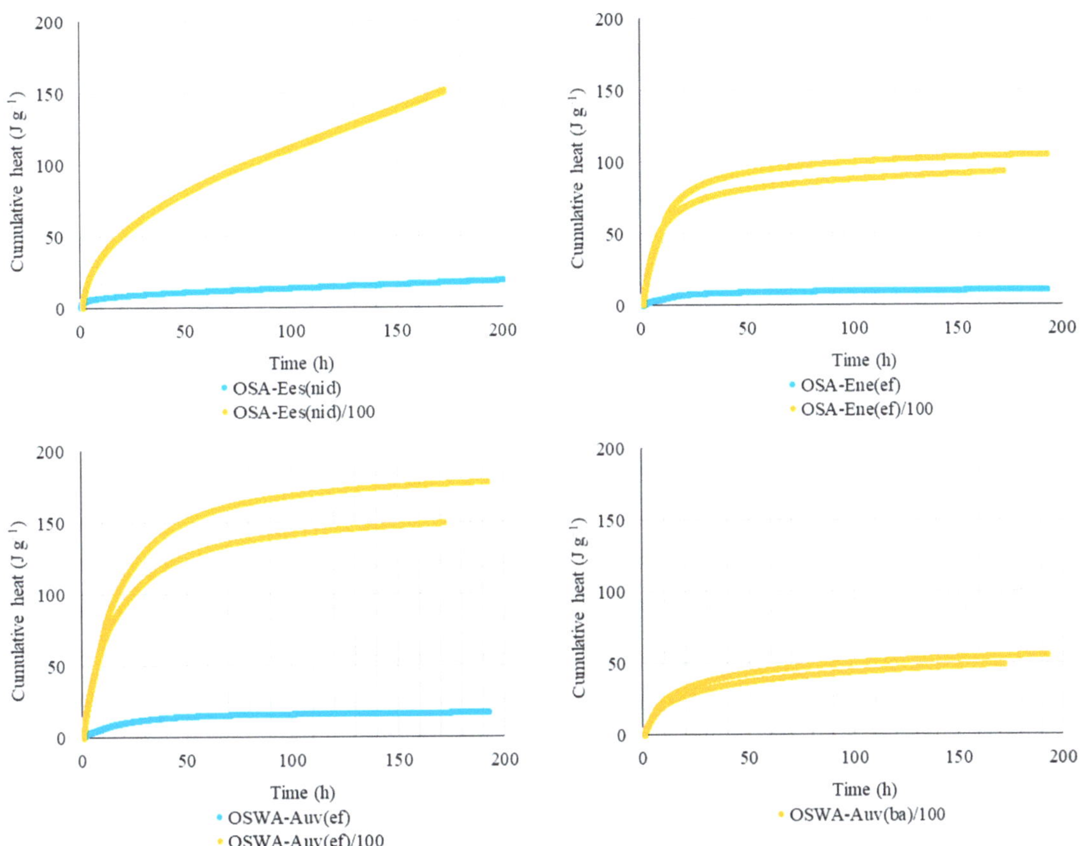

Figure 8. Chemical reactivity determined by isothermal calorimetry. Notation: ash: OSA—oil shale ash, OSWA—oil shale + wood ash; power plants: Ees—*Eesti*, Ene—*Enefit280*, and Auv—*Auvere*; ash extraction method: nid—novel integrated desulfurizer, ef—electrostatic filters, ba—bottom ash; collision milling frequency on a *Desi-11* disintegrator: 100 Hz.

The reactivity of OSWA-Auv(ba) was measured only on ash collision milled at 100 Hz since unmilled ash is too coarse for the isothermal calorimetry test. In comparison to the three types of milled fly ash, the milled bottom ash exhibits two to three times lower reactivity with a cumulative heat of ~50 J g^{-1}.

The results of the qualitative XRD phase analysis are summarized in Table 7. XRD phase analysis established that all the investigated types of ash contained calcite, quartz, and anhydrite. The reactive β-dicalcium silicate (larnite) was found in all samples, except for OSWA-Auv(ba). Its γ polymorph was identified in OSA-Ees(nid) and OSWA-Auv(ef), suggesting that during cooling, part of the β modification transformed to the γ polymorph. Calcite is most abundant in OSWA-Auv(ba), while its quantity is the lowest in OSWA-Auv(ef) and OSA-Ees(nid).

Table 7. Mineralogical composition established with X-ray diffraction (XRD) where a dot (•) marks the presence of a mineral. Notation: ash: OSA—oil shale ash, OSWA—oil shale + wood ash; power plants: Ees—*Eesti*, Ene—*Enefit280*, and Auv—*Auvere*; ash extraction method: nid—novel integrated desulfurizer, ef—electrostatic filters, ba—bottom ash.

Ash Designation	Anhydrite	β-Dicalcium Silicate	Calcite	γ-Dicalcium Silicate	Dolomite	Feldspar	Hematite	Lime	Mellilite	Muscovite	Periclase	Portlandite	Quartz	Rokühnite	Spinel	Sylvite
OSA-Ees(nid)	•	•	•	•				•			•	•	•	•	•	•
OSA-Ene(ef)	•	•	•		•	•	•		•	•			•			
OSWA-Auv(ef)	•	•	•	•		•	•	•	•		•	•	•			
OSWA-Auv(ba)	•		•		•			•	•			•	•			•

Lime content was the highest in OSWA-Auv(ef), and smaller amounts were also found in OSA-Ees(nid) and OSWA-Auv(ba). In OSA-Ene(ef), all lime and portlandite are fully hydrated and carbonated to form calcite. The highest amount of portlandite was detected in OSA-Ees(nid), whereas in the other ashes, portlandite was already transformed into calcite. Sulfate, present as anhydrite, was found in all ashes. In OSA-Ees(nid) and OSWA-Auv(ef), anhydrite was more abundant compared to OSWA-Auv(ba). Chlorine-bearing phases, such as sylvite, were identified in OSA-Ees(nid) and OSWA-Auv(ba). Additionally, a Fe-bearing chlorine phase was detected in OSA-Ees(nid).

3.2. Properties of Printable Concrete Containing OSA

The results of the compression and flexural tests conducted on the reference formulation and formulations with OSA are summarized in Figure 9. Generally, it is expected that replacing cement with OSA shall result in reduced strength, and, overall, this stipulation is confirmed. The 56-day compressive and flexural strength of the reference formulation was ~60 and ~7.5 MPa, respectively. With the exception of OSA-Ees(nid), these values were between ~30–45 and ~6.0–7.0 MPa, respectively, for ash-containing formulations. The OSA-Ees(nid) formulation, on the other hand, exhibits a similar strength range as the reference, especially when the intersample variability observed on two reference sample groups is taken into account. In terms of early strength, the results indicate that after 7 days, the reference and the OSA-Ees(nid) formulation reach ~60% of 56-day compressive strength, while for the other ash-containing formulations, this value is ~70%. Considering the intra- and intersample variability, it can be concluded that neither of the tested formulations exhibits any significant strength gain after 28 days.

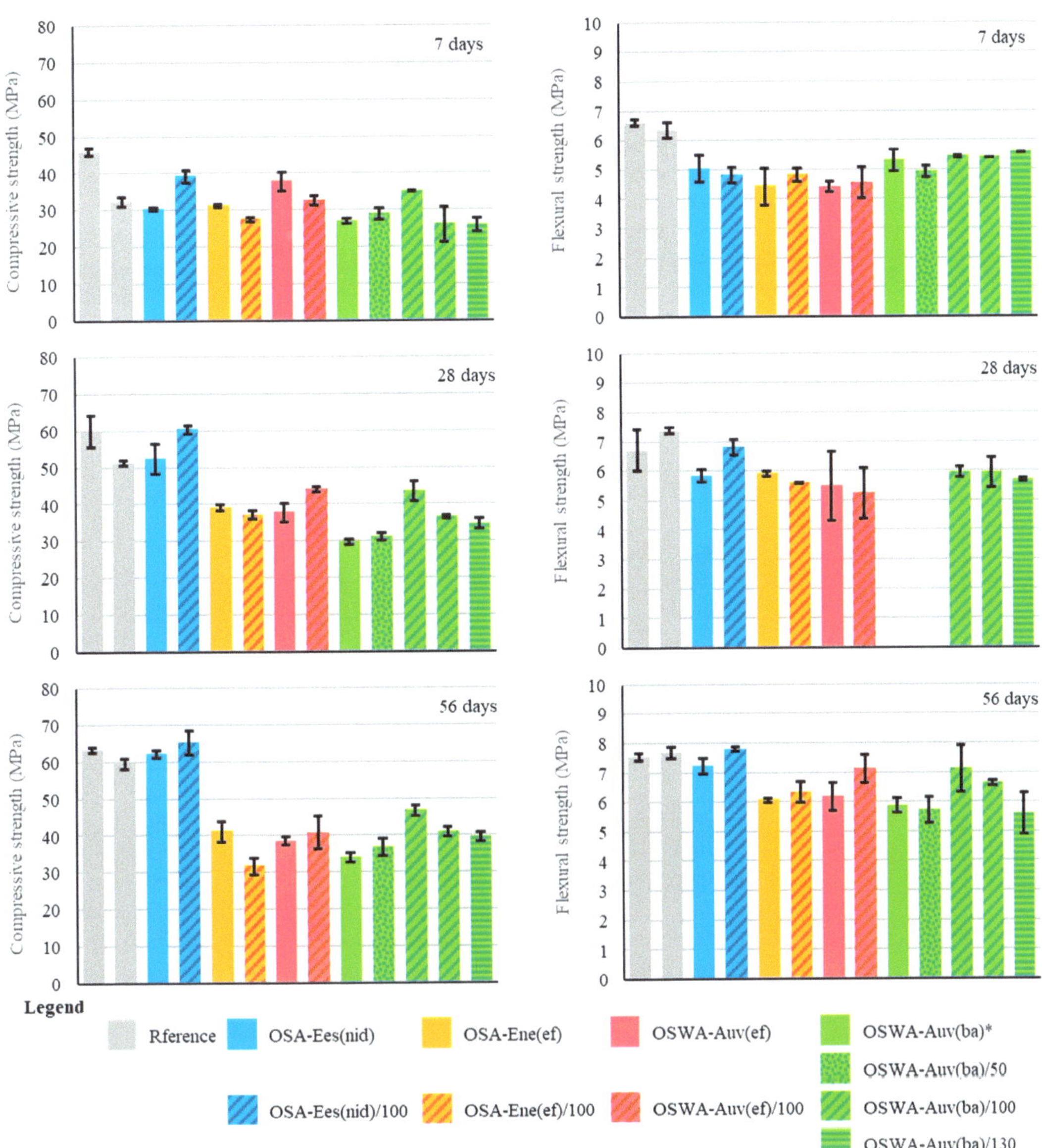

Figure 9. Compressive and flexural strength of printable concrete 7, 28, and 56 days after casting. Notation: ash: OSA—oil shale ash, OSWA—oil shale + wood ash; power plants: Ees—*Eesti*, Ene—*Enefit280*, and Auv—*Auvere*; ash extraction method: nid—novel integrated desulfurizer, ef—electrostatic filters, ba—bottom ash; collision milling frequency on a *Desi-11* disintegrator: 50, 100, 130 Hz; *—ash in original form was too coarse to be used without processing; thus, portion passing through a 1 mm sieve was used.

The compression test results obtained on the OSWA-Auv(ba) series consistently show strength gain when milling frequency increases from 50 to 100 Hz, whereas further frequency increase causes strength drop. Nevertheless, overall, it may be concluded that formulations with milled bottom ash performed better than those with unmilled fine bottom ash (sieved fraction of < 1 mm). This indicates that bottom ash, usually unsuitable for direct cement replacement due to its large particle size, may become a viable SCM after pretreatment with collision milling, which is most effective at around 100 Hz.

Overall, it may be concluded that collision milling of OSA enhances the mechanical properties of ash-containing concrete formulations, although the effectiveness varies by the type of ash. Neither the PSD nor the chemical reactivity of the ash is in a direct correlation with strength as far as the results of this study are concerned; however, the results pool is not sufficient to draw firm conclusions.

Based on the mechanical test results, two ash-containing formulations, namely, OSA-Ees(nid) and OSA-Ees(nid)/100, were selected for the printability assessment tests. The results, including the reference formulation, are collated in Table 8. These results show that the density of the reference and OSA-Ees(nid) formulation does not differ significantly (~15 g per 1 L of material), whereas the difference is more pronounced for OSA-Ees(nid)/100 (~100 g per 1 L of material). Since the milling of this ash increased the proportion of particles around the 10 μm mark, it may have improved particle packing.

Table 8. Properties measured on fresh concrete mixtures for printability assessment. Values for density, yield stress, and compressive stress at plastic collapse are rounded to the nearest 5. Notation: ash: OSA—oil shale ash; power plant: Ees—*Eesti*; ash extraction method: nid—novel integrated desulfurizer; collision milling frequency on a *Desi-11* disintegrator: 100 Hz.

Ash Designation	Density (kg m^{-3})	Slug Test	Buildability Test—Values at Plastic Collapse		
		Yield Stress τ_y (Pa)	Number of Layers	Compression Stress σ_y (Pa)	Yield Stress τ_y (Pa)
Reference	2140	1005	15	3150	1820
OSA-Ees(nid)	2125	1020	12	2500	1440
OSA-Ees(nid)/100	2230	1180	13	2840	1670

The yield stress measured with the slug test, despite being somewhat higher in the OSA-Ees(nid)/100 formulation (see Table 8), is in fact in a rather narrow range compared to the results reported by Ducoulombier et al. [39], who recorded values between ~250 and 1500 Pa. The buildability results show that the addition of ash reduces the number of layers at which the plastic collapse occurs. Comparing the yield stress results from the slug test to the results obtained in the buildability test, we find that the latter are consistently higher, which may be related to the time elapsed between the two tests. Yield stress from the buildability test is 1.8 times higher than observed in the slug test in the case of the reference formulation, while for the two ash formulations, this factor amounts to 1.4. These results may imply that the behavior of the reference formulation is more susceptible to elapsed time, and while its buildability is better than in the two ash formulations, pumpability might be impaired. This observation, however, requires further investigation.

Figure 10 shows the printed cylinders before and after the plastic collapse. Visual inspection reveals a small degree of tears in the filament of all three formulations. These were observed to be limited to the surface and not detrimental to the buildability. Adjustments to the extrusion rate and/or the speed of the printhead could fix the issue for these formulations, while fine tuning the composition is a matter of further investigation.

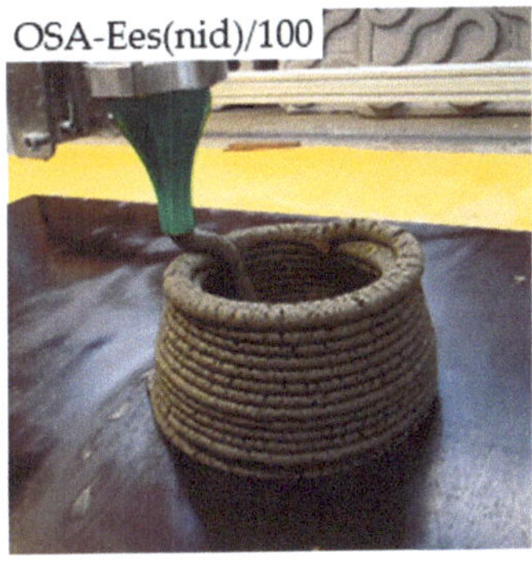

Figure 10. Direct buildability test showing printed objects before (**top** row) and after plastic collapse (**bottom** row). Notation: ash: OSA—oil shale ash; power plant: Ees—Eesti; ash extraction method: nid—novel integrated desulfurizer; collision milling frequency on a Desi-11 disintegrator: 100 Hz.

4. Discussion

OSA-Ees(nid), which is extracted from the flue gasses by the novel integrated desulfurizer, has the finest particles, with more than 50 wt% of them being smaller than 10 μm. This ash is the only one out of the four analyzed types of ash whose particles are mostly globular with a smooth surface. This ash contains ~35 wt% of CaO and ~20 wt% of SiO_2, which, as found in mineralogical analysis, form dicalcium silicate (C_2S). C_2S is present as β and γ polymorphs, with β being the reactive form and γ a rather inert form [44], which most likely precipitated from the β polymorph during cooling. The β polymorph of dicalcium silicate is also one of the principal minerals found in cement. On the other hand, ~5 wt% of CaO is found in a free form, but it does not cause an excessive expansion, which was tested with the soundness test. Furthermore, the equivalent alkali content and LOI are both within limits otherwise set for fly ash from coal power plants. Overall, these characteristics indicate that OSA-Ees(nid) could be suitable for use in concrete.

The reactivity of OSA-Ees(nid) measured with the isothermal calorimetry is, nevertheless low, with only ~20 J g^{-1} of released heat. While collision milling increased the released heat to ~150 J g^{-1}, this was not transformed into a significant improvement in mechanical properties. Namely, the 56-day compressive and flexural strength was found to be ~60 and ~7.5 MPa, respectively, for both OSA-Ees(nid) and OSA-Ees(nid)/100. These results, however, are approximately the same as measured on the reference concrete formulation. The printability of concrete formulations containing either milled or unmilled ash was also very similar. Hence, it should be concluded that while OSA-Ees(nid) is the most suitable ash out of the four tested, collision milling does not significantly improve its performance as an SCM, and such pretreatment is not likely to be viable.

The most significant impact of collision milling is observed in bottom ash OSWA-Auv(ba). In its original form, its particles are up to 2 cm big, and reducing the particle size is necessary to consider this ash as a potential SCM. Increasing the milling frequency from 30 to 130 Hz results in a decrease in average particle size, with a 50 wt% demarcation diameter being ~50 μm at 30 Hz and 15 μm at 130 Hz. The optimal frequency, however, appears to be ~100 Hz since the concrete formulation containing OSWA-Auv(ba)/100 exhibits the highest strength. Nevertheless, despite milling at 100 Hz, the reactivity of OSWA-Auv(ba) remains low with the cumulative heat release of ~50 J g^{-1}. This is probably related to the absence of any C_2S phases. On one hand, the SiO_2 content of ~5 wt% is low compared to ~50 wt% of CaO, and on the other hand, the temperature to which the particles are exposed during combustion is probably too low to promote the formation of reactive phases. This supposition is supported by the high LOI value of ~30 wt%.

The mechanical performance of the other two tested types of ash, namely, OSA-Ene(ef) and OSA-Auv(ef), is in about the same range as OSWA-Auv(ba). Their 56-day compressive and flexural strength is ~30–45 and ~6.0–7.0 MPa, respectively. However, in contrast to bottom ash, the former two both contain the β polymorph of dicalcium silicate, part of which is transformed to the γ polymorph solely in OSA-Auv(ef). In both types of electrostatic filter ash, milling increased reactivity measured with isothermal calorimetry from ~10 to ~100 J g^{-1} in OSA-Ene(ef) and from ~20 to ~150 J g^{-1} in OSA-Auv(ef); however, no significant impact on the mechanical properties of concrete was observed.

5. Conclusions

The principal aim of this study was to assess the feasibility of using OSA in concrete formulations for 3D printing and assess the effectiveness of collision milling as pretreatment. It was hypothesized that collision milling may activate the ash and that reduced particle size may be of benefit to printability.

Four types of OSA were collected, three extracted from the flue gasses (i.e., OSA-Ees(nid), OSA-Ene(ef), and OSA-Auv(ef)) and one from the bottom of the furnace (OSWA-Auv(ba)). The former were collision milled at 100 Hz, while the latter was milled at 30, 50, 100, and 130 Hz. The physical, chemical, and mineralogical characterization of ash was conducted. Furthermore, concrete formulations containing ash were derived from the reference formulation by replacing part of the cement with OSA and were tested for their mechanical performance. In terms of strength, two best-performing ash formulations, namely, OSA-Ees(nid) and its milled counterpart OSA-Ees(nid)/100, were also assessed for printability.

Overall, it can be concluded that the most suitable ash for use in concrete is OSA-Ees(nid); nevertheless, the results obtained on other types of ash also warrant further investigation. The second conclusion is that collision milling is probably a viable option only for bottom ash, which is too coarse for direct use as an SCM.

The first conclusion is supported by the following observations:

1. The particles of the ash extracted from the novel integrated desulfurizer, namely, OSA-Ees(nid), are mostly globular and smooth, resembling the cenospheres found in the FA from coal power plants. Such particles are formed at high combustion temperatures. The particles of the other two types of ash extracted from flue gasses and the bottom ash particles are predominantly angular;

2. The LOI value of OSA-Ees(nid) is <10 wt%, while in other types of ash, this value is significantly higher (~15–30 wt%). This indicates that in the case of the former, the combustion temperature is sufficiently high to burn the majority of kerogen and decompose the minerals;

3. The primary components of all tested ashes are SiO_2 and CaO. In ash extracted from flue gasses, they constitute ~20–30 and ~30–40 wt%, respectively, while in the bottom ash, they amount to ~5 and 50 wt% accordingly. In OSA-Ees(nid), the majority of these two oxides are consumed to form C_2S, which is found as an active β polymorph as well as a rather inert γ polymorph;

4. The compressive and flexural strength of the concrete formulation containing OSA-Ees(nid) was approximately the same as found in the reference formulation, amounting to ~60 and ~7 MPa at 56 days, respectively.

The second conclusion is supported by the following observations:

5. The size of the OSA-Ees(nid) particles is, on average, somewhat <10 μm and, overall, this ash is finer than CEM I, even before the pretreatment. Milling at 100 Hz only reduces the particles >10 μm, thus increasing the peak in the PSD curve at ~6 μm from ~6 to ~8 wt%;

6. In the case of the other three types of ash, which are coarser than CEM I, the milling shifts the PSD curve peak towards 10 μm, so that the overall results indicate that milling is effective only for particles >10 μm. The resulting change is significant only in the case of OSWA-Auv(ba);

7. Collision milling of the three types of flue gas ash resulted in a more than seven-fold increase of reactivity measured with the isothermal calorimetry. Namely, the released heat increased from ~10–20 J g^{-1} measured before milling to ~100–150 J g^{-1} after milling at 100 Hz;

8. Despite the increased reactivity measured by isothermal calorimetry, the mechanical properties of concrete formulations did not significantly improve when milled ash was used compared to their untreated counterparts. The only significant difference was found in OSWA-Auv(ba);

9. Concrete formulations incorporating either OSA-Ees(nid) or OSA-Ees(nid)/100 were checked for printability on a laboratory-scale gantry printer. No significant differences between the two were observed in the slug and the direct buildability tests. The yield stress calculated from the slug test was in a narrow range between ~1000 and 1200 Pa, including the reference formulation, while the yield stress calculated from the direct buildability test of the three formulations was between ~1500 and 2000 Pa.

Based on the findings of this study, future work will focus on using OSA-Ees(nid) in concrete formulations for 3D printing. Collision milling will not be further pursued in combination with this ash. First, the optimization of printable concrete formulations will be performed using screening methods such as measuring mechanical properties on cast specimens and optimizing the workability using the flow table, slug, and direct buildability tests. The optimized formulation will be assessed in detail for its mechanical properties, durability, and printability by adopting the methodologies developed by the technical committees (TCs) of the *International Union of Laboratories and Experts in Construction Materials, Systems, and Structures* (RILEM) [45], namely the TC 303-PFC *performance requirements and testing of fresh printable cement-based materials*, and the TC 304-ADC *assessment of additively manufactured concrete materials and structures*. One set of the results has already been submitted for publication [46]. Furthermore, the Life Cycle Assessment (LCA) and the Life Cycle Cost Assessment (LCCA) will be conducted for the optimized formulation.

In conclusion, the importance of this research for infrastructural projects should also be outlined since these projects consume a significant proportion of concrete and, in contrast to residential buildings, are expected to have a longer service life often under harsher conditions. Therefore, the challenge of using new types of cement and non-conventional SCMs in concrete for infrastructures will have to be addressed. This includes the nuclear power plants where concrete features heavily and is thus addressed by several research projects dealing with nuclear materials, such as Orient-NM [47], Connect-NM [48], and Aces [49]. On the other hand, the main benefit of 3D printing structural elements is in the optimization of geometry, leading to large material savings. Two full-scale showcase infrastructural projects have already been executed. Namely, 4.5 m high water tanks with a diameter of 7 m were printed in Kuwait with 25 % material savings compared to tanks built with conventional technology [50]. The other showcase project is the prototype of the wind turbine tower printed in Denmark [51]. Furthermore, shortages in the labor market and the scale of demand for new infrastructures will necessitate automation and digital fabrication.

Author Contributions: Conceptualization, L.H., M.Š. (Mateja Štefančič), M.Š. (Māris Šinka), E.Š. and L.K.B.; methodology, L.H., M.Š. (Mateja Štefančič), M.Š. (Māris Šinka), E.Š. and B.M.; formal analysis, L.H., V.Z.S., M.Š. (Māris Šinka) and A.S.; investigation, M.Š. (Mateja Štefančič), K.Š., V.Z.S., A.S., G.Š., B.M. and L.K.B.; writing—original draft preparation, L.H. and A.S.; visualization, L.H. and A.S.; writing—review and editing, L.K.B.; data curation, K.Š.; project administration, M.Š. (Māris Šinka), E.Š. and L.K.B.; funding acquisition, L.H., M.Š. (Māris Šinka) and E.Š. All authors have read and agreed to the published version of the manuscript.

Funding: This research was funded through the transnational M-ERA.NET Joint Call 2022, project title *"Transforming waste into high-performance 3D printable cementitious composite,"* acronym: Transition, funding period 2023–2026, https://www.m-era.net/home (accessed on 29 December 2024). Authors affiliated with the *Slovenian National Building and Civil Engineering Institute* acknowledge the co-funding of the Slovenian part of the transnational project by the *Ministry of Higher Education, Science and Innovation of the Republic of Slovenia,* under contract number C3360-25-452009.

Data Availability Statement: The data set is available in the OSF repository https://doi.org/10.176 05/OSF.IO/MWF9N.

Acknowledgments: The authors acknowledge the guidance kindly provided by Diana Bajare and the keen support of Eva Dzene.

Conflicts of Interest: The authors declare no conflicts of interest. The funders had no role in the design of the study; in the collection, analyses, or interpretation of data; in the writing of the manuscript; or in the decision to publish the results.

References

1. Wang, N.; Xu, Z.; Liu, Z. Innovation in the construction sector: Bibliometric analysis and research agenda. *J. Eng. Technol. Manag.* **2023**, *68*, 101747. [CrossRef]
2. Reichstein, T.; Salter, A.J.; Gann, D.M. Last among equals: A comparison of innovation in construction, services and manufacturing in the UK. *Constr. Manag. Econ.* **2005**, *23*, 631–644. [CrossRef]
3. Sawhney, A.; Riley, M.; Irizarry, J. *Construction 4.0: An Innovation Platform for the Built Environment*, 1st ed.; Sawhney, A., Riley, M., Irizarry, J., Eds.; Routledge: Abingdon, OX, UK; New York, NY, USA, 2020.
4. Sivam, A.; Trasente, T.; Karuppannan, S.; Chileshe, N. The impact of an ageing workforce on the construction industry in Australia. In *Valuing People in Construction*, 1st ed.; Emuze, F., Smallwood, J., Eds.; Routledge: Abingdon, OX, UK; New York, NY, USA, 2017.
5. Clarke, L.; Michielsens, E.; Snijders, S.; Wall, C. *'No More Softly, Softly': Review of Women in the Construction Workforce*; University of Westminster: London, UK, 2015.
6. French, E.; Strachan, G. Women in the construction industry: Still the outsiders. In *Valuing People in Construction*, 1st ed.; Emuze, F., Smallwood, J., Eds.; Routledge: Abingdon, OX, UK; New York, NY, USA, 2017.
7. Wierzbicki, M.; de Silva, C.W.; Krug, D.H. BIM—History and trends. In Proceedings of the International Conference on Construction Applications of Virtual Reality, Weimar, Germany, 3–4 November 2011.
8. Borkowski, A.S. Evolution of BIM: Epistemology, genesis and division into periods. *J. Inf. Technol. Constr.* **2023**, *28*, 646–661. [CrossRef]
9. Ma, G.; Buswell, R.; Leal da Silva, W.R.; Wang, L.; Xu, J.; Jones, S.Z. Technology readiness: A global snapshot of 3D concrete printing and the frontiers for development. *Cem. Concr. Res.* **2022**, *156*, 106774. [CrossRef]
10. Wangler, T.; Lloret, E.; Reiter, L.; Hack, N.; Gramazio, F.; Kohler, M.; Bernhard, M.; Dillenburger, B.; Buchli, J.; Roussel, N.; et al. Digital concrete: Opportunities and challenges. *RILEM Tech. Lett.* **2016**, *1*, 67–75. [CrossRef]

11. Buswell, R.A.; Leal de Silva, W.R.; Jones, S.Z.; Dirrenberger, J. 3D printing using concrete extrusion: A roadmap for research. *Cem. Concr. Res.* **2018**, *112*, 37–49. [CrossRef]
12. Wangler, T.; Pileggi, R.; Gürel, S.; Flatt, R.J. A chemical process engineering look at digital concrete processes: Critical step design, inline mixing, and scaleup. *Cem. Concr. Res.* **2022**, *155*, 106782. [CrossRef]
13. Bos, F.P.; Menna, C.; Pradena, M.; Kreiger, E.; da Silva, W.R.L.; Rehman, A.U.; Weger, D.; Wolfs, R.J.M.; Zhang, Y.; Ferrara, L.; et al. The realities of additively manufactured concrete structures in practice. *Cem. Concr. Res.* **2022**, *156*, 106746. [CrossRef]
14. Kreiger, M.; Kreiger, E.; Mansour, S.; Monkman, S.; Delavar, M.A.; Sideris, P.; Roberts, C.; Friedell, M.; Platt, S.; Jones, S. Additive construction in practice—Realities of acceptance criteria. *Cem. Concr. Res.* **2024**, *186*, 107652. [CrossRef]
15. Scrivener, K.; Martirena, F.; Bishnoi, S.; Maity, S. Calcined clay limestone cements (LC3). *Cem. Concr. Res.* **2018**, *114*, 49–56. [CrossRef]
16. Scrivener, K.L.; John, V.M.; Gartner, E.M. Eco-efficient cements: Potential economically viable solutions for a low-CO_2 cement-based materials industry. *Cem. Concr. Res* **2018**, *114*, 2–26. [CrossRef]
17. Kahru, A.; Põllumaa, L. Environmental hazard of the waste streams of Estonian oil shale industry: An ecotoxicological review. *Oil Shale* **2006**, *23*, 53. [CrossRef]
18. Velts, O.; Uibu, M.; Kallas, J.; Kuusik, R. Waste oil shale ash as a novel source of calcium for precipitated calcium carbonate: Carbonation mechanism, modeling, and product characterization. *J. Hazard. Mater.* **2011**, *195*, 139–146. [CrossRef]
19. Usta, M.C.; Yörük, C.R.; Hain, T.; Paaver, P.; Snellings, R.; Rozov, E.; Gregor, A.; Kuusik, R.; Trikkel, A.; Uibu, M. Evaluation of new applications of oil shale ashes in building materials. *Minerals* **2020**, *10*, 765. [CrossRef]
20. Smadi, M.M.; Haddad, R.H. The use of oil shale ash in Portland cement concrete. *Cem. Concr. Compos.* **2003**, *25*, 43–50. [CrossRef]
21. Al-Hamaiedh, H.; Maaitah, O.; Mahadin, S. Using oil shale ash in concrete binder. *Electron. J. Geotech. Eng.* **2010**, *15*, 601–608.
22. Raado, L.-M.; Hain, T.; Liisma, E.; Kuusik, R. Composition and properties of oil shale ash concrete. *Oil Shale* **2014**, *31*, 147–160. [CrossRef]
23. *EVS 927:2018*; Burnt Shale for Building Materials. Specification, Performance and Conformity. Estonian Centre for Standardisation and Accreditation: Tallinn, Estonia, 2018.
24. Sobolev, K.; Gutiérrez, M.F. How nanotechnology can change the concrete world (part two): Successfully mimicking nature's bottom-up construction processes is one of the most promising directions. *Am. Ceram. Soc. Bull.* **2005**, *84*, 16–20.
25. Hanžič, L.; Šter, K.; Štefančič, M. Implementing the excess paste concept for a systematic mix design of printable concrete. In Proceedings of the Supplementary Conference Proceedings: Fourth RILEM International Conference on Concrete and Digital Fabrication—Digital Concrete 2024, Munchen, Germany, 4–6 September 2024.
26. Tümanok, A.; Kulu, P.; Tümanok, A. Treatment of different materials by disintegrator systems. *Proc. Est. Acad. Sci.* **1999**, *5*, 222–242.
27. Bumanis, G.; Bajare, D. Compressive strength of cement mortar affected by sand microfiller obtained with collision milling in disintegrator. *Procedia Eng.* **2017**, *172*, 149–156. [CrossRef]
28. *SIST EN 1097-5:2008*; Tests for Mechanical and Physical Properties of Aggregates—Part 5: Determination of the Water Content by Drying in a Ventilated Oven. Slovenian Institute for Standardization: Ljubljana, Slovenia, 2008.
29. *SIST EN 1097-7:2023*; Tests for Mechanical and Physical Properties of Aggregates—Part 7: Determination of the Particle Density of Filler—Pyknometer Method. Slovenian Institute for Standardization: Ljubljana, Slovenia, 2023.
30. *ISO 13320:2020*; Particle Size Analysis—Laser Diffraction Methods. International Organization for Standardization: Vernier (Geneva), Switzerland, 2020.
31. *EN 196-2:2013*; Methods of Testing Cement—Part 2: Chemical Analysis of Cement. European Committee for Standardization: Brussels, Belgium, 2013.
32. *SIST EN 451-1:2017*; Method for Testing Fly Ash—Part 1: Determination of Free Calcium Oxide Content. Slovenian Institute for Standardization: Ljubljana, Slovenia, 2017.
33. *EN 196-3:2016*; Methods of Testing Cement—Part 3: Determination of Setting Times and Soundness. European Committee for Standardization: Brussels, Belgium, 2016.
34. *EN 450-1:2012*; Fly Ash for Concrete Definition, Specifications and Conformity Criteria. European Committee for Standardization: Brussels, Belgium, 2012.
35. *ASTM C1897-20*; Standard Test Methods for Measuring the Reactivity of Supplementary Cementitious Materials by Isothermal Calorimetry and Bound Water Measurement. ASTM International: Philadelphia, PA USA, 2020.
36. Stefanidou, M.; Kesikidou, F.; Konopisi, S.; Vasiadis, T. Investigating the suitability of waste glass as a supplementary binder and aggregate for cement and concrete. *Sustainability* **2023**, *15*, 3796. [CrossRef]
37. *EN 1015-3:1999*; Methods of Test for Mortar for Masonry. Determination of Consistence of Fresh Mortar (by Flow Table). European Committee for Standardization: Brussels, Belgium, 1999.
38. *EN 1015-11:2019*; Methods of Test for Mortar for Masonry. Determination of Flexural and Compressive Strength of Hardened Mortar. European Committee for Standardization: Brussels, Belgium, 2019.

39. Ducoulombier, N.; Carneau, P.; Mesnil, R.; Demont, L.; Caron, J.-F.; Roussel, N. 'The Slug Test': Inline assessment of yield stress for extrusion-based additive manufacturing. In *Second RILEM International Conference on Concrete and Digital Fabrication*, 1st ed.; Bos, F.P., Lucas, S.S., Wolfs, R.J.M., Salet, T.A.M., Eds.; Springer: Cham, Switzerland, 2020; pp. 216–224. [CrossRef]
40. Suiker, A.S.J.; Wolfs, R.J.M.; Lucas, S.M.; Salet, T.A.M. Elastic buckling and plastic collapse during 3D concrete printing. *Cem. Concr. Res.* **2020**, *135*, 106016. [CrossRef]
41. Tao, Y.; Ren, Q.; Lesage, K.; Van Tittelboom, K.; Yuan, Y.; De Schutter, G. Shape stability of 3D printable concrete with river and manufactured sand characterized by squeeze flow. *Cem. Concr. Compos.* **2022**, *133*, 104674. [CrossRef]
42. *EN 12350-6:2019*; Testing Fresh Concrete—Part 6: Density. European Committee for Standardization: Brussels, Belgium, 2019.
43. Labus, M.; Matyasik, I.; Ziemianin, K. Thermal decomposition processes in relation to the type of organic matter, mineral and maceral composition of menilite shales. *Energies* **2023**, *16*, 4500. [CrossRef]
44. Wang, Q.; Li, F.; Shen, X.; Shi, W.; Li, X.; Guo, Y.; Xiong, S.; Zhu, Q. Relation between reactivity and electronic structure for α'L-, β- and γ-dicalcium silicate: A first-principles study. *Cem. Concr. Res.* **2014**, *57*, 28–32. [CrossRef]
45. RILEM. International Union of Laboratories and Experts in Construction Materials, Systems and Structures. Available online: https://www.rilem.net/ (accessed on 15 November 2024).
46. Sapata, A.; Sinka, M.; Šahmenko, G.; Korat Bensa, L.; Hanžič, L.; Šter, K.; Rucevskis, S.; Bajare, D.; Bos, F.P. Establishing benchmark properties for 3d-printed concrete: A study of printability, strength, and durability. *J. Compos. Sci.* 2024, *submitted*.
47. Orient-NM. Organisation of the European Research Community on Nuclear Materials, Euratom Research and Training Programme. Available online: https://www.eera-jpnm.com/orient-nm/ (accessed on 29 December 2024).
48. Connect-NM. Coordination of the European Nuclear Materials Community for Energy Innovation, Euratom Research and Training Programme. Available online: https://indico.scc.kit.edu/event/3644/attachments/7119/11266/5_Innumat_WS_23 _Connect-nminnumatb_Malerba.pdf (accessed on 17 December 2024).
49. Aces. Towards Improved Assessment of Safety Performance for LTO of Nuclear Civil Engineering Structures, Euratom Research and Training Programme. Available online: https://aces-h2020.eu/ (accessed on 15 November 2024).
50. Cobod. World's First 3D Printed Large Concrete Tanks Made with Materials Savings of 25%, Press release 05/03/2024. Available online: https://cobod.com/worlds-first-3d-printed-large-concrete-tanks-made-with-materials-savings-of-25/ (accessed on 23 December 2024).
51. Cobod. The Revolutionary Impact of 3D Printing on Industries, Blog 27/04/2023. Available online: https://cobod.com/the-revolutionary-impact-of-3d-printing-on-industries/ (accessed on 23 December 2024).

MDPI AG
Grosspeteranlage 5
4052 Basel
Switzerland
Tel.: +41 61 683 77 34

Infrastructures Editorial Office
E-mail: infrastructures@mdpi.com
www.mdpi.com/journal/infrastructures

www.ingramcontent.com/pod-product-compliance
Lightning Source LLC
LaVergne TN
LVHW071934160726
843515LV00011B/2611